Electromagnetic Theory

Electromagnetic Theory

ATTAY KOVETZ

School of Physics and Astronomy
Tel Aviv University

OXFORD
UNIVERSITY PRESS

OXFORD
UNIVERSITY PRESS

Great Clarendon Street, Oxford OX2 6DP

Oxford University Press is a department of the University of Oxford.
It furthers the University's objective of excellence in research, scholarship,
and education by publishing worldwide in

Oxford New York

Athens Auckland Bangkok Bogotá Buenos Aires Calcutta
Cape Town Chennai Dar es Salaam Delhi Florence Hong Kong Istanbul
Karachi Kuala Lumpur Madrid Melbourne Mexico City Mumbai
Nairobi Paris São Paulo Singapore Taipei Tokyo Toronto Warsaw

with associated companies in Berlin Ibadan

Published in the United States
by Oxford University Press Inc., New York

A catalogue record for this book is available from the British Library

Library of Congress Cataloging in Publication Data
Kovetz, Attay.
Electromagnetic theory / Attay Kovetz.
Includes bibliographical references and index.
1. Electromagnetic theory. I. Title.
QC670 .K693 2000   537'.12 21–dc21   99–044818

ISBN 0 19 850604 X (Hbk acid-free paper)
ISBN 0 19 850603 1 (Pbk acid-free paper)

Typeset by
Newgen Imaging Systems (P) Ltd., Chennai, India

Printed in Great Britain by
Biddles Ltd, Guildford and Kings Lynn

In memoriam

Dr. Akiva Stefan Kovács 1904–1988
Martha Kovács (née Spindel) 1905–1996

Preface

This book is the outcome of a course entitled 'Analytical Electromagnetism', which I have taught on and off during the last 20 years at Tel Aviv University. The course is attended by physics students during the latter half of their second year, or at the beginning of their third year, undergraduate studies. It is at an advanced level: the students have all been exposed to electromagnetism, as well as to special relativity, through several courses (both in the class and in the laboratory) on classical and modern physics.

Obviously, then, the students who attend such an advanced undergraduate course have all seen Maxwell's equations in one form or another. They have all heard of Poynting's vector, electric displacement, Joule heating, Lorentz force and electromagnetic waves. But, despite this familiarity, the average student is uncertain about the concepts and laws of electromagnetic theory. And this uncertainty stands in remarkable contrast to the confidence with which students regard the concepts and laws of mechanics. Unless there are some gaping faults in the logical structure of electromagnetic theory, the reason for these difficulties must lie in the way in which this theory is usually presented.

I believe that an advanced undergraduate course affords a good opportunity for presenting electromagnetism as a coherent and logical theory. This is the purpose of this book. Its level is roughly that of other texts on electromagnetism that start from Maxwell's equations. In particular, it assumes that the reader requires little or no motivation for the introduction of various electromagnetic concepts and definitions. But it differs from most other textbooks in presenting electromagnetism as a *classical* theory, based on principles that are independent of the atomic constitution of matter. Whereas the concepts and principles of classical mechanics are never claimed to depend on the fact that ordinary (terrestrial) matter consists of atoms and molecules, textbooks on electromagnetism abound with discussions of stretched molecules forming tiny dipoles and of Ampèrian currents associated with atomic electrons, all dating back to Lorentz's 90-year-old *Theory of electrons*. Not only are the students asked to believe that the average value of a quantity can be found in the absence of any knowledge regarding its distribution, they are also expected to forget that the existence of atoms can only be established on the basis of quantum mechanics, using the very Hamiltonian operators that are suggested by electromagnetism.

Obviously, then, contemporary authors, who no longer believe the classical theory of electrons—and who look to the solid state physicists for an explanation of its well-known successes—are not being quite frank with their readers. Clearly, a presentation of electromagnetic theory that avoids the atomic constitution of matter is to be preferred on both didactic and logical grounds.

The concepts of force and energy are a major source of confusion for students of electromagnetism. They know that force and energy are subject to the laws of mechanics and thermodynamics, and they had no special difficulties in mastering and applying these concepts in their studies of those two disciplines. But in electromagnetism force and energy reappear in strange and apparently conflicting forms. There is a reason for this: the discussion of force and energy requires the establishment of a definite relationship between mechanics, thermodynamics and electromagnetism. Textbooks are only too often strangely vague about this relationship. The treatment is usually wordy, in the worst thermodynamic tradition; indeed the laws of mechanics and thermodynamics are seldom stated in mathematical form. And the manner in which these laws must be modified in order to account for both electromagnetic momentum *and* energy flux remains unclear. Special attention is therefore paid in this book to the precise link between the three major disciplines of classical physics.

The outline of the book is as follows: starting from the law of charge conservation, Faraday's laws and the Maxwell–Lorentz aether relations, the principles of electromagnetism are stated, in Chapters 1–5, in tensor form. Since undergraduate students regard tensors with apprehension, there is a short chapter on antisymmetric tensors which, I hope, will provide not only a gentle introduction, but also the necessary motivation for their use. Besides taking care of questions regarding transformations and invariance, the tensorial formulation has the added advantage that students will not have to relearn electromagnetism when they take a course in general relativity. For, although curved space-time is not mentioned in these chapters, the possibility is nowhere denied.

Chapter 6 deals with the motion of charged particles. Since this subject is closely connected with measurement (the oscilloscope), it is a good place for introducing Système International (SI) electromagnetic units.

Chapters 7–11 cover what is roughly the subject matter of pre-Maxwellian electromagnetism—polarization and magnetization, electrostatics, slowly varying currents and their magnetic effects. But the treatment is modern: relativistic effects are not always ignored, and the student is encouraged to develop considerable skill in dealing with moving bodies. I have left out the treatment of alternating-current circuits because they are adequately treated in other courses (but there is a section on the skin effect).

Electromagnetic radiation and wave propagation are treated in Chapters 12 and 13.

Chapter 14 is a review of continuum mechanics, and introduces Coleman and Noll's method in order to establish the link between mechanics and thermodynamics on a rational basis. Everything comes together in Chapter 15, where the complete set of laws of electromagnetism, mechanics and thermodynamics is treated in a manner which is a direct generalization of Coleman and Noll's method. Maxwell stress tensors, ponderomotive forces, equations of motion and of energy and many other results all follow by a process of mathematical, and therefore logical, deduction.

From a strictly logical point of view, Chapters 14 and 15 should precede Chapter 8. But this would entail a serious drawback: students become impatient when they are presented with all the principles and theoretical deductions before encountering any real application.

The results of Chapter 15 are applied in Chapter 16, on magnetohydrodynamics, and in Chapters 17 and 18, which deal with electric and magnetic properties of materials, such as piezoelectricity, ferroelectricity, ferromagnetism and superconductivity.

To some degree, the choice of subjects invariably reflects a personal preference. I have tried to convince the reader that electromagnetism is a theory, not only of phenomena, but also of materials, and that it can be readily applied to moving, as well as to stationary, bodies. I have also attempted to dispel some common prejudices (such as the belief, which most students share, that electromagnetism is a linear theory) and misconceptions (for example, that the theory needs to be supplemented by a principle of 'causality').

Appendix A introduces the Gaussian system of electromagnetic units and deals with the relations (conversion, or rather correspondence, factors) between Gaussian and SI units.

Fifty solved example problems are scattered throughout the book. In addition, there are 175 exercises. Some of these are really parts of a mathematical proof or derivation that have been left to the reader. The remaining exercises, representing applications of the theory, are of the kind usually found at the end of a textbook chapter. Appendix B provides detailed solutions of all the exercises, but students are urged to try and solve the exercises before looking at these solutions.

The prerequisites for a student who intends to use this book as a text for an intermediate or advanced course on electromagnetic theory should present no problem. The mathematical tools required are no more than vector calculus and Legendre polynomials. Tensors are introduced for the purpose of formulating the basic laws of electromagnetism in the most general manner, and for dealing efficiently with changes of frame. But a prior familiarity with tensor calculus is not required; and beyond the first five chapters tensors are hardly ever mentioned. Of the physics curriculum, besides an introductory course on electromagnetism, some familiarity with the basic concepts of analytical mechanics, special relativity and thermodynamics is assumed, but any of these can be studied concurrently with a reading of this book.

I would like to thank Leon Mestel for a critical reading of the manuscript, and for many helpful comments. I am very grateful to my wife Dina Prialnik for numerous comments and suggestions, and for her constant encouragement during the writing of this book. Finally, it gives me special pleasure to thank my daughter Michal Semo, Head of Tel Aviv University's Graphic Design Office, for all the graphical work.

Tel Aviv 1999   A. K.

Contents

Notation

In some cases I have found it necessary to depart from the notation that has become standard in treatments of electromagnetism. Of these departures, the most important are:

Electric charge	$Q,\ e$
Charge density	q
Mass density	ρ
Stress tensor	T
Temperature	$\vartheta.$

In each section figures, problems, exercises and equations—or groups of equations—are numbered consecutively. Thus, for example, (10.6) denotes the sixth equation of Section 10; and if this is a group of equations, $(10.6)_2$ denotes its second member. In a few cases simple, temporary labels—(1), (2) etc.—are used to denote intermediate results within a problem or exercise solution.

In the Index, page numbers are <u>underlined</u> when they represent the main source of information about an index entry; for example, its definition. A page number is given in italics (e.g. *324*) when that page contains an instructive example of the use of an index entry.

1

Electric charges and currents

1. Charge conservation

Electric charge is a property of bodies in nature. Logically speaking, it is a primitive: its status in electromagnetism is like that of point or plane in geometry, mass or force in mechanics, and energy or temperature in thermodynamics. For the time being we shall be concerned with those cases in which the charge Q is smoothly distributed over volume; more precisely, those situations in which it is possible to define a charge density q such that

$$Q = \int q \, dV, \tag{1.1}$$

where Q is the charge residing in some region and the integral is over the volume of that region. Of course, Q and q may each be positive, negative or zero.

The fundamental assumption we shall make regarding electric charge is that it is conserved. By that we mean that, if the charge within a certain region has changed, it is because charge has passed out of the region, or entered into it, through the bounding surface:

$$\Delta Q = \Delta \int q \, dV = - \int_{t_1}^{t_2} dt \oint \alpha \, dS. \tag{1.2}$$

Here $\Delta Q = Q(t_2) - Q(t_1)$ is the change in charge over the time interval from t_1 to t_2. The last integral, which extends over the complete surface, is the net rate at which electric charge leaves the region. The function α, the existence of which is part of the principle (or axiom) of charge conservation, depends on position $\mathbf{x}$ and time t, and on properties of the surface. But a theorem of Cauchy in continuum mechanics imposes severe restrictions on the latter dependence. We shall assume (with Cauchy) that α has a common value for all surface elements dS having a common tangent plane. It is therefore a function only of the orientation of the tangent plane to the surface, which in turn is determined by the components of the unit outward normal $\mathbf{n}$.

Thus α is a function of $\mathbf{x}$, t and $\mathbf{n}$. Cauchy's theorem states that the dependence of α on the Cartesian components n_i of $\mathbf{n}$ is linear and homogeneous:

$$\alpha(\mathbf{x}, t, \mathbf{n}) = j_1(\mathbf{x}, t)n_1 + j_2(\mathbf{x}, t)n_2 + j_3(\mathbf{x}, t)n_3. \tag{1.3}$$

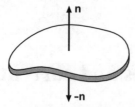

Fig. 1.1

In order to prove the theorem, we first apply the law of charge conservation (1.2) to a thin plane slab of thickness ϵ (Fig. 1.1). The rate of change of the charge Q in the slab will be proportional to the volume, and will therefore be of the first order in ϵ. The rate of outflow through the bottom (1) and top (2) faces will be $\int \alpha \, dS_1 + \int \alpha \, dS_2$, and the contribution from the sides will again be of order ϵ. In the limit of vanishing thickness, we shall have

$$\int \alpha \, dS_1 + \int \alpha \, dS_2 = 0,$$

with the top and bottom coinciding, except that their orientations are opposite. Since this holds for a slab of any size, we have *Cauchy's lemma*

$$\alpha(\mathbf{x}, t, -\mathbf{n}) = -\alpha(\mathbf{x}, t, \mathbf{n}). \tag{1.4}$$

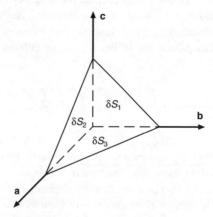

Fig. 1.2

Next, we apply the law of charge conservation to a tetrahedron with three orthogonal faces (Fig. 1.2). Let these orthogonal faces have areas $\delta S_1, \delta S_2, \delta S_3$ and unit outward normals $-\mathbf{a}$, $-\mathbf{b}$, $-\mathbf{c}$, and let the fourth inclined face have area δS and unit outward normal $\mathbf{n}$. The total rate of outflow of charge is

$$\alpha(\mathbf{n})\delta S + \alpha(-\mathbf{a})\delta S_1 + \alpha(-\mathbf{b})\delta S_2 + \alpha(-\mathbf{c})\delta S_3,$$

where, by the mean value theorem of the integral calculus, each α is evaluated at an appropriate point on the corresponding face. Since the δS_i are projections of δS, we have three relations like

$$\delta S_1 = \mathbf{a} \cdot \mathbf{n}\,\delta S$$

for the three orthogonal faces. With the use of Cauchy's lemma, the rate of outflow becomes

$$[\alpha(\mathbf{n}) - \alpha(\mathbf{a})\mathbf{a} \cdot \mathbf{n} - \alpha(\mathbf{b})\mathbf{b} \cdot \mathbf{n} - \alpha(\mathbf{c})\mathbf{c} \cdot \mathbf{n}]\delta S. \tag{1.5}$$

If we now let the linear dimensions of the tetrahedron vanish without changing its shape, ΔQ will vanish like the volume, whereas the outflow, according to (1.5), will vanish like δS. Hence, in the limit,

$$\alpha(\mathbf{n}) = [\alpha(\mathbf{a})\mathbf{a} + \alpha(\mathbf{b})\mathbf{b} + \alpha(\mathbf{c})\mathbf{c}] \cdot \mathbf{n}, \tag{1.6}$$

where all the α's now refer to the same point $\mathbf{x}$ and time t. If we regard the three orthogonal vectors $\mathbf{a}$, $\mathbf{b}$ and $\mathbf{c}$ as a Cartesian frame, eqn (1.6), with a change of notation, becomes

$$\alpha(\mathbf{n}) = j_1 n_1 + j_2 n_2 + j_3 n_3, \tag{1.7}$$

which is Cauchy's theorem (1.3).

Cauchy's basic assumption, that α for any point on a surface is the same as for the tangent plane through that point, can be greatly relaxed. This has been done by Hamel and Noll, but we shall not pause to provide the proof.

In matrix notation, we can write the last equation in the concise form

$$\alpha = j^T n, \tag{1.8}$$

where j^T (a row) is the transpose of the column j. If we now rotate the axes, the new components of $\mathbf{n}$ will be given by $n' = An$, where A is an orthogonal matrix. In the new frame, the *same* α will be given, according to Cauchy's theorem, by

$$\alpha = (j')^T n' = (j')^T An = (A^T j')^T n. \tag{1.9}$$

Comparison with (1.8) gives $A^T j' = j$, or, since A is orthogonal,

$$j' = Aj. \tag{1.10}$$

Thus j has the transformation properties of a vector. We shall therefore denote it by $\mathbf{j}$, and write (1.7) in the form

$$\alpha = \mathbf{j} \cdot \mathbf{n} = j_n(\mathbf{x}, t).\tag{1.11}$$

According to (1.2), the amount of charge passing (in the direction of $\mathbf{n}$), per unit time, through the surface element dS is $\mathbf{j} \cdot \mathbf{n}\, dS$. The vector $\mathbf{j}$ is called the *current density*.

We can now write the principle of charge conservation in the form

$$\Delta \int q\, dV + \int_{t_1}^{t_2} dt \oint \mathbf{j} \cdot \mathbf{n}\, dS = 0.\tag{1.12}$$

If q and $\mathbf{j}$ are differentiable, this becomes

$$\int_{t_1}^{t_2} dt \int dV \left(\frac{\partial q}{\partial t} + \operatorname{div} \mathbf{j} \right) = 0,\tag{1.13}$$

where we have used Gauss's theorem[†] in order to replace $\oint \mathbf{j} \cdot \mathbf{n}\, dS$ by $\int \operatorname{div} \mathbf{j}\, dV$. Since (1.13) holds for any volume, and for any time interval, we have the local form

$$\frac{\partial q}{\partial t} + \operatorname{div} \mathbf{j} = 0.\tag{1.14}$$

If we denote the dimensions of length, time and charge by L, T and C, respectively, the dimensions of charge density q are C/L^3, and the dimensions of current density $\mathbf{j}$ are $C/(TL^2)$.

The integral, or global, eqn (1.12) and the differential, or local, eqn (1.14) are equivalent only when $\mathbf{j}$ is differentiable and the time derivative of q exists. Consider, for example, a charged body in motion. When its boundary reaches a point P, both $\operatorname{div} \mathbf{j}$ and $\partial q/\partial t$ cease to exist at P. Similar circumstances may obtain when a boundary between two adjoining bodies is moving. In such cases, when $\mathbf{j}$ or q suffer discontinuities, the local eqn (1.14) becomes meaningless, but the global eqn (1.12) remains valid, and provides a restriction on the discontinuities.

Problem 1.1   Let S be a surface across which q and $\mathbf{j}$ are discontinuous. If $\mathbf{n}$ is a unit normal on S (chosen in one of the two possible ways), we call the side of S to which $\mathbf{n}$ points the positive side of S, and the other side the negative one. Show that the discontinuities are constrained by

$$\mathbf{n} \cdot (\mathbf{j}_+ - \mathbf{j}_-) - v_n(q_+ - q_-) = 0,\tag{1.15}$$

where v_n is the normal velocity of the surface.

[†]The correct attribution is to Green.

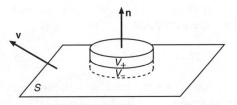

Fig. 1.3

Solution   Around a piece dS of the surface we erect a *fixed* pillbox with generators parallel to **n** (Fig. 1.3). The charge contained in the pillbox is

$$Q = q_+ V_+ + q_- V_-.$$

During a short time δt, the normal displacement $v_n \delta t$ of the surface element decreases V_+ by $v_n \delta t dS$, and increases V_- by the same amount, since the pillbox is fixed (v_n may of course be negative). Thus

$$\delta Q = q_+ \delta V_+ + q_- \delta V_- = -(q_+ - q_-) v_n \delta t dS.$$

According to the principle of charge conservation,

$$\delta Q + \delta t \oint j_n \, dS = 0.$$

In the limit of vanishing height of the pillbox, the contribution to $\oint j_n \, dS$ from the sides vanishes, and

$$\delta t \oint j_n \, dS = \delta t (\mathbf{j}_+ - \mathbf{j}_-) \cdot \mathbf{n} \, dS.$$

The minus sign is due to **n** pointing inward at the bottom of the pillbox. Thus, in the limit, we obtain

$$\mathbf{n} \cdot (\mathbf{j}_+ - \mathbf{j}_-) - v_n (q_+ - q_-) = 0.$$

Note that this equation does not depend on the choice of **n**: the alternative choice will reverse the sign of **n** *along with* the definitions of 'positive' and 'negative' sides of S, and will also reverse the sign of v_n.

---

2. Charge conservation in four dimensions

The local form (1.14) of the principle of charge conservation is

$$\frac{\partial q}{\partial t} + \operatorname{div} \mathbf{j} = 0. \tag{2.1}$$

In Cartesian coordinates, the notation $\operatorname{div} \mathbf{j}$ stands for $\partial j_i / \partial x^i$, where summation over the repeated index i is understood. We shall now write the principle of charge conservation in four-dimensional notation. First, we add the time as a fourth coordinate, x^4, to the three space coordinates $x^i, i = 1, 2, 3$:

$$x = (x^\alpha) = (x^1, x^2, x^3, x^4) = (x^i, x^4) = (\mathbf{x}, t); \qquad (2.2)$$

we use Latin indices (like i) for the three space coordinates and Greek indices (like α) for the four space-time coordinates. Accordingly, a repeated Latin index implies summation from 1 to 3, and a Greek index, from 1 to 4. A point $x = (\mathbf{x}, t)$ in space-time is called an *event*. We shall take the time-axis to be orthogonal to the space axes: its positive (future) direction is $(0, 0, 0, 1)$; its negative (past) direction is $(0, 0, 0, -1)$. A space-time frame constructed in this way is called a *Euclidean frame*. It is the four-dimensional analogue of a three-dimensional Cartesian frame.

We now define

$$s = (s^\alpha) = (s^i, s^4) = (\mathbf{j}, q) \qquad (2.3)$$

and call the collection s of s^α's the *charge-current density*. With this notation, $\partial q / \partial t = \partial s^4 / \partial x^4$ and $\operatorname{div} \mathbf{j} = \partial s^i / \partial x^i$, so that (2.1) becomes

$$\partial_\alpha s^\alpha = 0, \qquad (2.4)$$

where ∂_α stands for $\partial / \partial x^\alpha$. The positioning of indices, as in j_i or s^α, need not concern us for the moment. We shall soon derive formulae which will make the reasons for the choice of an upper or a lower index apparent. We do, however, admit to a definite preference—purely aesthetic for the time being—for placing a summation index, like α in (2.4), in upper and lower positions respectively. We may regard the sum over α in (2.4) as a four-dimensional divergence of the charge-current density s, and write

$$\operatorname{div} s = 0. \qquad (2.5)$$

The four-dimensional forms (2.4) or (2.5) of the principle of charge conservation are not general enough, because they are local, whereas the general principle of charge conservation has the global, integral form (cf. (1.12))

$$\int_{(t_2)} q \, dV - \int_{(t_1)} q \, dV + \int_{t_1}^{t_2} dt \oint \mathbf{j} \cdot \mathbf{n} \, dS = 0. \qquad (2.6)$$

In order to cast this general form of the principle of charge conservation in four-dimensional notation, we make use of a four-dimensional diagram. The cylinder in Fig. 2.1 is a three-dimensional hypersurface in space-time. Its top, a 3-space in the hyper-plane $t = t_2$, with element $d_3v = dV$ and four-dimensional outward-pointing normal $n = (0, 0, 0, 1)$ in the upward (future) direction, represents the spatial region (volume) of (2.6) at the time t_2. Over this region, the four-dimensional scalar product

$s \cdot n = (\mathbf{j}, q) \cdot (0, 0, 0, 1) = q$, so that $s \cdot n \, d_3 v = q(\mathbf{x}, t_2) \, dV$. Hence the first term of (2.6) is simply $\int s \cdot n \, d_3 v$, taken over the top of the cylinder.

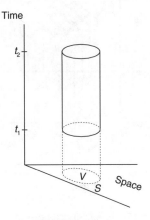

Fig. 2.1

The bottom of the cylinder, a 3-space in the hyperplane $t = t_1$, similarly represents the volume at time t_1. Over this region, the outward-pointing 4-normal is in the downward (past) direction, and $s \cdot n = (\mathbf{j}, q) \cdot (0, 0, 0, -1) = -q$. Thus the second term of (2.6), together with its sign, is simply $\int s \cdot n \, d_3 v$, taken over the bottom of the cylinder.

The side of the cylinder represents the *history*, between t_1 and t_2, of the surface in (2.6). It is a 3-space with element $d_3 v = dS \, dt$ and horizontal outward-pointing 4-normal $n = (\mathbf{n}, 0)$. Over this region, $s \cdot n = (\mathbf{j}, q) \cdot (\mathbf{n}, 0) = \mathbf{j} \cdot \mathbf{n}$. Thus the third term of (2.6) is simply $\int s \cdot n \, d_3 v$, taken over the side of the cylinder.

Collecting all terms, the three-dimensional law of charge conservation (2.6) becomes

$$\oint s \cdot n \, d_3 v = 0, \tag{2.7}$$

where the integral extends over any closed 3-space, with 'flat' bottom and top, and side parallel to the time axis, like the cylinder of Fig. 2.1.

In fact, (2.7) applies to a closed 3-space of any shape, as in Fig. 2.2. This is because we can approximate the integral over any region by a sum of integrals over vertical cylinders with horizontal ends and rectangular cross-sections. The contributions from *adjoining* sides will cancel in pairs, because their outward pointing normals are opposite. By the definition of the Riemann integral, this is sufficient for proving that (2.7) holds for a general closed 3-space in space-time.

Wherever $\operatorname{div} s$ exists, the local eqn (2.5) now follows from the four-dimensional analogue of Gauss's formula, which equates the triple integral $\oint s \cdot n \, d_3 v$ to a quadruple integral over $\operatorname{div} s$, just as the local three-dimensional eqn (1.14) followed from the global three-dimensional eqn (1.12). The proof of the

Time

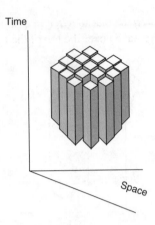

Space

Fig. 2.2

four-dimensional Gauss formula is based on the four-dimensional Stokes theorem, which we shall state in Section 4.

The principle of charge conservation as originally formulated refers to 'charge in a region', $\int q\, dV$, and 'charge that has crossed the boundary of a region during a time interval', $\int dt \oint \mathbf{j} \cdot \mathbf{n}\, dS$. In Fig. 2.1 these correspond to the top–bottom and to the side of the cylinder. The final generalization to a three-dimensional boundary of a four-dimensional region of any shape, as in Fig. 2.2, fuses the two kinds of charge into one—'charge in a three-dimensional subspace of four-dimensional space-time', namely

$$s \cdot n\, d_3 v. \tag{2.8}$$

It is only when the 4-normal in (2.8) is 'time-like' (vertical) or 'space-like' (horizontal) that this generalized charge has one of the original meanings.

Our next object is to require that the new, generalized, concept of charge have a meaning which transcends any particular choice of space-time coordinates, like the Euclidean frame in which (2.7) holds. In other words, we shall insist on the *invariance* of the generalized charge (2.8). This will, of course, guarantee the frame-invariance of the law of charge conservation (2.7).

We are thus led to consider coordinate transformations and, in particular, the consequences of an assumed invariance of an *integral*, or a piece of an integral, such as (2.8), under arbitrary transformations. This is necessary because, as shall see later, one of the principles of electromagnetism concerns a special coordinate frame. Without an understanding of the transformation properties of the electric quantities—such as charge and current densities (the components of s)—this principle cannot be appreciated. It cannot even be meaningfully stated.

In the next chapter we shall present a calculus which is an extension of three-dimensional vector analysis to spaces of an arbitrary number n of dimensions. In this calculus, there are differential operators called rot, curl and div, which are generalizations of the **grad, curl** and div of standard vector analysis. Finally, n-dimensional extensions will be given of the familiar Gauss and Stokes integration formulae, as well as of the theorem on the existence of a potential for a conservative field.

Problem 2.1   Let $\Sigma(x)$ and $\Sigma'(x')$ be two Cartesian frames connected by the *Galilean transformation* (Fig. 2.3)

$$\mathbf{x}' = \mathbf{x} - \mathbf{u}t, \qquad t' = t, \tag{2.9}$$

$\mathbf{u}$ being the *constant* velocity of the primed with respect to the unprimed frame. If (for any $\mathbf{u}$) observers in Σ and Σ' agree on any 'charge in a region' and any 'charge that has crossed a surface during a time interval', how are $s = (\mathbf{j}, q)$ and $s' = (\mathbf{j}', q')$ connected?

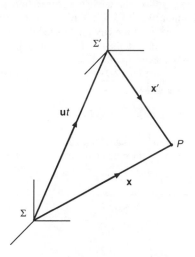

Fig. 2.3

Solution   Agreement on any 'charge in a volume' means $\int q \, dV = \int q' \, dV'$, where the integrals extend over the same spatial domain. But the Galilean transformation (2.9) preserves distances, and hence volumes. Thus, if we insist on agreement for any domain, however small, the two equations $q' \, dV' = q \, dV$ and $dV' = dV$ result in $q' = q$.

Agreement on the second kind of charge, namely 'charge crossing a surface during a time interval', is not so trivial. Although the transformation preserves areas ($dS' = dS$), angles ($\mathbf{n}' = \mathbf{n}$) and time intervals ($dt' = dt$), it would be wrong to use $\delta t' \mathbf{j}' \cdot \mathbf{n}' \, dS' = \delta t \mathbf{j} \cdot \mathbf{n} \, dS$ in order to deduce $\mathbf{j}' = \mathbf{j}$. This is because the same surface cannot be stationary in both frames, and motion of a surface through a charged region involves a transfer of charge across the surface, beside the one due to $\mathbf{j}$. We must therefore allow for this motion-induced current, at least in one of the frames.

Let $\mathbf{v}$ be the velocity of the surface element in the unprimed frame. During δt, dS sweeps (Fig. 2.4) over a volume $\delta V = \mathbf{n} \, dS \cdot \mathbf{v} \delta t$. This volume, which has the sign of $v_n = \mathbf{v} \cdot \mathbf{n}$, is transferred from the positive side of dS (the side to which $\mathbf{n}$ points) to the negative side. In order to calculate the net amount of charge crossing dS in the direction $\mathbf{n}$, a charge $q \delta V$

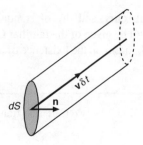

Fig. 2.4

must therefore be subtracted from $\delta t \mathbf{j} \cdot \mathbf{n}\,dS$. The net amount of charge crossing, in the unprimed frame, the directed surface element $\mathbf{n}\,dS$ during δt is thus $\delta t\,(\mathbf{j} - q\mathbf{v}) \cdot \mathbf{n}\,dS$. If $\mathbf{v}'$ is the velocity of the same surface element in the primed frame, the net amount of charge crossing in that frame is, similarly, $\delta t'\,(\mathbf{j}' - q'\mathbf{v}') \cdot \mathbf{n}'\,dS'$. The correct deduction is therefore $\mathbf{j}' - q'\mathbf{v}' = \mathbf{j} - q\mathbf{v}$ or, since we have already established that $q' = q$, $\mathbf{j}' = \mathbf{j} - q(\mathbf{v} - \mathbf{v}')$. But according to (2.9) the velocities in the two frames are related by the Galilean velocity composition rule $\mathbf{v}' = \mathbf{v} - \mathbf{u}$. Hence, finally, $\mathbf{j}' = \mathbf{j} - q\mathbf{u}$.

   Under the Galilean transformation (2.9), then, charge invariance leads to the following transformation formulae for q and $\mathbf{j}$:

$$q' = q, \qquad \mathbf{j}' = \mathbf{j} - q\mathbf{u}. \tag{2.10}$$

Thus charge density is a Galilean invariant, but current density is not. In Chapter 3 we shall derive these formulae (cf. (5.8)) by the much more efficient method of tensor transformation rules.

2

The calculus of antisymmetric tensors

3. Transformations and tensors

A point in n-dimensional space is specified by the n-dimensional array $(x^i) = (x^1, x^2, \ldots, x^n)$. In this chapter we shall assume that all Latin superscripts and subscripts range over the integers from 1 to n. We shall also use the summation convention: a repeated index implies summation—from 1 to n—over that index.

In n dimensions, a line is defined by n functions $x^r(u)$ $(r = 1, \ldots, n)$ of a parameter u. Each point on the line has n coordinates x^r, but the line is a one-dimensional subspace—a 1-*space*—because these coordinates are determined by the values of a single parameter. The line element corresponding to a change du of u is the displacement dx, with components

$$dx^r = \frac{dx^r}{du} du. \tag{3.1}$$

Let A be a collection $A_r(x)$ of n functions (the 'components of A'). Consider the line integral

$$\int A_r \, dx^r. \tag{3.2}$$

It is easily seen that the value of this integral is independent of the choice of parameter along the line. But it does usually depend on the line, and on the coordinates used. We wish to arrange things so that it may depend on the former, but not on the latter. Let

$$x^{r'} = x^{r'}(x^1, \ldots, x^n) \tag{3.3}$$

be an analytic transformation from the coordinates x^r to a new set of coordinates $x^{r'}$. The first-order changes in the coordinates will then be related by the chain rule

$$dx^{r'} = \frac{\partial x^{r'}}{\partial x^r} dx^r. \tag{3.4}$$

If the mapping from the unprimed to the primed coordinates is to be one-to-one, the matrix of the coefficients on the right-hand side must not be singular. Its determinant J, called the *Jacobian*, must not be allowed to vanish. Thus J must be positive everywhere, or negative everywhere, in the region under consideration.

Let $\epsilon_{i_1 i_2 \cdots i_n}$ denote the permutation symbol, defined as follows:

$$\epsilon_{i_1 i_2 \cdots i_n} = 0 \quad \text{unless } \{i_1, i_2, \ldots, i_n\} \text{ is a permutation of } \{1, 2, \ldots, n\};$$
$$\epsilon_{i_1 i_2 \cdots i_n} = +1 \quad \text{if } \{i_1, i_2, \ldots, i_n\} \text{ is an even permutation of } \{1, 2, \ldots, n\};$$
$$\epsilon_{i_1 i_2 \cdots i_n} = -1 \quad \text{if } \{i_1, i_2, \ldots, i_n\} \text{ is an odd permutation of } \{1, 2, \ldots, n\}.$$

By recalling the manner in which a determinant is formed from the matrix elements we then have

$$\epsilon_{i'_1 i'_2 \cdots i'_n} \frac{\partial x^{i'_1}}{\partial x^{i_1}} \frac{\partial x^{i'_2}}{\partial x^{i_2}} \cdots \frac{\partial x^{i'_n}}{\partial x^{i_n}} = J \epsilon_{i_1 i_2 \cdots i_n}. \tag{3.5}$$

Exercises

3.1. The permutation symbol $\epsilon^{i_1 i_2 \cdots i_n}$ is defined as numerically equal to $\epsilon_{i_1 i_2 \cdots i_n}$. Show that

$$\epsilon^{i_1 i_2 \cdots i_n} \epsilon_{i_1 i_2 \cdots i_n} = n!.$$

3.2. Show that, in three dimensions, the components of the vector product $\mathbf{c} = \mathbf{a} \times \mathbf{b}$ are given by

$$c^r = \epsilon^{rst} a_s b_t.$$

3.3. Show that, in three dimensions, the components of $\mathbf{a} = \mathbf{curl}\, \mathbf{b}$ are given by

$$a^r = \epsilon^{rst} \partial_s b_t.$$

Let us now return to the line integral (3.2). The question is, what are the $A_{r'}(x')$ in the 'primed frame', if the line integral is to be invariant? Since we require invariance for any line, however short, it is clear that $A_r dx^r$ must be invariant. Now, since $dx^r = (\partial x^r / \partial x^{r'}) dx^{r'}$,

$$A_r dx^r = A_r \frac{\partial x^r}{\partial x^{r'}} dx^{r'},$$

and if this is to equal $A_{r'} dx^{r'}$ for any $dx^{r'}$ (any line element), we must have

$$A_{r'}(x') = \frac{\partial x^r}{\partial x^{r'}} A_r(x) \tag{3.6}$$

as the transformation law for A.

The reader who is familiar with Lagrangian mechanics may be reminded at this point of the transformation formula for the components Q_r of the generalized force: if the sum $Q_r dq^r$, which is the work done in a virtual displacement of the generalized coordinates q^r, is to be invariant under an analytical transformation from the q^r to a new set of generalized coordinates $q^{r'}$, then the new $Q_{r'}$ are related to the Q_r by an equation similar to (3.6), $Q_{r'} = (\partial q^r / \partial q^{r'})Q_r$, and for the same reason.

A surface is a 2-*space*: the coordinates of points on the surface are functions $x^r(u^1, u^2)$ of two parameters. If we change u^1, keeping u^2 fixed, we obtain a displacement $d_1 x$ with components $d_1 x^r = (\partial x^r / \partial u^1)du^1$. Similarly, if we change u^2, keeping u^1 fixed, we have a displacement $d_2 x$ with components $d_2 x^r = (\partial x^r / \partial u^2)du^2$. The two displacements $d_1 x$ and $d_2 x$ 'lie on the surface', because each one connects two points on the surface. They are not parallel: if they were, the two parameters would describe a *line* on the surface (a 1-space). The parallelogram formed by the two displacements has the 'area projections'

$$d\tau_{(2)}^{rs} = d_1 x^r d_2 x^s - d_1 x^s d_2 x^r = \left(\frac{\partial x^r}{\partial u^1}\frac{\partial x^s}{\partial u^2} - \frac{\partial x^s}{\partial u^1}\frac{\partial x^r}{\partial u^2}\right)du^1 du^2 \qquad (3.7)$$

on the 'coordinate planes' (in a three-dimensional Cartesian frame these are the components of the vector product $d_1 \mathbf{x} \times d_2 \mathbf{x}$). Each of these 'projections' is a determinant. There are $n \times n$ such $d\tau_{(2)}^{rs}$, but since this double array is antisymmetric, it has only $\frac{1}{2}n(n-1)$ independent members.

For any array $a^{rsu\cdots}$ with M running indices, we shall indicate complete antisymmetrization by square brackets: $a^{[rsu\cdots]}$. This is done by permuting the indices in all possible ways, attaching a positive (negative) sign to the terms corresponding to even (odd) permutations, adding these terms and dividing by the number of terms $M!$. Parentheses, $a^{(rsu\cdots)}$, indicate complete symmetrization with respect to the enclosed indices. This is done in the same way, except that the positive sign is attached to all permutations. For example,

$$d_1 x^r d_2 x^s = d_1 x^{[r} d_2 x^{s]} + d_1 x^{(r} d_2 x^{s)}, \qquad (3.8)$$

$$d\tau_{(2)}^{rs} = 2! \, d_1 x^{[r} d_2 x^{s]} = 2! \frac{\partial x^{[r}}{\partial u^1}\frac{\partial x^{s]}}{\partial u^2} du^1 du^2. \qquad (3.9)$$

Let A_{rs} now be an array of $n \times n$ functions, and consider the surface integral

$$\int \frac{1}{2} A_{rs} \, d\tau_{(2)}^{rs}, \qquad (3.10)$$

taken over a surface. Since $d\tau_{(2)}$ is antisymmetric, the integral depends only on the antisymmetric part $A_{[rs]}$ of A. We lose nothing by considering only antisymmetric A's in (3.10). Since, in this case, $A_{rs} \, d\tau_{(2)}^{rs}$ and $A_{sr} \, d\tau_{(2)}^{sr}$ are equal, it makes sense to introduce the factor $\frac{1}{2}$. Again, we ask: if the integral (3.10) is to be invariant, what is the transformation law for the antisymmetric A_{rs}? We must first establish the transformation law for the $d\tau_{(2)}$'s, but this is easy: according to (3.7), $d\tau_{(2)}^{rs}$ is a difference of products of differentials; hence (cf. (3.4))

$$d\tau_{(2)}^{r's'} = \frac{\partial x^{r'}}{\partial x^r}\frac{\partial x^{s'}}{\partial x^s} d\tau_{(2)}^{rs}. \tag{3.11}$$

It now follows, by an argument which is similar to the one that has led to (3.6), that the surface integral (3.10) will be invariant (for any surface element) if

$$A_{r's'} = \frac{\partial x^r}{\partial x^{r'}}\frac{\partial x^s}{\partial x^{s'}} A_{rs}. \tag{3.12}$$

In an n-dimensional space with $n > 3$ we have, in addition to lines and surfaces, *hypersurfaces*, which are M-dimensional $(2 < M < n)$, and n-spaces, which are regions of the n-dimensional space itself. Generally, for an M-space $(M \leq n)$, we consider an M-cell with edges $d_1 x, d_2 x, \ldots, d_M x$ and define its *extension*[†] by

$$d\tau_{(M)}^{i_1 i_2 \cdots i_M} = M!\, d_1 x^{[i_1} d_2 x^{i_2} \cdots d_M x^{i_M]}$$

$$= M!\, \frac{\partial x^{[i_1}}{\partial u^1}\frac{\partial x^{i_2}}{\partial u^2} \cdots \frac{\partial x^{i_M]}}{\partial u^2} du^1 du^2 \cdots du^M. \tag{3.13}$$

Of course $d\tau_{(1)}^i = dx^i$ for a 1-space (a line). The extension of an M-cell depends on the order $d_1 x d_2 x \cdots d_M x$. An exchange of u^1 and u^2 (say) will change the signs of all components of $d\tau_{(M)}$. The ordering of the parameters is called the *orientation* of the M-cell. Thus the extension depends on the size of the M-cell (the magnitudes of the du's) and on its orientation.

As a sum of products of differentials, the extension obeys the transformation law

$$d\tau_{(M)}^{i'_1 \cdots i'_M} = \frac{\partial x^{i'_1}}{\partial x^{i_1}} \cdots \frac{\partial x^{i'_M}}{\partial x^{i_M}} d\tau_{(M)}^{i_1 \cdots i_M}. \tag{3.14}$$

For an antisymmetric $A_{i_1 i_2 \cdots i_M}$, the integral

$$\int \frac{1}{M!} A_{i_1 \cdots i_M}\, d\tau_{(M)}^{i_1 \cdots i_M}, \tag{3.15}$$

taken over an M-space, will be invariant if

$$A_{i'_1 \cdots i'_M} = \frac{\partial x^{i_1}}{\partial x^{i'_1}} \cdots \frac{\partial x^{i_M}}{\partial x^{i'_M}} A_{i_1 \cdots i_M}. \tag{3.16}$$

Quantities that transform like a product of differentials, according to the transformation law (3.14), are called *contravariant tensors*. Those that transform like the A of

---

[†]The concept is of some age: it was introduced, for a space with any number of dimensions, by Grassmann, in a book entitled *Ausdehnungslehre*, in 1844.

(3.16) are called *covariant tensors*. The transformation formulae have a neatness about them that makes them easier to remember than the names 'contravariant' (for upper indices) and 'covariant' (for lower indices). Of course our choice of upper indices for the coordinates, which might have appeared idiosyncratic at first, is now seen to be consistent with the chain rule (3.4).

A function $g(x)$, with no indices, which transforms according to $g' = g$ (this means $g(x') = g(x)$), is called a *scalar*. The temperature at an event is an example of a scalar, because it does not depend on the coordinate labels of the event.

The number of indices of a tensor is its *rank*. The A_r of (3.6) is a covariant tensor of rank one, or of the first rank; the A_{rs} of (3.12) is a second rank tensor. A scalar is a tensor of rank zero.

The tensor transformations (3.14) and (3.16) are linear and homogeneous in the tensor components. It follows from the homogeneity that, if the components of a tensor all vanish in one frame, they will do so in any other frame. If A and B are tensors of the same type—both contravariant, or both covariant, and with the same number of indices (same rank)—then, by the linearity of the tensor transformations (3.14) or (3.16), the sum or difference $A \pm B$ is again a tensor of the same type. Thus a tensor equation, that is, an equality like $A^{rs\cdots} = B^{rs\cdots}$, or like $A_{rs\cdots} = B_{rs\cdots}$, between two tensors of the same type, is either true in all frames, or in none.

Exercises

The following exercises contain some easy theorems on tensors.

3.4. If $A_{r'} = (\partial x^r / \partial x^{r'}) A_r$ and $A_{r''} = (\partial x^{r'} / \partial x^{r''}) A_{r'}$, then $A_{r''} = (\partial x^r / \partial x^{r''}) A_r$. This is the transitive property of the tensor transformation law.

3.5. If A_{rs} is a tensor then the symmetric part $A_{(rs)}$ and the antisymmetric part $A_{[rs]}$ are both tensors. Thus the tensor transformation preserves symmetry.

3.6. If a_r is covariant and b^s is contravariant, then the sum $a_r b^r$ is a scalar.

3.7. If the sum $A_s X^s$ is a scalar for arbitrary contravariant X^r, then A_r is a covariant tensor. This statement is *the quotient theorem* for tensors.

3.8. If u^r and v^r are contravariant tensors, then $A^{rs} = u^r v^s$ is a second rank, contravariant tensor.

3.9. If ϕ is a scalar, then $A_r = \partial_r \phi$ is a covariant tensor. This is no more than a restatement of the chain rule of the differential calculus.

An *antisymmetric* tensor with M indices is called an *M-vector*. For any covariant M-vector A and any contravariant M-vector B, we shall denote

$$A \cdot B = \frac{1}{M!} A_{i_1 \cdots i_M} B^{i_1 \cdots i_M}. \tag{3.17}$$

With this notation, the integral (3.15) is $\int A \cdot d\tau_{(M)}$.

The standard three-dimensional vector calculus does not distinguish between covariant and contravariant vectors. That is because it deals with Cartesian frames, connected by orthogonal transformations of the form $\mathbf{x}' = A(\mathbf{x} - \mathbf{a})$, where A is a constant, orthogonal matrix and $\mathbf{a}$ is a constant translation vector. The inverse transformation is $\mathbf{x} = A^T\mathbf{x}' + \mathbf{a}$, where A^T denotes the transpose of A. For such transformations, $\partial x^{i'}/\partial x^i$ and $\partial x^i/\partial x^{i'}$ are equal (to the matrix element $A_{i'i}$), so that the distinction between contravariant and covariant tensors disappears.

4. Integral theorems and dual tensors

If F is a covariant M-vector, its *rotation* is defined by

$$(\text{rot } F)_{rs_1...s_M} = (M+1)\, \partial_{[r} F_{s_1...s_M]}. \tag{4.1}$$

Note that rot rot $F = 0$, because $\partial_{r_1}\partial_{r_2} - \partial_{r_2}\partial_{r_1} = 0$. The *curl* of an M-vector will be defined later; it is related, but not identical, to the rotation. The rotation of a scalar χ (a 0-vector, with no indices), is $(\text{rot }\chi)_r = \partial_r\chi$, an n-dimensional generalization of the gradient of a scalar.

The *generalized Stokes theorem* states: let R_M be the complete closed boundary of the $(M+1)$-dimensional region R_{M+1}; then if F is any covariant M-vector which is continuously differentiable in R_{M+1}, we have

$$\oint_{R_M} F \cdot d\tau_{(M)} = \int_{R_{M+1}} \text{rot } F \cdot d\tau_{(M+1)}, \tag{4.2}$$

provided the extension $d\tau_{(M+1)}$ is constructed from the $d\tau_{(M)}$ of (3.13) by adding a $d_{M+1}x$, which points *out* of R_M, as a *first* factor. Otherwise a minus sign will appear on one side of (4.2). We need not be concerned by this question of orientation, because the conclusions we shall draw from (4.2) will not depend on this.

Stokes's generalized theorem[†] is really a form of the formula for integration by parts . It holds independently of the invariance of the integral on the left-hand side of (4.2), and therefore does not require F to be an M-vector. In (4.2) F may be any antisymmetric M-component symbol, with rot F defined by (4.1). But if F is an M-vector, the left-hand side is invariant and so, by the theorem, is the integral on the right-hand side. Hence if F is an M-vector then rot F is an $(M+1)$-vector.

A covariant M-vector F which satisfies

$$\oint F \cdot d\tau_{(M)} = 0, \tag{4.3}$$

where the integral is taken over the complete boundary of *any* $(M+1)$-dimensional region, is called a *conservative M-vector field*. For such fields the following theorem

[†] Due to Poincaré (1877).

holds: there exists at least one $(M-1)$-vector field V, called *a potential* of F, such that

$$\int_{R_M} F \cdot d\tau_{(M)} = \oint_{R_{M-1}} V \cdot d\tau_{(M-1)} \qquad (4.4)$$

for every R_M, where R_{M-1} is the complete boundary of R_M. The field V is not uniquely determined by F and (4.4): any conservative $(M-1)$-vector field can be added to V, and the new potential will still satisfy (4.4). For $M = 1$ the theorem (4.4) takes the form

$$\int_{x_1}^{x_2} F \cdot d\tau_{(1)} = V(x_2) - V(x_1).$$

Wherever the conservative field F and its potential V are smooth we can use Stokes's theorem (4.2) to obtain the local forms of the foregoing statements: a conservative M-vector field is irrotational, that is, it satisfies

$$\operatorname{rot} F = 0; \qquad (4.5)$$

a potential V, an $(M-1)$-vector field, then exists such that

$$F = \operatorname{rot} V; \qquad (4.6)$$

for any smooth $(M-2)$-vector S,

$$\tilde{V} = V + \operatorname{rot} S \qquad (4.7)$$

is another potential of F, and we have $F = \operatorname{rot} \tilde{V}$ because $\operatorname{rot} \operatorname{rot} S = 0$.

Since $\operatorname{rot} \operatorname{rot} V = 0$, any $F = \operatorname{rot} V$ is irrotational, and thus conservative. The theorem on the existence of a potential tells us that, conversely, if $\operatorname{rot} F = 0$, there exists a V such that $F = \operatorname{rot} V$.

In three-dimensional space, the surface extension (cf. (3.7)) is

$$d\tau_{(2)}^{ij} = d_1 x^i d_2 x^j - d_1 x^j d_2 x^i. \qquad (4.8)$$

In a Cartesian frame we can form, out of the parallelogram defined by the two surface displacements $d_1 \mathbf{x}$ and $d_2 \mathbf{x}$, a directed surface element $\mathbf{n}\,dS$ by constructing the vector product $d_1 \mathbf{x} \times d_2 \mathbf{x}$. For any two vectors $\mathbf{a}$ and $\mathbf{b}$, the i'th component of $\mathbf{a} \times \mathbf{b}$ is $a^j b^k \epsilon_{jki}$, where ϵ_{jki} is the three-dimensional permutation symbol (cf. Exercise 3.2). Thus

$$n_i\,dS = d_1 x^j d_2 x^k \epsilon_{jki} = \tfrac{1}{2} d\tau_{(2)}^{jk} \epsilon_{jki}. \qquad (4.9)$$

This motivates (in n dimensions) the definition of the *dual* of a contravariant M-vector F as the antisymmetric $(n-M)$-symbol given by

$$(\text{dual } F)_{r_1 r_2 \dots r_{n-M}} = \frac{1}{M!} F^{s_1 s_2 \dots s_M} \epsilon_{s_1 s_2 \dots s_M r_1 r_2 \dots r_{n-M}}; \qquad (4.10)$$

and of the dual of a covariant M-vector G, as the antisymmetric $(n - M)$-symbol given by

$$(\text{dual } G)^{r_1 r_2 \dots r_{n-M}} = \frac{1}{M!} \epsilon^{r_1 r_2 \dots r_{n-M} s_1 \dots s_M} G_{s_1 s_2 \dots s_M}. \tag{4.11}$$

The permutation symbol $\epsilon^{i_1 i_2 \dots}$, with upper indices, is defined as numerically equal to $\epsilon_{i_1 i_2 \dots}$. It follows that for any type of M-vector we have

$$\text{dual dual } F = F. \tag{4.12}$$

We shall also use the notation

$$\hat{F} = \text{dual } F. \tag{4.13}$$

An M-vector and its $(n - M)$-dual have the same number of independent components, only differently arranged. For example, for $M = 3$ in four dimensions, $\hat{F}_1 = F^{243}$, $\hat{F}_2 = F^{341}$, $\hat{F}_3 = F^{142}$, $\hat{F}_4 = F^{123}$. Equation (4.12) assures us that nothing is lost by passing from an antisymmetric symbol to its dual, because the process is reversible. In choosing between an M-vector and its dual we usually prefer the symbol with the smaller number of indices.

If F is a contravariant M-vector, and G a covariant M-vector, the following identity holds:

$$\text{dual } F \cdot \text{dual } G = (-)^{(n-M)M} F \cdot G. \tag{4.14}$$

The sign factor is due to the different positions of the summation indices $s_1 s_2 \dots s_M$ in the permutation symbols in (4.10)–(4.11). In a space with an odd number n of dimensions (e.g. $n = 3$) it is always $+1$.

According to (4.14), the dual of the integral (3.15) is

$$(-)^{(n-M)M} \int \hat{A} \cdot d\hat{\tau}_{(M)}. \tag{4.15}$$

We have called the duals defined by (4.10)–(4.11) *symbols* because their transformation properties must still be determined. In order to do this, we recall the formula (3.5) for the Jacobian, which has the form of a transformation formula for the permutation symbol, except that it contains the factor J. It is a straightforward matter to use (3.5) in order to establish the transformation formulae for the duals defined by (4.10)–(4.11). In the case of $d\hat{\tau}_{(M)}$, the result is

$$d\hat{\tau}_{(M)i'_1 \dots i'_{n-M}} = J \frac{\partial x^{i_1}}{\partial x^{i'_1}} \cdots \frac{\partial x^{i_{n-M}}}{\partial x^{i'_{n-M}}} d\hat{\tau}_{(M)i_1 \dots i_{n-M}}. \tag{4.16}$$

For the integral (4.15) to be invariant, it is thus necessary that

$$\hat{A}^{i'_1 \dots i'_{n-M}} = J^{-1} \frac{\partial x^{i'_1}}{\partial x^{i_1}} \cdots \frac{\partial x^{i'_{n-M}}}{\partial x^{i_{n-M}}} \hat{A}^{i_1 \dots i_{n-M}}. \tag{4.17}$$

The difference between (4.16) and (3.16), or between (4.17) and (3.14), is in the common J factor, or its reciprocal. Quantities that transform in accordance with (4.16)–(4.17) are called *relative* covariant or contravariant M-tensors (relative M-vectors, if they are skew-symmetric). In order to emphasize the difference, tensors that are *not* relative, which transform according to (3.14) or (3.16), are also called *absolute*.

Since the common J or J^{-1} factor does not alter the linearity or the homogeneity of the tensor transformations, the remarks following (3.16) on tensor equations, between tensors *of the same type*, apply to relative tensors as well: any equality such as $A^{rs\cdots} = B^{rs\cdots}$, or $A_{rs\cdots} = B_{rs\cdots}$, between two tensors of the same type—the word 'type' now including 'absolute' or 'relative', in addition to 'contravariant' or 'covariant' and 'rank'—is either true in all frames, or in none.

If F is a covariant M-vector, its *curl* is defined as

$$\text{curl } F = \text{dual rot } F. \tag{4.18}$$

The identity curl rot $F = $ dual rot rot $F = 0$ is a consequence of rot rot $= 0$. For a scalar χ, curl rot $\chi = 0$ is a generalization of the three-dimensional identity **curl grad** $\chi = 0$ (cf. the remarks following (4.1)).

According to (4.15) and (4.18), the dual form of Stokes's formula (4.2) is

$$\oint_{R_M} \hat{F} \cdot d\hat{\tau}_{(M)} = (-)^{n+1} \int_{R_{M+1}} \text{curl } F \cdot d\hat{\tau}_{(M+1)}. \tag{4.19}$$

Thus, if F is a covariant M-vector, then curl F is a relative $(n - M - 1)$-vector.

If G is a relative contravariant M-vector, its *divergence* is defined by

$$(\text{div } G)^{r_1 \cdots r_{M-1}} = \partial_s G^{r_1 \cdots r_{M-1} s}. \tag{4.20}$$

Equation (2.5) is a special case for $M = 1$. Since an M-vector is antisymmetric, the identity div div $G = 0$ holds. From the definitions (4.11), (4.18) and (4.20) it is easy to prove the identity

$$\text{curl } F = \text{div dual } F. \tag{4.21}$$

Thus, the identity div curl $F = 0$ is a consequence of div $\overset{\cdot}{\text{div}} = 0$.

According to (4.21), the dual (4.19) of Stokes's formula can also be written in the form

$$\oint_{R_M} \hat{F} \cdot d\hat{\tau}_{(M)} = (-)^{n+1} \int_{R_{M+1}} \text{div } \hat{F} \cdot d\hat{\tau}_{(M+1)}. \tag{4.22}$$

Thus, if G ($= \hat{F}$) is a relative contravariant M-vector, then div G is a relative $(M - 1)$-vector.

Exercises

4.1. Prove the identity (4.14).

4.2. Using the transitive property of Jacobian determinants, formulate appropriate extensions to relative tensors of the theorems in Exercises 3.4–3.8.

4.3. Using the three-dimensional analogues of (4.10)–(4.11), show that, for a 1-vector F, the curl F of (4.18) has the same components as the **curl** usually defined in three-dimensional vector analysis.

4.4. Show that (4.18) and (4.21) are equivalent.

4.5. By forming the duals of (4.3) and (4.4), show that if G is (dual-) conservative in the sense that

$$\oint_{R_M} G \cdot d\hat{\tau}_{(M)} = 0, \tag{4.23}$$

where the integral extends over the complete boundary R_M of *any* R_{M+1}, then a W exists such that

$$\int_{R_M} G \cdot d\hat{\tau}_{(M)} = (-)^{n+1} \oint_{R_{M-1}} W \cdot d\hat{\tau}_{(M-1)}. \tag{4.24}$$

Of course W is not uniquely determined by the last equation.

Wherever G and W are smooth, the dual (4.22) of Stokes's theorem can be used to obtain local forms of these statements: a G that is (dual-) conservative in the sense of (4.23) is solenoidal, that is,

$$\text{div } G = 0; \tag{4.25}$$

a (dual-) potential W, an $[n - (M - 1)]$-vector, then exists such that

$$G = \text{div } W; \tag{4.26}$$

for any $[n - (M - 2)]$-vector T,

$$\tilde{W} = W + \text{div } T \tag{4.27}$$

is another (dual-) potential for G, and we have $G = \text{div } \tilde{W}$ because div div $T = 0$.

Since div div $W = 0$, any $G = \text{div } W$ is solenoidal, and thus (dual-) conservative in the sense of (4.23). The theorem (4.24) tells us that, conversely, if div $G = 0$, there exists a W such that $G = \text{div } W$.

3

The first pair of Maxwell's equations

5. Invariance of the principle of charge conservation

Our motivation for introducing the duals was the expression (4.9) for the directed surface element $\mathbf{n}dS$ in a Cartesian three-dimensional space. According to (4.9), $\mathbf{n}dS$ is equal to the three-dimensional dual of $d\tau_{(2)}$. The analogous relation, in Euclidean space-time, is

$$n d_3 v = d\hat{\tau}_{(3)}, \tag{5.1}$$

where $d_3 v$ is the volume of an element of hypersurface and n is its unit normal 4-vector. Actually, in order to ensure that n be outward-pointing, and that the 3-volume element $d_3 v$ be positive, we must choose a 'correct' orientation for $d\tau_{(3)}$. Otherwise, eqn (5.1) will hold with a minus sign before one of the members.

We have already shown that the law of charge conservation in Euclidean space-time is (cf. (2.7))

$$\oint s \cdot n \, d_3 v = 0. \tag{5.2}$$

We can now use (5.1) in order to arrive at the general form[†] of the first principle of electromagnetism: for any closed 3-space,

$$\oint_{R_3} s \cdot d\hat{\tau}_{(3)} = 0. \tag{5.3}$$

Alternatively: the charge-current density s is conservative. In (5.3) the choice of orientation does not matter, because minus zero equals zero.

If s is continuously differentiable, it must be solenoidal (cf. (4.25)), that is, it must satisfy the equation $\operatorname{div} s = 0$ (cf. (2.5)). On a surface across which $s = (\mathbf{j}, q)$ is discontinuous, the integral laws (5.2) or (5.3) lead to the jump condition $\mathbf{n} \cdot (\mathbf{j}_+ - \mathbf{j}_-) - v_n(q_+ - q_-) = 0$ (cf. (1.15)).

[†]Due to Bateman (1910).

In order to free ourselves of the shackles of Euclidean space-time, we simply note that extensions and their duals are defined for any type of coordinates—Euclidean or otherwise. Let us therefore require that the product-sum

$$s \cdot d\hat{\tau}_{(3)} \tag{5.4}$$

be an *invariant*. This requires the charge-current density s to be a relative 1-vector, with the transformation law (4.17), and guarantees the invariance of the law (5.3). But a problem arises in connection with transformations that have a negative Jacobian; for example, transformations that involve an odd number of reflections. Such transformations change the orientation of $d\tau_{(3)}$, and destroy (5.1). The quantity which we really wish to keep invariant is the infinitesimal, generalized charge (cf. (2.8)),

$$s \cdot n \, d_3 v. \tag{5.5}$$

The remedy is to require that

$$s' \cdot d\hat{\tau}'_{(3)} = \pm s \cdot d\hat{\tau}_{(3)},$$

the sign being that of the Jacobian of the transformation. Since the product of J and its sign is the absolute value $|J|$, the result is the replacement of J^{-1} in (4.17) by $|J|^{-1}$. The invariance of the charge (5.5), and hence of the law of charge conservation (5.2), thus leads to the transformation formula

$$s^{\alpha'} = |J|^{-1} \frac{\partial x^{\alpha'}}{\partial x^{\alpha}} s^{\alpha}. \tag{5.6}$$

Quantities with the transformation law (5.6), which differs from (4.17) by having the Jacobian in absolute value, are called *tensor densities*.

The name 'density' derives from the mass density ρ, defined such that ρdV is the mass contained in a volume element dV. If we transform the spatial coordinates, $dV' = |J| dV$. Therefore the invariance of the mass $\rho' dV' = \rho dV$ requires $\rho' = |J|^{-1} \rho$. Thus ρ transforms as a *scalar density*.

Exercise

5.1. Formulate the appropriate extensions involving tensor densities of the theorems in Exercises 3.4–3.8.

As an example, we consider the *Galilean transformation* which, according to classical mechanics, connects any two inertial, Cartesian frames:

$$x^{r'} = A_r^{r'} (x^r - u^r x^4), \quad x^{4'} = x^4, \tag{5.7}$$

u being the *constant* velocity of the primed with respect to the unprimed frame, and A a constant orthogonal matrix. This transformation corresponds to Fig. 2.3, except that the primed frame Σ' is no longer parallel to Σ: it has undergone a rotation, and perhaps some reflections, all of which are determined by the orthogonal matrix A. For this simple transformation, $|J| = |\det A| = 1$. The formulae (5.6) give

$$s^{r'} = A^{r'}_r (s^r - u^r s^4), \qquad s^{4'} = s^4.$$

The first three of these equations state that the components $j^{r'} = s^{r'}$ of $\mathbf{j}'$ are those of $\mathbf{j} - q\mathbf{u}$, referred to the primed axes. In the standard notation of vector calculus, the transformation formulae for the charge and current densities under a Galilean transformation are therefore

$$q' = q, \qquad \mathbf{j}' = \mathbf{j} - q\mathbf{u}. \tag{5.8}$$

Thus charge density is a Galilean invariant, but current density is not. When charge-carrying material is moving with a velocity $\mathbf{v}$, there is a *convection current density* $q\mathbf{v}$. If this is subtracted from the total current density, the result,

$$\mathcal{J} = \mathbf{j} - q\mathbf{v}, \tag{5.9}$$

is called the *conduction current density*. This is a Galilean invariant, for $\mathcal{J}' - \mathcal{J} = q(\mathbf{v} - \mathbf{v}' - \mathbf{u})$, which vanishes because $\mathbf{v}' = \mathbf{v} - \mathbf{u}$ is the Galilean transformation law for velocities (the classical 'velocity composition' formula).

Finally, we observe that if the word 'type' is now extended to include tensor densities, in addition to absolute and relative tensors, then it is still true that a tensor equation between tensors of the same type either holds in all frames, or in none.

6. The charge-current potential

According to the principle of charge conservation (5.3) the 1-vector density s is conservative (cf. (4.23)). According to (4.24) a 2-vector potential f exists such that

$$\int_{R_3} s \cdot d\hat{\tau}_{(3)} = -\oint_{R_2} f \cdot d\hat{\tau}_{(2)}. \tag{6.1}$$

We call f a *charge-current potential*. It is not uniquely determined by s. Wherever f is smooth (cf. (4.26)),

$$s = \operatorname{div} f. \tag{6.2}$$

In a Euclidean space-time, it is customary to name the components of the charge-current potential in accordance with the following scheme:

$$f = \begin{pmatrix} 0 & H_3 & -H_2 & -D^1 \\ -H_3 & 0 & H_1 & -D^2 \\ H_2 & -H_1 & 0 & -D^3 \\ D^1 & D^2 & D^3 & 0 \end{pmatrix}. \tag{6.3}$$

Alternatively,

$$f^{4r} = D^r, \qquad f^{rs} = \epsilon^{rst} H_t. \tag{6.4}$$

Of course the mere fact that we denote the components of f in this way does not mean that D^r is a contravariant tensor, or that H_r is a covariant tensor. A subset of the components of a tensor is not itself a tensor, but it may transform as a tensor (of lower rank) with respect to transformations in a subspace. It is then said to constitute a *subtensor*. Problem 6.2 shows that, for a restricted class of transformations, D^r and H_r are indeed subtensors.

With the names (6.4) for the Euclidean components of f, and $s = (\mathbf{j}, q)$, the eqns (6.2) become

$$q = \operatorname{div} \mathbf{D},$$
$$\mathbf{j} = \operatorname{curl} \mathbf{H} - \frac{\partial \mathbf{D}}{\partial t}. \tag{6.5}$$

These constitute the first pair of Maxwell's equations. We call $\mathbf{D}$ the *charge potential*, and $\mathbf{H}$ the *current potential*. It is clear that the dimensions of $\mathbf{D}$ and $\mathbf{H}$ are, respectively, C/L^2 and $C/(LT)$.

We have already mentioned that the potential f is not uniquely determined by the charge-current density s. It is indeed evident from (6.5) that, for any vector field $\mathbf{a}(\mathbf{x}, t)$ and any function $\psi(\mathbf{x}, t)$, the potentials $\mathbf{D}$ and $\mathbf{H}$ and the potentials

$$\tilde{\mathbf{D}} = \mathbf{D} + \operatorname{curl} \mathbf{a}, \qquad \tilde{\mathbf{H}} = \mathbf{H} + \frac{\partial \mathbf{a}}{\partial t} + \operatorname{grad} \psi \tag{6.6}$$

correspond to the same charge and current densities q and $\mathbf{j}$.

The non-uniqueness presents no problem when we use Maxwell's equations (6.5) to calculate the charge and current densities q and $\mathbf{j}$ from *given* potentials $\mathbf{D}$ and $\mathbf{H}$. This is indeed the way in which these equations are used in material media. As we shall see in Section 20, the response functions of materials specify the charge-current potentials, and the equations (6.5) are then used to calculate the so-called polarization and magnetization charge and current densities.

Since we have required $s' \cdot d\hat{\tau}'_{(3)} = \pm s \cdot d\hat{\tau}_{(3)}$ under transformations, we must, according to (6.1), require $f' \cdot d\hat{\tau}'_{(2)} = \pm f \cdot d\hat{\tau}_{(2)}$. Thus f must be a 2-vector density, with the transformation formula

$$f^{\alpha'\beta'} = |J|^{-1} \frac{\partial x^{\alpha'}}{\partial x^\alpha} \frac{\partial x^{\beta'}}{\partial x^\beta} f^{\alpha\beta}. \tag{6.7}$$

Problem 6.1 Show that, with respect to the subset of space transformations $x^{r'} = x^{r'}(x^r)$, $x^{4'} = x^4$, D^r is a three-dimensional tensor density and H_r is an axial covariant three-dimensional tensor (an axial tensor undergoes a sign change whenever J is negative).

Solution From the transformation law (6.7) of the tensor density f,

$$D^{r'} = f^{4'r'} = |J|^{-1}\frac{\partial x^{4'}}{\partial x^4}\frac{\partial x^{r'}}{\partial x^r}f^{4r} = |J|^{-1}\frac{\partial x^{r'}}{\partial x^r}D^r,$$

$$\epsilon^{r's't'}H_{t'} = f^{r's'} = |J|^{-1}\frac{\partial x^{r'}}{\partial x^r}\frac{\partial x^{s'}}{\partial x^s}f^{rs} = |J|^{-1}\frac{\partial x^{r'}}{\partial x^r}\frac{\partial x^{s'}}{\partial x^s}\epsilon^{rst}H_t.$$

Use of $\epsilon^{rst} = \epsilon^{rsu}\delta_u^t$, the chain rule $(\partial x^{t'}/\partial x^u)(\partial x^t/\partial x^{t'}) = \delta_u^t$ and the determinant formula (3.5) gives

$$H_{t'} = \mathrm{sign}(J)\frac{\partial x^t}{\partial x^{t'}}H_t.$$

Problem 6.2 Derive the Galilean transformation formulae for **D** and **H**.

Solution For the Galilean transformation (5.7), $|J| = 1$, and

$$D^{r'} = f^{4'r'} = \frac{\partial x^{4'}}{\partial x^\alpha}\frac{\partial x^{r'}}{\partial x^\beta}f^{\alpha\beta} = \frac{\partial x^{4'}}{\partial x^4}\frac{\partial x^{r'}}{\partial x^r}f^{4r} = A_r^{r'}D^r.$$

Thus the components of $\mathbf{D}'$ are those of $\mathbf{D}$, referred to the primed axes. Next,

$$\epsilon^{r's't'}H_{t'} = f^{r's'}$$

$$= \frac{\partial x^{r'}}{\partial x^\alpha}\frac{\partial x^{s'}}{\partial x^\beta}f^{\alpha\beta}$$

$$= A_r^{r'}A_s^{s'}(\epsilon^{rst}H_t - u^r D^s + u^s D^r)$$

$$= A_r^{r'}A_s^{s'}\epsilon^{rst}(\mathbf{H} - \mathbf{u}\times\mathbf{D})_t$$

$$= A_r^{r'}A_s^{s'}\epsilon^{rsu}\delta_u^t(\mathbf{H} - \mathbf{u}\times\mathbf{D})_t.$$

Use of the orthogonality property $AA^T = I$ (δ_u^t is the element of the unit matrix I) and the determinant formula (cf. (3.5)) $A_r^{r'}A_s^{s'}A_u^{u'}\epsilon^{rsu} = (\det A)\epsilon^{r's'u'}$ results in

$$H_{t'} = (\det A)(A^T)_{t'}^t(\mathbf{H} - \mathbf{u}\times\mathbf{D})_t.$$

Thus the components of $\mathbf{H}'$ are those of $\mathbf{H} - \mathbf{u}\times\mathbf{D}$, referred to the primed axes and multiplied by the common factor $\det A$. In the standard notation of vector calculus,

$$\mathbf{D}' = \mathbf{D}, \qquad \mathbf{H}' = (\det A)(\mathbf{H} - \mathbf{u}\times\mathbf{D}).$$

For a *proper* Galilean transformation $\det A = +1$, and the transformation formulae become

$$\mathbf{D}' = \mathbf{D}, \qquad \mathbf{H}' = \mathbf{H} - \mathbf{u}\times\mathbf{D}. \tag{6.8}$$

According to (6.8) the charge potential $\mathbf{D}$ is a Galilean invariant. The current potential $\mathbf{H}$ is not, but if $\mathbf{v}$ is a material velocity, a Galilean invariant can be constructed (cf. (5.9)) as follows:

$$\mathcal{H} = \mathbf{H} - \mathbf{v}\times\mathbf{D}. \tag{6.9}$$

Exercises

6.1. Show that the first pair (6.5) of Maxwell's equations results when the relations (6.4) are substituted in (6.2).

6.2. Show that

$$\tfrac{1}{2} f \cdot \text{dual } f = -\mathbf{D} \cdot \mathbf{H}.$$

In many cases electric charge may be distributed over a surface, and electric current may flow on a surface. Formally, such cases correspond to infinite q or $\mathbf{j}$, and the local differential equations (6.2) or (6.5) then become meaningless. If, however, these infinite q or $\mathbf{j}$ are *integrable*, we can return to the integral equation (6.1), which remains meaningful. It relates an integral of s over a 3-space to the integral of f over its complete 2-boundary, and leads to *two* kinds of integral formulae, according to whether the 3-space lies in a hyperplane of four-dimensional space (Fig. 6.1), or not (Fig. 6.2).

It is easier, however, to obtain the required integral formulae by re-integrating the local Maxwell's equations (6.5), using the familiar three-dimensional integration theorems of Gauss and Stokes. If we integrate $(6.5)_1$ over some volume and use Gauss's theorem to transform the right-hand side to an integral over the bounding surface, we obtain

$$\int q \, dV = \oint D_n \, dS, \tag{6.10}$$

where D_n is the component of $\mathbf{D}$ in the direction of the outward unit normal $\mathbf{n}$. The charge in any volume is therefore equal to the flux of $\mathbf{D}$ through the boundary surface.

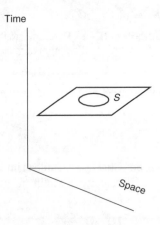

Fig. 6.1

Equation (6.10) could also have been obtained by direct application of (6.1) to a 3-space, with closed boundary S, lying in the hyperplane $t = $ const., as in Fig. 6.1.

We now calculate the flux of either side of $(6.5)_2$ through a fixed, open, surface. Use of Stokes's theorem[†] gives

$$\int j_n \, dS = \oint_c \mathbf{H} \cdot \mathbf{ds} - \frac{d}{dt} \int D_n \, dS. \qquad (6.11)$$

The line integral on the right is taken along the curve c that constitutes the boundary of the open surface, in the sense that bears to the direction of the unit normal $\mathbf{n}$ the same relation that a rotation of the x-axis into the y-axis bears to the z-axis; otherwise a minus sign has to be inserted before the line integral in (6.11). This line integral is called the *circulation* of $\mathbf{H}$ around c. In the second term on the right we have used the fact that the surface is fixed in order to place the time differentiation in front of the integral.

Another integration, over time from t_1 to t_2, gives

$$\int_{t_1}^{t_2} dt \int j_n \, dS = \int_{t_1}^{t_2} dt \oint_c \mathbf{H} \cdot \mathbf{ds} - \int_{(t_2)} D_n \, dS + \int_{(t_1)} D_n \, dS. \qquad (6.12)$$

Equation (6.12) could also have been obtained by applying (6.1) to the *history* of an open surface S, with boundary c, lying in the hyperplane $t = $ const., as in Fig. 6.2. This history of a 2-space S is a 3-space, bounded at the bottom and top by 2-spaces $S(t_1)$ and $S(t_2)$, and on the side by the history of c, a 2-space as well.

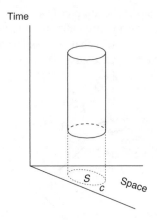

Fig. 6.2

[†]The correct attribution is to Kelvin.

7. The jump conditions

We shall now use the integral laws[†] (6.10) and (6.12) in order to deduce the jump conditions that hold on a surface of discontinuity.

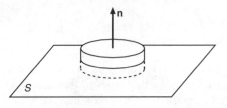

Fig. 7.1

In Fig. 7.1, S is a surface of discontinuity of $\mathbf{D}$. Let $\mathbf{n}$ be a unit normal to the surface, chosen in any one of the two possible ways (up in Fig. 7.1). We shall refer to the side into which $\mathbf{n}$ is pointing as the positive side of S, and to the other side as the negative one. Accordingly, we denote the value of $\mathbf{D}$ on the positive side by $\mathbf{D}_+$, and on the negative side by $\mathbf{D}_-$. Around a piece δS of the surface we erect a pillbox with generators parallel to $\mathbf{n}$. We now apply (6.10) to the pillbox and assume that the surface carries a *surface charge density* (charge per unit area) σ. If we then pass to the limit in which the height of the pillbox vanishes, the contributions from space charges (if any) on either side of S will vanish, and we shall obtain

$$\mathbf{n} \cdot [\![\mathbf{D}]\!] = \sigma, \tag{7.1}$$

where $[\![\mathbf{D}]\!] = \mathbf{D}_+ - \mathbf{D}_-$ denotes the jump of $\mathbf{D}$ across S.

It should be noted that the jump condition (7.1) holds independently of the choice of $\mathbf{n}$: the alternative choice will change the signs of both $\mathbf{n}$ and $[\![\mathbf{D}]\!]$, and will therefore not affect their product. Of course, $\mathbf{D}_n$ will be continuous if, and only if, the surface charge density σ vanishes.

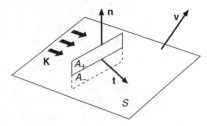

Fig. 7.2

[†]We refrain at this stage from identifying (6.10) with Gauss's law, or (6.12) with Ampère's law. That is because $\mathbf{D}$ and $\mathbf{H}$—the components of the charge-current potential f—are as yet unrelated to the electric and magnetic fields $\mathbf{E}$ and $\mathbf{B}$. These fields and this relation are the subjects of two further principles of electromagnetism, which we shall introduce in the next two chapters.

In Fig. 7.2 we have a *fixed* rectangle in a plane perpendicular to S. The orientation of the plane is defined by its normal $\mathbf{t}$, which is, of course, parallel to the surface. We apply (6.12) to the rectangle and pass to the limit in which it collapses on to S.

Consider, first, the current $\int \mathbf{j} \cdot \mathbf{t} \, dS$ passing through the rectangle as the latter collapses on to a segment Δl on S. We define the *surface current density* $\mathbf{K}$ in such a way that the amount of charge flowing per unit time through the *line segment* Δl, in the direction of the unit normal $\mathbf{t}$, is $\mathbf{K} \cdot \mathbf{t} \Delta l$. Thus $\mathbf{K} \cdot \mathbf{t} \Delta l$ is the surface analogue of $\mathbf{j} \cdot \mathbf{n} \, dS$. In the limit, during a time interval δt, we shall obtain $\delta t (\mathbf{K} \cdot \mathbf{t}) \Delta l$ on the left-hand side of (6.12). On the right-hand side, the first term gives $\delta t (\mathbf{t} \times \mathbf{n}) \cdot [\![\mathbf{H}]\!] \Delta l = \delta t \mathbf{n} \times [\![\mathbf{H}]\!] \cdot \mathbf{t} \Delta l$. The last two terms on the right-hand side of (6.12) will cancel, unless the surface S is moving through the fixed rectangle. Each of these terms is a sum

$$\int_{(t)} \mathbf{D} \cdot \mathbf{t} \, dA = (\mathbf{D}_+ \cdot \mathbf{t}) A_+ + (\mathbf{D}_- \cdot \mathbf{t}) A_-, \tag{7.2}$$

where A_+ and A_- are the areas of the rectangle on the positive and negative sides of S. Let $\mathbf{v}$ be the velocity of the surface, so that during a short time δt it is displaced a distance $v_n \delta t$ in the direction of $\mathbf{n}$. Of course v_n may be negative. This displacement reduces A_+ by $v_n \delta t \Delta l$, and increases A_- by the same amount. Hence

$$\int_{(t+\delta t)} \mathbf{D} \cdot \mathbf{t} \, dA = (\mathbf{D}_+ \cdot \mathbf{t})(A_+ - v_n \delta t \Delta l) + (\mathbf{D}_- \cdot \mathbf{t})(A_- + v_n \delta t \Delta l). \tag{7.3}$$

For the two last terms of (6.12) we need the difference between (7.3) and (7.2), which is $v_n \delta t [\![\mathbf{D}]\!] \cdot \mathbf{t} \Delta l$. Collecting all terms, and noting that $\mathbf{t}$ is arbitrary, we get

$$\mathbf{n} \times [\![\mathbf{H}]\!] + v_n [\![\mathbf{D}]\!] = \mathbf{K}. \tag{7.4}$$

Again, this jump condition holds independently of the choice of $\mathbf{n}$, because the jumps, as well as v_n, will reverse their signs along with $\mathbf{n}$.

From now on we shall assume that all relevant quantities are piecewise smooth. By this we mean that they are continuously differentiable up to any desired order everywhere, except on surfaces of discontinuity; and that the number of such surfaces in any closed subdomain of space is finite. For piecewise smooth quantities the integral laws (6.10) and (6.12) are everywhere meaningful. Wherever the integrands are smooth ('between' surfaces of discontinuity), the laws are correctly given by their local forms (6.5). On a surface of discontinuity, we have the conditions (7.1) and (7.4), which provide restrictions on the jumps that the charge and current potentials may suffer.

To sum up, the first principle of electromagnetism—the principle of electric charge conservation—has the invariant form

$$\oint_{R_3} s \cdot d\hat{\tau}_{(3)} = 0. \tag{7.5}$$

That is, the charge-current density s is conservative. Therefore a charge-current potential f exists such that the invariant equation

$$\int_{R_3} s \cdot d\hat{\tau}_{(3)} = -\oint_{R_2} f \cdot d\hat{\tau}_{(2)} \tag{7.6}$$

holds. For piecewise smooth potentials these equations are equivalent to Maxwell's equations

$$\text{div } \mathbf{D} = q,$$
$$\mathbf{curl\ H} = \mathbf{j} + \frac{\partial \mathbf{D}}{\partial t}, \tag{7.7}$$

together with the jump conditions

$$\mathbf{n} \cdot [\![\mathbf{D}]\!] = \sigma,$$
$$\mathbf{n} \times [\![\mathbf{H}]\!] + v_n [\![\mathbf{D}]\!] = \mathbf{K}. \tag{7.8}$$

The student need only memorize the pair (7.7) of Maxwell's equations. The more general integral laws can be easily constructed from them by use of Gauss's and Stokes's theorems; and the jump conditions follow directly from the integral laws. The law of charge conservation, too, is an immediate consequence of Maxwell's equations: take the divergence of $(7.7)_2$ and eliminate div $\mathbf{D}$. The result, eqn (1.14), can be integrated over a volume to yield the original statement (1.12).

It is only when we wish to transform to another frame that we need to recall that $\mathbf{j}$ and q are components of a single four-dimensional vector density s, and that $\mathbf{D}$ and $\mathbf{H}$ are components of a single four-dimensional 2-vector density f.

Obviously, the first pair of Maxwell's equations is equivalent to the local form of the law of charge conservation. What has been gained by replacing the simple equation div $s = 0$, involving the four functions s^α, with the equation $s = \text{div } f$, which requires the additional six components of the 2-vector f? The answer is, nothing at *this* stage. The charge-current potential f owes its importance to the second principle of electromagnetism (Chapter 4), which introduces another 2-vector F, and to the third principle (Chapter 5), which connects f and F.

Equations that have the form div $s = 0$ will appear in this book in two other contexts (cf. (45.1) and (51.12)). Each one of them can of course be replaced by a 'first pair of Maxwell's equations' $s = \text{div } f$, where f is a suitable 2-vector 'potential'. But in the absence of any additional principle regarding these 'potentials', their introduction would be for purely mathematical purposes.

4

The second pair of Maxwell's equations

8. The electromagnetic field

The first principle of electromagnetism was based on a primitive—the electric charge—which was subject to a conservation law. Its formulation did not take account of any other phenomena that we commonly associate with charges and currents: there was no mention of like charges repelling, or of unlike charges attracting each other, or of wires heating up whenever a current passes through them. Historically, of course, these phenomena led to the discovery of charge conservation.

The second principle of electromagnetism, similarly, involves another primitive concept—the electromagnetic field. It postulates the existence of this field and lays down the equations that it satisfies, but it does not tell us what the field is supposed to do. In particular, it does *not* identify the electromagnetic field as a force field. That aspect, as we shall see, not only requires a separate postulate concerning the link between electromagnetism and mechanics; it may also involve the properties of materials.

We assume the existence of two (primitive) vector fields, the electric field $\mathbf{E}$ and the magnetic field $\mathbf{B}$, which satisfy the following two integral laws:

$$\oint B_n \, dS = 0,$$

$$\int_{(t_2)} B_n \, dS - \int_{(t_1)} B_n \, dS = -\int_{t_1}^{t_2} dt \oint_c \mathbf{E} \cdot \mathbf{ds}. \tag{8.1}$$

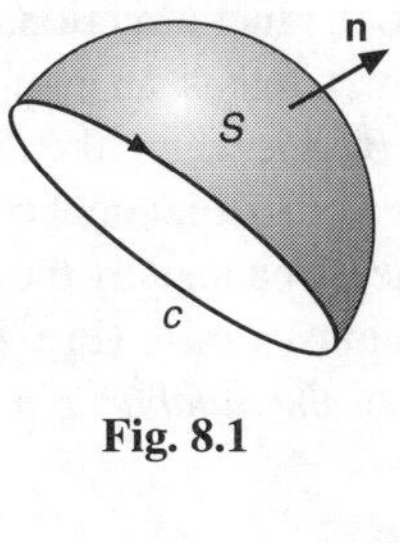

Fig. 8.1

The second of these equations, in which the relation between **n** and the sense of integration along c is the same as in (6.11), has the form of a conservation law for the flux of **B** through a *fixed, open* surface, as in Fig. 8.1: this flux changes when something happens on the boundary c of the surface, that 'something' being minus the circulation of **E**.

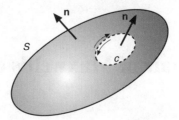

Fig. 8.2

We can always divide any *closed* surface S into two parts by drawing an arbitrary closed curve c on the surface, as in Fig. 8.2. If we apply the conservation law $(8.1)_2$ to the parts of S on either side of c, the two circulations, which have opposite senses, cancel exactly. Thus, according to $(8.1)_2$, the flux of **B** through a fixed, closed surface is constant. The first equation, $(8.1)_1$, tells us that this constant is zero. These considerations suggest that it is quite natural to regard the *pair* of eqns (8.1) as a single (second) principle of electromagnetism, which expresses the conservation of magnetic flux.

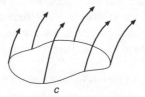

Fig. 8.3

Let us now consider any closed curve c in a region in which **B** does not vanish. We draw, as in Fig. 8.3, the field lines of **B** that pass through every point of c. These lines form a tube, which Faraday called a *tube of force*. The tube is everywhere parallel to the direction of **B**. It is clear that two field lines, or two tubes of force, can never meet, for that would involve two distinct directions of **B** at the common point.

If S is any surface spanned by c, that is, an open surface which has the boundary c, we can form the flux $\Phi = \int B_n \, dS$, where the direction of **n** is chosen such that $B_n = \mathbf{B} \cdot \mathbf{n} > 0$. This positive flux is independent of S, since the difference between the fluxes through any two surfaces spanned by the same curve c is equal to the flux through a closed surface, which must vanish (Fig. 8.2). Thus Φ depends only on c, and is called *the flux through c*, or *the number of magnetic field lines through c*.

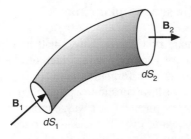

Fig. 8.4

Let S_1 and S_2 be two sections (not necessarily plane) along a tube of force, as in Fig. 8.4. Since the total flux of **B** out of the volume formed in this way must vanish, and since the wall of the tube, where $B_n = 0$, does not contribute to this outward flux, we conclude that $\int \mathbf{n} \cdot \mathbf{B}_1 \, dS_1 = \int \mathbf{n} \cdot \mathbf{B}_2 \, dS_2$. Thus the flux is the same through any closed curve c drawn around the tube, and remains constant along the tube. It is called *the strength of the tube*. Faraday's tubes of force are nowadays called *flux tubes*.

The laws (8.1) were discovered by Faraday, who of course used various effects of **E** and **B** that we are deliberately ignoring at this stage. We shall therefore refer to them as *Faraday's laws*. Their mathematical statement in the form (8.1) is due to Kelvin and Maxwell.

The dimensions of **E** and **B** are of course connected by $(8.1)_2$. If we denote the dimension of magnetic flux $\int B_n \, dS$ by Φ, the dimensions of **B** are Φ/L^2 and those of **E** are $\Phi/(LT)$.

If the fields are smooth (differentiable), the laws (8.1) have the local forms

$$\operatorname{div} \mathbf{B} = 0,$$
$$\frac{\partial \mathbf{B}}{\partial t} = -\operatorname{\mathbf{curl}} \mathbf{E}. \tag{8.2}$$

This is the second pair of Maxwell's equations. Whenever the fields suffer discontinuities, we expect restrictions in the form of jump conditions. We could derive these by following the method which led us from the integral laws (6.10) and (6.12) to the jump conditions (7.8). It is simpler, however, to note that Faraday's laws can be (formally) obtained from the integral laws (6.10) and (6.12) by replacing **H** by **E**, and **D** by $-\mathbf{B}$, and setting $q = \mathbf{j} = 0$. By making the same substitutions in (7.8)—this also entails the omission of the surface charge and surface current—we obtain

$$\mathbf{n} \cdot [\![\mathbf{B}]\!] = 0,$$
$$\mathbf{n} \times [\![\mathbf{E}]\!] - v_n [\![\mathbf{B}]\!] = 0. \tag{8.3}$$

9. Invariance of the second principle

Following the programme of Chapter 3, we shall now seek an invariant four-dimensional formulation of Faraday's laws, which of course refer to a Euclidean frame. This will lead to the transformation laws for **B** and **E**. Since, as we have already observed, Faraday's laws constitute a principle of conservation of magnetic flux, we shall also discover a potential. In a way, our present task is simpler, because the laws (8.1) involve integrals over 2-spaces—a surface S or the history of a line c. The law of electric charge conservation involved a 3-space.

We begin by constructing an antisymmetric 2-index symbol F according to the following scheme:

$$F = \begin{pmatrix} 0 & B^3 & -B^2 & E_1 \\ -B^3 & 0 & B^1 & E_2 \\ B^2 & -B^1 & 0 & E_3 \\ -E_1 & -E_2 & -E_3 & 0 \end{pmatrix}. \tag{9.1}$$

Alternatively,

$$F_{r4} = E_r, \qquad F_{rs} = \epsilon_{rst} B^t. \tag{9.2}$$

The remarks following (6.4) apply to the positions of the indices of **B** and **E** in these relations. Note the difference in the orders of the indices r and 4 in (9.2) and (6.4).

We shall now consider the double integral (cf. (3.10))

$$\int_{R_2} F \cdot d\tau_{(2)} = \int_{R_2} \tfrac{1}{2} F_{\alpha\beta} \, d\tau_{(2)}^{\alpha\beta}, \tag{9.3}$$

taken over suitable 2-spaces R_2 in Euclidean space-time. First, let R_2 be a *closed* surface in the hyperplane $t = $ const., as in Fig. 9.1 (cf. Fig. 6.1). Since all points on this surface have the same x^4, the displacements $d_m x$ have vanishing fourth

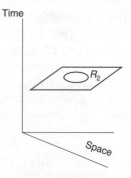

Fig. 9.1

components. For this R_2, then, the $d\tau_{(2)}^{\alpha\beta}$ with $\alpha = 4$ or $\beta = 4$ will all vanish, and the double sum in (9.3) is actually over Latin indices:

$$\tfrac{1}{2} F_{ij} \, d\tau_{(2)}^{ij} = \tfrac{1}{2} \epsilon_{ijk} B^k \, 2! d_1 x^i d_2 x^j = \epsilon_{ijk} B^k \, d_1 x^i d_2 x^j, \qquad (9.4)$$

but this is simply the triple product $\mathbf{B} \cdot (d_1\mathbf{x} \times d_2\mathbf{x})$, or $\mathbf{B} \cdot \mathbf{n} dS$. From Faraday's law $(8.1)_1$, we conclude that, for any closed R_2 in the hyperplane $t = \text{const.}$,

$$\oint_{R_2} F \cdot d\tau_{(2)} = 0. \qquad (9.5)$$

Now let R_2 be the closed 2-space corresponding to the history of an *open* surface S, with boundary curve c, as in Fig. 9.2 (cf. Fig. 6.2). The contributions to the integral (9.3) from top and bottom (hyperplanes $t = t_2$ and $t = t_1$) will be the first two terms of Faraday's law $(8.1)_2$; the minus sign before the second term is due to the $\mathbf{n}$ in $(8.1)_2$, which is out of the cylinder at the top, and *into* the cylinder at the bottom.

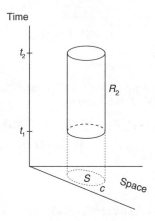

Fig. 9.2

As for the side, we may choose, as the two parameters for surface integration, the parameter u^1 along the curve c, and $u^2 = t$. In Fig. 9.2, u^1 runs horizontally, and u^2 vertically. In the antisymmetric surface extension $d\tau_{(2)}^{\alpha\beta}$, one of the indices *must* now be 4, and therefore the second index Latin. The integrand of (9.3) for the side of the cylinder is therefore

$$\tfrac{1}{2} \left(F_{r4} d\tau_{(2)}^{r4} + F_{4r} d\tau_{(2)}^{4r} \right) = F_{r4} d\tau_{(2)}^{r4};$$

both F and $d\tau_{(2)}$ change sign when their indices are interchanged. According to the definition (3.9) of the surface extension,

$$d\tau_{(2)}^{r4} = (dx^r / du^1) du^1 du^2 = dx^r dt.$$

Since, by (9.2), $F_{r4} = E_r$, we get precisely the integral on the right-hand side of $(8.1)_2$. Thus, (9.5) holds for our cylinder. This is sufficient for stating that (9.5) holds for *any* closed 2-space in space-time (cf. Fig. 2.2). We shall regard this result[†] as the general statement of the second principle of electromagnetism: F is conservative.

Since the extension can be defined in any coordinate system, we can guarantee the general invariance of the principle (9.5) by requiring that the product-sum

$$F \cdot d\tau_{(2)} \tag{9.6}$$

be an *invariant*. Then F—which we shall from now on call *the electromagnetic field*—becomes an absolute, covariant 2-vector, with the transformation law (cf. (3.12))

$$F_{\alpha'\beta'} = \frac{\partial x^\alpha}{\partial x^{\alpha'}} \frac{\partial x^\beta}{\partial x^{\beta'}} F_{\alpha\beta}. \tag{9.7}$$

Problem 9.1   Prove that

$$f \cdot F = \mathbf{H} \cdot \mathbf{B} - \mathbf{D} \cdot \mathbf{E}.$$

This is a scalar density.

Solution   The charge-current potential f is a tensor density, and the electromagnetic field F an absolute tensor. Hence $f \cdot F$ is a scalar density and

$$f \cdot F = \tfrac{1}{2} f^{\alpha\beta} F_{\alpha\beta} = \tfrac{1}{2} \left(f^{rs} F_{rs} + f^{4r} F_{4r} + f^{r4} F_{r4} \right) = \tfrac{1}{2} f^{rs} F_{rs} - f^{4r} F_{r4}$$
$$= \tfrac{1}{2} \epsilon^{rst} H_t \epsilon_{rsu} B^u - D^r E_r = \mathbf{H} \cdot \mathbf{B} - \mathbf{D} \cdot \mathbf{E}.$$

Problem 9.2   Derive the Galilean transformation formulae for $\mathbf{E}$ and $\mathbf{B}$.

Solution   In matrix notation, the Galilean transformation (5.7) is $x' = A(x - ux^4)$, $x^{4'} = x^4$. Since A is orthogonal, the unprimed coordinates are given, in terms of the primed ones, by $x = A^T x' + ux^{4'}$, $x^4 = x^{4'}$. Now the transformation formulae can be obtained directly from those of the covariant tensor F. It is much easier, however, to start from the result of the last problem. Since $\mathbf{H} \cdot \mathbf{B} - \mathbf{D} \cdot \mathbf{E}$ is a scalar density, and the Galilean transformation has $|J| = 1$,

$$\mathbf{H'} \cdot \mathbf{B'} - \mathbf{D'} \cdot \mathbf{E'} = (\det A)(\mathbf{H} - \mathbf{u} \times \mathbf{D}) \cdot \mathbf{B'} - \mathbf{D} \cdot \mathbf{E'} = \mathbf{H} \cdot \mathbf{B} - \mathbf{D} \cdot \mathbf{E},$$

where, in the first equality, we have used the Galilean transformation formulae (Problem 6.2) for $\mathbf{H}$ and $\mathbf{D}$. The last equality must hold identically, that is, for all $\mathbf{H}$ and $\mathbf{D}$. This yields the two equations

$$(\det A)\mathbf{B'} = \mathbf{B}, \qquad (\det A)\mathbf{u} \times \mathbf{B'} - \mathbf{E'} = -\mathbf{E}.$$

[†] Due to Hargreaves (1908).

Solving for $\mathbf{B}'$ and $\mathbf{E}'$, and remembering that det $A = \pm 1$, we get

$$\mathbf{B}' = (\det A)\mathbf{B}, \qquad \mathbf{E}' = \mathbf{E} + \mathbf{u} \times \mathbf{B}.$$

For a *proper* Galilean transformation det $A = +1$, and the transformation formulae become

$$\mathbf{B}' = \mathbf{B}, \qquad \mathbf{E}' = \mathbf{E} + \mathbf{u} \times \mathbf{B}. \tag{9.8}$$

The magnetic field $\mathbf{B}$ is a Galilean invariant (under proper Galilean transformations). The electric field $\mathbf{E}$ is not, but if $\mathbf{v}$ is a material velocity, a Galilean invariant can be constructed (cf. (6.9)) as follows:

$$\mathcal{E} = \mathbf{E} + \mathbf{v} \times \mathbf{B}. \tag{9.9}$$

This is called the *electromotive intensity*.

According to the generalized Stokes theorem (4.2), if F is smooth,

$$\oint_{R_2} F \cdot d\tau_{(2)} = \int_{R_3} \text{rot } F \cdot d\tau_{(3)}. \tag{9.10}$$

We conclude that F must satisfy the tensor equation

$$\text{rot } F = 0. \tag{9.11}$$

In a Euclidean frame, with the identifications (9.2), the four independent components of (9.11) become Maxwell's second pair (8.2).

Exercises

9.1. Show that

$$\tfrac{1}{2} F \cdot \text{dual } F = \mathbf{B} \cdot \mathbf{E}.$$

9.2. Show that in a Euclidean frame, in which the components of F are given by (9.2), rot $F = 0$ reduces to the second pair of Maxwell's equations.

---

10. The electromagnetic potential

According to the theorem (4.3)–(4.5), since the 2-vector F is conservative, there exists a covariant 1-vector A, an *electromagnetic potential*, such that for every R_2 and its complete boundary R_1

$$\int_{R_2} F \cdot d\tau_{(2)} = \oint_{R_1} A \cdot d\tau_{(1)} = \oint_{R_1} A_\alpha \, d\tau_{(1)}^\alpha. \tag{10.1}$$

Wherever A is smooth,

$$F = \text{rot } A, \quad \text{or} \quad F_{\alpha\beta} = \partial_\alpha A_\beta - \partial_\beta A_\alpha. \tag{10.2}$$

The electromagnetic potential A is not uniquely determined by F and (10.1). For any smooth function χ,

$$\tilde{A} = A + \text{rot } \chi, \quad \text{or} \quad \tilde{A}_\alpha = A_\alpha + \partial_\alpha \chi \tag{10.3}$$

is another potential for the same electromagnetic field F, because $\text{rot}^2 \chi$ vanishes identically. Equation (10.3) is called a *gauge transformation* of A. The 1-vector A is also called a *four-potential*.

Since A is a covariant 1-vector, its transformation law is (3.6). For example, under a Galilean transformation,

$$A_{r'} = A_r, \quad A_{4'} = A_4 + u^r A_r. \tag{10.4}$$

In a Euclidean frame, it is customary to denote

$$(A_r) = \mathbf{A}, \quad A_4 = -V. \tag{10.5}$$

The three-dimensional vector field $\mathbf{A}$ is called the *magnetic potential*, or the *vector potential*, and the function V, the *electric potential*. In terms of these, and the fields $\mathbf{B}$ and $\mathbf{E}$, eqns (10.2) are

$$\mathbf{B} = \text{curl } \mathbf{A},$$

$$\mathbf{E} = -\frac{\partial \mathbf{A}}{\partial t} - \text{grad } V. \tag{10.6}$$

The fields given by these equations evidently satisfy Maxwell's equations (8.2). With the notation (10.5), the gauge transformation (10.3) is

$$\tilde{\mathbf{A}} = \mathbf{A} + \text{grad } \chi, \quad \tilde{V} = V - \frac{\partial \chi}{\partial t}. \tag{10.7}$$

Again, it is clear that the fields (10.6) are unchanged when $\mathbf{A}$ and V are replaced by $\tilde{\mathbf{A}}$ and $\tilde{V}$.

Consider, for example, the potentials

$$\mathbf{A} = 0, \quad V(\mathbf{r}) = \frac{1}{4\pi \epsilon_0} \frac{e}{r}, \tag{10.8}$$

where ϵ_0 and e are constants. According to (10.6), these correspond to the fields

$$\mathbf{B} = 0, \quad \mathbf{E}(\mathbf{r}) = \frac{1}{4\pi \epsilon_0} \frac{e}{r^3} \mathbf{r}. \tag{10.9}$$

A gauge transformation (10.7) with the function $\chi(\mathbf{r}, t) = Vt$ leads to the potentials

$$\tilde{\mathbf{A}}(\mathbf{r}, t) = -\frac{1}{4\pi\epsilon_0}\frac{et}{r^3}\mathbf{r}, \qquad \tilde{V} = 0. \tag{10.10}$$

Although this pair looks quite different from the pair (10.8), it corresponds to the same fields (10.9), as one can easily verify by using the potentials (10.10) in (10.6).

Finally, in terms of the magnetic potential $\mathbf{A}$ and the electric potential V, the Galilean transformation (10.4) becomes

$$\mathbf{A}' = \mathbf{A}, \qquad V' = V - \mathbf{u} \cdot \mathbf{A}. \tag{10.11}$$

Thus V is not a Galilean invariant. But if $\mathbf{v}$ is a material velocity, the difference $V - \mathbf{v} \cdot \mathbf{A}$ is a Galilean invariant (cf. (5.9)). It will be used for constructing a Lagrangian for a charged particle.

Problem 10.1   Let (r, ϕ, z) be cylindrical coordinates. An axisymmetric field $\mathbf{a}$ is called *toroidal* if it has a ϕ component only, that is, $\mathbf{a} = (0, a_\phi(r, z), 0)$. It is called *poloidal*, or *meridional*, if it has no ϕ component, that is, $\mathbf{a} = (a_r(r, z), 0, a_z(r, z))$. Show that, if the magnetic field $\mathbf{B}$ is toroidal, the vector potential $\mathbf{A}$ is poloidal; and that, for any poloidal $\mathbf{B}$, a gauge can be chosen such that $\mathbf{A}$ is toroidal.

Solution   In cylindrical coordinates

$$\mathbf{curl\ A} = \left(\frac{1}{r}\frac{\partial A_z}{\partial\phi} - \frac{\partial A_\phi}{\partial z}, \frac{\partial A_r}{\partial z} - \frac{\partial A_z}{\partial r}, \frac{1}{r}\frac{\partial}{\partial r}(rA_\phi) - \frac{1}{r}\frac{\partial A_r}{\partial\phi}\right).$$

For an axisymmetric $\mathbf{A}$

$$\mathbf{curl\ A} = \left(-\frac{\partial A_\phi}{\partial z}, \frac{\partial A_r}{\partial z} - \frac{\partial A_z}{\partial r}, \frac{1}{r}\frac{\partial}{\partial r}(rA_\phi)\right).$$

If $\mathbf{B} = \mathbf{curl\ A}$ is toroidal then

$$\frac{\partial A_\phi}{\partial z} = 0 \quad \text{and} \quad \frac{1}{r}\frac{\partial}{\partial r}(rA_\phi) = 0.$$

Unless A_ϕ is singular on the z axis, it must vanish. Thus $\mathbf{A}$ is poloidal.

If $\mathbf{B} = \mathbf{curl\ A}$ is poloidal then

$$\frac{\partial A_r}{\partial z} - \frac{\partial A_z}{\partial r} = 0.$$

This is the integrability condition for the existence of a single-valued function

$$\chi(P) = -\int_O^P (A_r\, dr + A_z\, dz),$$

defined by a line integral from an arbitrary, fixed point O. The gauge transformation $\tilde{\mathbf{A}} = \mathbf{A} + \mathbf{grad}\,\chi$ now leads to a toroidal $\tilde{\mathbf{A}}$.

Problem 10.2   According to the last problem an axisymmetric meridional (poloidal) magnetic field is derivable from a toroidal vector potential which, in cylindrical coordinates (r, ϕ, z), has the form $\mathbf{A} = (0, A_\phi(r, z), 0)$. In the plane $z = $ const. let c be the circle of radius r with centre on the axis. Show that the magnetic flux through c is $2\pi r A_\phi(r, z)$. Deduce that $r A_\phi(r, z) = $ const. is the equation of the magnetic field lines.

Solution   The flux through c is

$$\int_0^r B_z(r', z) 2\pi r' \, dr' = \int_0^r \frac{1}{r'} \frac{\partial}{\partial r'} [r' A_\phi(r', z)] 2\pi r' \, dr' = 2\pi r A_\phi(r, z).$$

The equation $2\pi r A_\phi(r, z) = $ const. therefore determines the locus of circles c around the z axis that enclose the same flux, that is, the boundary of a flux tube (Fig. 10.1). This boundary is a field line.

Fig. 10.1

To sum up, the second principle of electromagnetism has the invariant form

$$\oint_{R_2} F \cdot d\tau_{(2)} = 0. \tag{10.12}$$

That is, the electromagnetic field F is conservative. Therefore an electromagnetic potential A exists such that the invariant equation

$$\int_{R_2} F \cdot d\tau_{(2)} = \oint_{R_1} A \cdot d\tau_{(1)} \tag{10.13}$$

holds. For piecewise smooth potentials these equations are equivalent to the second pair of Maxwell's equations

$$\text{div } \mathbf{B} = 0,$$

$$\text{curl } \mathbf{E} = -\frac{\partial \mathbf{B}}{\partial t}, \tag{10.14}$$

together with the jump conditions

$$\mathbf{n} \cdot [\![\mathbf{B}]\!] = 0,$$
$$\mathbf{n} \times [\![\mathbf{E}]\!] - v_n [\![\mathbf{B}]\!] = 0,$$

(10.15)

and the equations relating the fields to the potentials,

$$\mathbf{B} = \mathbf{curl\ A},$$
$$\mathbf{E} = -\frac{\partial \mathbf{A}}{\partial t} - \mathbf{grad}\ V.$$

(10.16)

The student need only memorize the pair (10.14) of Maxwell's equations—along with the first pair (7.7), which constituted the concise expression of the first principle. The more general integral laws (8.1) can be easily constructed from (10.14) by use of Gauss's and Stokes's theorems; and the jump conditions follow directly from the integral laws.

It is only when one wishes to transform to another frame that one needs to recall that **B** and **E** are components of a single four-dimensional covariant, absolute 2-vector F, and that **A** and $(-V)$ are components of a single four-dimensional covariant, absolute 1-vector A. Such changes of frame will concern us in the next two sections. By the end of Section 12 the tensors f and F will have served their purpose, and will not be mentioned any more. The advanced student will encounter them again when he studies the general theory of relativity.

Exercises

10.1. Let $f^{\alpha\beta}$ be an arbitrary antisymmetric tensor density, and s^α an arbitrary vector density. If the scalar density $s \cdot A - f \cdot F$ is integrated over a four-dimensional region in Euclidean space-time, the result is the invariant

$$I = \int (s \cdot A - f \cdot F)\, d^4 x.$$

If $F = \mathrm{rot}\, A$, which is equivalent to assuming the second pair of Maxwell's equations, I becomes a functional $I[A]$ of the components of A. Prove that, if the A_α have fixed values on the boundary, the first pair of Maxwell's equations results from the requirement that $I[A]$ be stationary.

10.2. Show that, at a surface of discontinuity, the jump conditions on the electromagnetic potential are

$$\mathbf{n} \times [\![\mathbf{A}]\!] = 0, \quad [\![V]\!] - v_n \mathbf{n} \cdot [\![\mathbf{A}]\!] = 0,$$

(10.17)

where v_n is the normal velocity of the surface.

5

The aether relations and the theory of relativity

11. The Maxwell–Lorentz aether relations

Maxwell's equations (we use the notation $f_t = \partial f / \partial t$),

$$\operatorname{div} \mathbf{D} = q,$$
$$\operatorname{curl} \mathbf{H} - \mathbf{D}_t = \mathbf{j},$$
$$\operatorname{div} \mathbf{B} = 0,$$
$$\operatorname{curl} \mathbf{E} + \mathbf{B}_t = 0,$$

(11.1)

which are the local (differential) expressions (div $f = s$, rot $F = 0$) of the first and second principles of electromagnetism in a Euclidean frame, consist of two disjoint pairs.

We shall now introduce a third principle of electromagnetism. Its aim is to connect the hitherto independent first and second principles, and to create a link between electromagnetism and mechanics. Oddly enough, it connects, not the charge-current densities $\{\mathbf{j}, q\}$ and the electromagnetic field $\{\mathbf{B}, \mathbf{E}\}$; nor their potentials $\{\mathbf{H}, \mathbf{D}\}$ and $\{\mathbf{A}, V\}$; but the *potentials* $\{\mathbf{H}, \mathbf{D}\}$ and the *fields* $\{\mathbf{B}, \mathbf{E}\}$. Not only are the charge-current potentials thereby raised from an auxiliary to a primary status, they also lose the somewhat indefinite character in which they were left by the first principle, and become uniquely defined fields. But they gain this unique and elevated status by giving up their independence. The electromagnetic potentials $\{\mathbf{A}, V\}$, on the other hand, retain their auxiliary status: often a very convenient tool, and often dispensable.

Any formulation of a principle connecting $\{\mathbf{H}, \mathbf{D}\}$ with $\{\mathbf{B}, \mathbf{E}\}$ must take account of the fact that the two pairs of vector fields have different laws of transformation: the former are components of the charge-current potential f, a contravariant tensor density, whereas the latter are components of the electromagnetic field F, an absolute covariant tensor.

The situation is not unlike the one underlying the second law of mechanics, which connects mass, acceleration and force. Since mass and force are absolute— the same in all frames—and acceleration is relative, we cannot expect a relation such as $m\mathbf{a} = \mathbf{F}$ to hold generally. Even the existence of a single frame, called *inertial*, in which $m\mathbf{a} = \mathbf{F}$ *does* hold constitutes an axiom, and this axiom *is* the second law. Once this is assumed, it is clear that, since acceleration has a common value in all frames that are moving relative to each other at constant velocity, the inertial frame is not unique. Furthermore, it is clear that $m\mathbf{a} = \mathbf{F}$ will *not* hold in any frame which is not inertial: in such frames we have $m(\mathbf{a} + \Delta\mathbf{a}) = \mathbf{F}$ or $m\mathbf{a} = \mathbf{F} - m\Delta\mathbf{a}$; the last term is usually called a 'fictitious force'.

The third principle of electromagnetism states: a Euclidean, inertial frame exists in which the relations

$$\mathbf{D} = \epsilon_0\mathbf{E}, \qquad \mathbf{H} = \mathbf{B}/\mu_0, \tag{11.2}$$

with ϵ_0 and μ_0 two positive, universal constants, hold everywhere and at all times— inside material bodies as well as in empty space. We emphasize that the charge-current potentials $\mathbf{D}$ and $\mathbf{H}$ in (11.2) refer to the *total* charges and currents, including those which we shall later associate with polarization and magnetization[†].

In accordance with the foregoing remarks, the restriction of the relations (11.2) to a special frame is necessary. The assumption that this frame is also an inertial one provides a link between electromagnetism and mechanics.

Now, according to classical mechanics, all inertial frames follow from any one of them by applying Galilean transformations. As we have seen, these transformations have the following effects on $\mathbf{D}$, $\mathbf{E}$, $\mathbf{H}$ and $\mathbf{B}$:

$$\begin{aligned} \mathbf{D}' &= \mathbf{D}, & \mathbf{E}' &= \mathbf{E} + \mathbf{u} \times \mathbf{B}, \\ \mathbf{H}' &= \mathbf{H} - \mathbf{u} \times \mathbf{D}, & \mathbf{B}' &= \mathbf{B}. \end{aligned} \tag{11.3}$$

Clearly, the relations (11.2) cannot hold in all inertial frames. On the contrary, the frame in which they do hold emerges as a special, preferred one. It is called *the aether frame*, and the relations (11.2) are called *the aether relations*. We shall presently comment on the reasons behind these names.

If we recall the dimensions of $\mathbf{D}$ and $\mathbf{E}$, which were C/L^2 and $\Phi/(LT)$ respectively, we see that the dimensions of the universal constant ϵ_0 are $(C/\Phi)/(L/T)$. Similarly, the dimensions of μ_0 are the dimensions Φ/L^2 of $\mathbf{B}$, divided by the dimensions $C/(LT)$ of $\mathbf{H}$, that is, $(\Phi/C)/(L/T)$. The dimensions of $(\epsilon_0\mu_0)^{-1}$ are therefore the dimensions of velocity squared. We shall now prove that this velocity is the speed of an electromagnetic disturbance in vacuum.

The jump conditions across a surface of discontinuity in vacuum ($\mathbf{K} = \sigma = 0$) which is moving with velocity $\mathbf{v}$ are

$$\begin{aligned} \mathbf{n} \cdot [\![\mathbf{D}]\!] &= 0, \\ \mathbf{n} \times [\![\mathbf{H}]\!] + v_n[\![\mathbf{D}]\!] &= 0, \\ \mathbf{n} \cdot [\![\mathbf{B}]\!] &= 0, \\ \mathbf{n} \times [\![\mathbf{E}]\!] - v_n[\![\mathbf{B}]\!] &= 0. \end{aligned} \tag{11.4}$$

[†]In its definitive form, the third principle is due to Lorentz.

In the aether frame, we also have the relations (11.2). It is clear that, if $v_n = 0$, $\mathbf{E}$ and $\mathbf{B}$ (and by the aether relations, $\mathbf{D}$ and $\mathbf{H}$ as well) are all continuous. A surface across which these fields are discontinuous—called an *electromagnetic shock*—must therefore be moving. But then, according to $(11.4)_2$, $[\![\mathbf{D}]\!]$ is perpendicular to $\mathbf{n}$, so that $(11.4)_1$ is a consequence of $(11.4)_2$. Similarly, if $v_n \neq 0$, $(11.4)_3$ follows from $(11.4)_4$. It is therefore sufficient to consider the two conditions

$$\mathbf{n} \times [\![\mathbf{B}]\!]/\mu_0 + \epsilon_0 v_n [\![\mathbf{E}]\!] = 0,$$
$$\mathbf{n} \times [\![\mathbf{E}]\!] - v_n [\![\mathbf{B}]\!] = 0. \tag{11.5}$$

According to (11.5), the jumps $[\![\mathbf{E}]\!]$ and $[\![\mathbf{B}]\!]$ are parallel to the surface and perpendicular to each other. If we eliminate one of the jumps between the two equations, we obtain for v_n the condition

$$v_n^2 = (\epsilon_0\mu_0)^{-1}. \tag{11.6}$$

Since electromagnetism associates propagating electromagnetic disturbances with light, we identify this velocity with the speed of light c:

$$c^2 = (\epsilon_0\mu_0)^{-1}. \tag{11.7}$$

Equation (11.7) provides another argument for the uniqueness of the aether frame in classical physics: if there were other aether frames—inertial frames in which the aether relations (11.2) hold—then the speed of a given electromagnetic shock (pulse of light) would have the common, *universal* value $c = (\epsilon_0\mu_0)^{-1/2}$ in each one of them. But this would be in contradiction with the Galilean law of composition of velocities, which requires the relation $\mathbf{c}' = \mathbf{c} - \mathbf{u}$ between the velocities $\mathbf{c}$ and $\mathbf{c}'$ of the same light pulse in any two inertial frames with relative velocity $\mathbf{u}$.

We have established the formula $c^2 = (\epsilon_0\mu_0)^{-1}$ by determining the speed, relative to the aether frame, of a shock in the absence of charges and currents. It is also possible to consider smooth fields (which must satisfy Maxwell's equations), and this leads to the same formula for the speed of light. In the aether frame, Maxwell's equations in the absence of charges and currents are

$$\text{div } \epsilon_0\mathbf{E} = 0,$$
$$\text{curl } \mathbf{B}/\mu_0 = \epsilon_0\mathbf{E}_t,$$
$$\text{div } \mathbf{B} = 0,$$
$$\text{curl } \mathbf{E} = -\mathbf{B}_t. \tag{11.8}$$

Taking the curl of the second equation, and substituting for **curl E** from the fourth, we obtain

$$\text{curl}^2 \mathbf{B} = -\epsilon_0\mu_0\mathbf{B}_{tt}. \tag{11.9}$$

Applying the identity $\mathbf{curl}^2\mathbf{a} = \mathbf{grad}\,\mathrm{div}\,\mathbf{a} - \Delta\mathbf{a}$, where the last term is the Laplacian of $\mathbf{a}$, and noting that, according to $(11.8)_3$, $\mathrm{div}\,\mathbf{B} = 0$, we have the result

$$\Delta\mathbf{B} - \epsilon_0\mu_0\mathbf{B}_{tt} = 0. \tag{11.10}$$

A similar procedure, starting with the curl of $(11.8)_4$, leads to

$$\Delta\mathbf{E} - \epsilon_0\mu_0\mathbf{E}_{tt} = 0. \tag{11.11}$$

We shall show in Chapter 12 that these equations for the magnetic and electric fields possess solutions in the form of waves, travelling at the speed (relative to the aether frame) $c = (\epsilon_0\mu_0)^{-1/2}$. Such solutions are (continuous) *light waves*.

The founders of electromagnetic theory believed that propagating disturbances required the presence of a medium. But, as the foregoing discussion has shown, an electromagnetic shock, or a light wave, can propagate inside, as well as outside, material bodies, the only condition being the absence of charges and currents. This has led to the idea of a medium, called the *aether*, which pervades all matter and fills the empty space between material bodies, but has none of the properties of ordinary matter—no mass, no charge, no temperature, etc.[†] Its only role is to provide a seat for electromagnetic phenomena, and the aether frame is the one in which the aether is at rest[‡].

In other frames, particularly in other inertial frames, we should be able, by a simultaneous application of the laws of mechanics and electromagnetism, to detect an *aether wind*, that is, a motion with respect to the aether. This means, for example, that if we measure the speed of light in an inertial frame which is not the aether frame, we shall find a value different from c, a result of the Galilean composition of the velocity c in the aether frame and the aether wind. Of course there may be other methods of detecting an aether wind; light is not the only phenomenon that follows from electromagnetic theory.

The speed of light has been measured with continually increasing accuracy over a long time. An approximate value of c in terms of the old platinum metre in Paris is $3 \times 10^8\,\mathrm{m\,s}^{-1}$. Since ϵ_0 and μ_0 are *universal constants*, so is $c = (\epsilon_0\mu_0)^{-1/2}$. We can therefore *assume* a value of c, together with a given unit of time, in order to fix the unit of length. Several international committees on physical data and constants have indeed recommended that we set

$$c = 299\,792\,458\,\mathrm{m\,s}^{-1} \tag{11.12}$$

as the definition of the metre in terms of the second. On this recommendation, which has gained wide acceptance, the value of c is, by definition, *exactly* the one given by (11.12), and any improved measurement of the speed of light is really an improved measurement of the metre.

[†]The aether was introduced into physics by Descartes

[‡]In a work entitled *Nova Theoria Lucis et Colorum*, published in 1746, Euler stated that 'light is in the aether the same thing as sound in air'.

12. Lorentz transformations

It is well known that all attempts, beginning more than a century ago, to detect an aether wind have failed. In particular, the velocity of light has turned out to be c in different frames which were obviously moving relative to each other. Perhaps, then, there is more than one aether frame. This question can be answered by the theory developed so far.

Let $(x) = (\mathbf{x}, t)$ stand for the Cartesian coordinates and time in one (or *the*) aether frame $\Sigma(x)$. Let $g_{\alpha\beta}$ be the symmetric, absolute, covariant tensor that in this frame has the constant components

$$(g_{\alpha\beta}) = \begin{pmatrix} \delta_{rs} & 0 \\ 0 & -c^2 \end{pmatrix}, \tag{12.1}$$

where $\delta_{rs} = 1$ if $r = s$; otherwise $\delta_{rs} = 0$. The components of $g_{\alpha\beta}$ in any other frame will be determined by the tensor transformation rule

$$g_{\alpha'\beta'}(x') = \frac{\partial x^\alpha}{\partial x^{\alpha'}} \frac{\partial x^\beta}{\partial x^{\beta'}} g_{\alpha\beta}(x). \tag{12.2}$$

Hence (12.1)–(12.2) define $g_{\alpha\beta}$ in any frame[†]. We may regard the right-hand side of (12.2) as a product of three matrices. Taking the determinants of both sides and denoting $g = \det(g_{\alpha\beta})$, we obtain

$$g'(x') = J^{-2} g(x), \tag{12.3}$$

which proves that g has the same sign in all frames. According to (12.1), $g = -c^2$ in the frame Σ. From (12.3) we therefore obtain

$$\sqrt{-g'} = |J|^{-1}\sqrt{-g}, \tag{12.4}$$

where the radicals denote the positive square roots. Equation (12.4) shows that $\sqrt{-g}$ is a scalar density (cf. the remarks following (5.6)).

In any frame, let $g^{\alpha\beta}$ denote the inverse matrix of the non-singular matrix $g_{\alpha\beta}$, defined by

$$g_{\alpha\gamma} g^{\gamma\beta} = \delta_\alpha^\beta, \tag{12.5}$$

where $\delta_\alpha^\beta = 1$ if $\alpha = \beta$; otherwise $\delta_\alpha^\beta = 0$. It is easy to show that $g^{\alpha\beta}$ is an absolute, symmetric, contravariant tensor (cf. Problem 12.1). In the frame Σ, $g^{\alpha\beta}$ has the same components as $g_{\alpha\beta}$, except for $g^{44} = -c^{-2} = -\epsilon_0\mu_0$.

[†]The electromagnetic tensor $g_{\alpha\beta}$, with its definition involving $c^2 = (\epsilon_0\mu_0)^{-1}$, is the fundamental tensor of Einstein's theory of gravitation (the general theory of relativity).

Consider now the equation

$$f^{\alpha\beta} = \sqrt{\frac{\epsilon_0}{\mu_0}} \sqrt{-g}\, g^{\alpha\gamma} g^{\beta\delta} F_{\gamma\delta}. \tag{12.6}$$

This is a tensor equation, because both sides are 2-vector densities. It is true in the aether frame Σ, because there (12.6) reduces to the aether relations (11.2), as can easily be verified. Hence it is true in every frame. We shall therefore take (12.6) to be the tensor expression of the aether relations.

Now, if there are any other aether frames $\Sigma'(x')$, in which the aether relations (11.2) hold, they must be those in which the components of

$$g^{\alpha'\beta'}(x') = \frac{\partial x^{\alpha'}}{\partial x^{\alpha}} \frac{\partial x^{\beta'}}{\partial x^{\beta}} g^{\alpha\beta}(x) \tag{12.7}$$

are equal to the constant components of $g^{\alpha\beta}(x)$ in $\Sigma(x)$. For in all such frames the tensor equation (12.6) will reduce to the aether relations (11.2). It is clear that the partial derivatives in (12.7), which connect the two sets of constants $g^{\alpha\beta}$ and $g^{\alpha'\beta'}$, must all be constants (a detailed proof is provided by Problem 12.2). Hence the transformations leading to these frames must be linear. They can all be obtained from the special transformation

$$x' = \frac{x - ut}{\sqrt{1 - u^2/c^2}}, \quad y' = y, \quad z' = z, \quad t' = \frac{t - ux/c^2}{\sqrt{1 - u^2/c^2}}, \tag{12.8}$$

by orthogonal transformations of the space coordinates and time inversions. Their general form is

$$x^{r'} = A_r^{r'} \left\{ \left[\delta_s^r + \frac{\gamma^2}{\gamma + 1} \frac{u^r u_s}{c^2} \right] x^s - \gamma u^r x^4 \right\},$$
$$x^{4'} = \pm \gamma \left(x^4 - \frac{u_r x^r}{c^2} \right), \tag{12.9}$$

where A is an orthogonal matrix, $\gamma = (1 - u^2/c^2)^{-1/2}$, $u_r = u^r$ and $u^2 = u_r u^r$ must be less than c^2 for the transformation to be real. In three-dimensional vector notation, the transformation formulae (12.9) take the form

$$\mathbf{x}' = \mathbf{x} + \frac{\gamma^2}{\gamma + 1} \mathbf{u}(\mathbf{u} \cdot \mathbf{x})/c^2 - \gamma \mathbf{u}t,$$
$$t' = \pm \gamma (t - \mathbf{u} \cdot \mathbf{x}/c^2). \tag{12.10}$$

The transformations (12.9) or (12.10) are called *Lorentz transformations*, and the frames to which they lead from the aether frame are called *Lorentz frames*. Of course they are all aether frames.

The theory of relativity identifies the Lorentz frames with the inertial frames and replaces the Galilean transformations with Lorentz transformations. This requires a major revision of all the laws of physics, except, of course, those of electromagnetism. Thus mechanics, thermodynamics, the theory of gravitation and quantum mechanics all have to be revised in order to conform to the new transformation laws: the basic laws of each theory must now be such as to retain their canonical forms under Lorentz, rather than Galilean, transformations. This revision is still underway. Since in the new theory all inertial frames are aether frames, the aether itself (which can no longer remain at rest in each of the Lorentz frames) has gone out of vogue and is hardly ever mentioned.

Problem 12.1   Show that the inverse $g^{\alpha\beta}$ of $g_{\alpha\beta}$ is an absolute, symmetric, contravariant tensor.

Solution   The inverse of a symmetric matrix is itself symmetric. Since (12.5) defines $g^{\alpha\beta}$ in any frame, and $g_{\alpha\beta}$ is a covariant tensor, we have

$$g_{\alpha'\gamma'}g^{\gamma'\beta'} = \frac{\partial x^\alpha}{\partial x^{\alpha'}}\frac{\partial x^\gamma}{\partial x^{\gamma'}}g_{\alpha\gamma}g^{\gamma'\beta'} = \delta_{\alpha'}^{\beta'}.$$

Multiply the last equality by $(\partial x^{\alpha'}/\partial x^\lambda)(\partial x^\beta/\partial x^{\beta'})$. The result is

$$\frac{\partial x^\gamma}{\partial x^{\gamma'}}\frac{\partial x^\beta}{\partial x^{\beta'}}g_{\lambda\gamma}g^{\gamma'\beta'} = \delta_\lambda^\beta.$$

But the inverse of the non-singular $g_{\lambda\gamma}$ is unique. Hence

$$\frac{\partial x^\gamma}{\partial x^{\gamma'}}\frac{\partial x^\beta}{\partial x^{\beta'}}g^{\gamma'\beta'} = g^{\gamma\beta},$$

which proves that the symmetric $g^{\alpha\beta}$ is an absolute contravariant tensor.

Problem 12.2   Prove that Lorentz transformations are linear.

Solution   Lorentz transformations were defined as those for which eqns (12.7),

$$g^{\alpha'\beta'} = \frac{\partial x^{\alpha'}}{\partial x^\alpha}\frac{\partial x^{\beta'}}{\partial x^\beta}g^{\alpha\beta},$$

hold with the canonical components for both $g^{\alpha'\beta'}$ and $g^{\alpha\beta}$. We may equivalently define them by eqns (12.2),

$$g_{\alpha'\beta'} = \frac{\partial x^\alpha}{\partial x^{\alpha'}}\frac{\partial x^\beta}{\partial x^{\beta'}}g_{\alpha\beta},$$

where both $g_{\alpha'\beta'}$ and $g_{\alpha\beta}$, the reciprocal matrices of $g^{\alpha'\beta'}$ and $g^{\alpha\beta}$, have their canonical components (12.1). Interchanging the primed and unprimed coordinates in the last equations, we have

$$g_{\alpha\beta} = \frac{\partial x^{\alpha'}}{\partial x^\alpha}\frac{\partial x^{\beta'}}{\partial x^\beta} g_{\alpha'\beta'}.$$

Since the $g_{\alpha\beta}$ and the $g_{\alpha'\beta'}$ are constants, indeed *the same* constants, the matrix $\partial x^{\alpha'}/\partial x^\alpha$ has the determinant $J = \pm 1$ (cf. (12.3)) and is therefore non-singular. Differentiation with respect to x^γ gives

$$0 = \frac{\partial^2 x^{\alpha'}}{\partial x^\gamma \partial x^\alpha}\frac{\partial x^{\beta'}}{\partial x^\beta} g_{\alpha'\beta'} + \frac{\partial x^{\alpha'}}{\partial x^\alpha}\frac{\partial^2 x^{\beta'}}{\partial x^\gamma \partial x^\beta} g_{\alpha'\beta'}.$$

We wish to prove that the second derivatives vanish. In order to do so we add to the last equation the same equation with α and γ interchanged, and subtract the same equation with β and γ interchanged:

$$0 = g_{\alpha'\beta'}\left(\frac{\partial^2 x^{\alpha'}}{\partial x^\gamma \partial x^\alpha}\frac{\partial x^{\beta'}}{\partial x^\beta} + \frac{\partial x^{\alpha'}}{\partial x^\alpha}\frac{\partial^2 x^{\beta'}}{\partial x^\gamma \partial x^\beta} \right.$$

$$+ \frac{\partial^2 x^{\alpha'}}{\partial x^\alpha \partial x^\gamma}\frac{\partial x^{\beta'}}{\partial x^\beta} + \frac{\partial x^{\alpha'}}{\partial x^\gamma}\frac{\partial^2 x^{\beta'}}{\partial x^\alpha \partial x^\beta}$$

$$\left. - \frac{\partial^2 x^{\alpha'}}{\partial x^\beta \partial x^\alpha}\frac{\partial x^{\beta'}}{\partial x^\gamma} - \frac{\partial x^{\alpha'}}{\partial x^\alpha}\frac{\partial^2 x^{\beta'}}{\partial x^\beta \partial x^\gamma} \right).$$

The last term on the right-hand side cancels the second one, and the last but one cancels the fourth (because $g_{\beta'\alpha'} = g_{\alpha'\beta'}$). The remaining first and third terms are equal. Thus

$$\frac{\partial^2 x^{\alpha'}}{\partial x^\gamma \partial x^\alpha} g_{\alpha'\beta'} \frac{\partial x^{\beta'}}{\partial x^\beta} = 0.$$

For each α and γ the four ($\alpha' = 1, 2, 3, 4$) second derivatives $X_2^{\alpha'}$ (say) satisfy a linear, homogeneous system of equations $A_{\beta\alpha'} X_2^{\alpha'} = 0$ with the coefficient matrix $A_{\beta\alpha'} = g_{\alpha'\beta'}\partial x^{\beta'}/\partial x^\beta$. As a product of non-singular matrices, A is itself non-singular. Thus the second derivatives must vanish, and the transformations must be linear.

Problem 12.3 Derive the following transformation formulae from the transformation laws of the tensors $f^{\alpha\beta}$ and $F_{\alpha\beta}$ under proper Lorentz transformations (det $A = 1$ in (12.9)$_1$, +sign in (12.9)$_2$):

$$\mathbf{E}' = \gamma(\mathbf{E} + \mathbf{u} \times \mathbf{B}) - \frac{\gamma^2}{\gamma + 1}\mathbf{u}(\mathbf{u} \cdot \mathbf{E})/c^2, \qquad (12.11)$$

$$\mathbf{B}' = \gamma(\mathbf{B} - \mathbf{u} \times \mathbf{E}/c^2) - \frac{\gamma^2}{\gamma + 1}\mathbf{u}(\mathbf{u} \cdot \mathbf{B})/c^2, \qquad (12.12)$$

$$\mathbf{D}' = \gamma(\mathbf{D} + \mathbf{u} \times \mathbf{H}/c^2) - \frac{\gamma^2}{\gamma+1}\mathbf{u}(\mathbf{u} \cdot \mathbf{D})/c^2, \tag{12.13}$$

$$\mathbf{H}' = \gamma(\mathbf{H} - \mathbf{u} \times \mathbf{D}) - \frac{\gamma^2}{\gamma+1}\mathbf{u}(\mathbf{u} \cdot \mathbf{H})/c^2, \tag{12.14}$$

$$\mathbf{A}' = \mathbf{A} + \left[\frac{\gamma^2}{\gamma+1}(\mathbf{u} \cdot \mathbf{A}) - \gamma V\right]\mathbf{u}/c^2, \tag{12.15}$$

$$V' = \gamma(V - \mathbf{u} \cdot \mathbf{A}), \tag{12.16}$$

$$\mathbf{j}' = \mathbf{j} + \left[\frac{\gamma^2}{\gamma+1}(\mathbf{u} \cdot \mathbf{j})/c^2 - \gamma q\right]\mathbf{u}, \tag{12.17}$$

$$q' = \gamma\left[q - (\mathbf{u} \cdot \mathbf{j})/c^2\right]. \tag{12.18}$$

Solution   The formulae for $\mathbf{D}'$ and $\mathbf{H}'$ are obtained directly from the transformation law of the tensor density $f^{\alpha\beta}$, as in Problem 6.2. The current and charge densities $\mathbf{j}$ and q are the components of s^α, which transforms as

$$s^{\alpha'} = \frac{\partial x^{\alpha'}}{\partial x^\alpha}s^\alpha,$$

since $|J| = 1$ for the Lorentz transformations. But since the transformations are linear, the equation states that the (s^α)'s transform as the coordinates. Thus $\mathbf{j}$ transforms like $\mathbf{x}$, and q like x^4. Hence, from (12.9),

$$j^{r'} = A_r^{r'}\left\{\left[\delta_s^r + \frac{\gamma^2}{\gamma+1}\frac{u^r u_s}{c^2}\right]j^s - \gamma u^r q\right\},$$

$$q' = \gamma(q - u_r j^r/c^2),$$

from which the formulae for $\mathbf{j}'$ and q' follow.

The electromagnetic fields $\mathbf{B}$ and $\mathbf{E}$, and their potentials $\mathbf{A}$ and V, are components of covariant tensors F and A. Their transformation laws involve derivatives of the unprimed, with respect to the primed, coordinates. It is not difficult to solve the Lorentz transformation for the x^α in terms of the $x^{\alpha'}$, but it is much easier to note that $f \cdot F = \mathbf{H} \cdot \mathbf{B} - \mathbf{D} \cdot \mathbf{E}$ is a scalar density, and to obtain the formulae for $\mathbf{B}'$ and $\mathbf{E}'$ as in Problem 9.2.

In order to obtain the formulae for $\mathbf{A}'$ and V', we note that $s \cdot A = \mathbf{j} \cdot \mathbf{A} - qV$ is a scalar density. Hence

$$\mathbf{j}' \cdot \mathbf{A}' - q'V' = \left[\mathbf{j} + \left(\frac{\gamma^2}{\gamma+1}\mathbf{u} \cdot \mathbf{j}/c^2 - \gamma q\right)\mathbf{u}\right] \cdot \mathbf{A}' - \gamma(q - \mathbf{u} \cdot \mathbf{j})V'$$
$$= \mathbf{j} \cdot \mathbf{A} - qV$$

must hold for all $\mathbf{j}$ and q. This yields the two equations

$$\mathbf{A}' + \frac{\gamma^2}{\gamma+1}(\mathbf{u} \cdot \mathbf{A}')\mathbf{u}/c^2 + \gamma V'\mathbf{u}/c^2 = \mathbf{A}, \tag{1}$$

$$\gamma(V' + \mathbf{u} \cdot \mathbf{A}') = V. \tag{2}$$

In order to solve for $\mathbf{A}'$ and V', multiply the first of these by $\mathbf{u}$, to obtain

$$\gamma\left(\mathbf{u}\cdot\mathbf{A}' + \frac{u^2}{c^2}V'\right) = \mathbf{u}\cdot\mathbf{A}. \tag{3}$$

Now solve eqns (2) and (3) for $\mathbf{u}\cdot\mathbf{A}'$ and V', and then use eqn (1) to obtain

$$V' = \gamma(V - \mathbf{u}\cdot\mathbf{A}),$$

$$\mathbf{A}' = \mathbf{A} + \left[\frac{\gamma^2}{\gamma+1}(\mathbf{u}\cdot\mathbf{A}) - \gamma V\right]\mathbf{u}/c^2.$$

In introducing the antisymmetric tensors f and F we were guided by the frame-invariance of integrals, which in turn guaranteed the frame-invariance of the first and second principles of electromagnetism. The motivation for introducing the symmetric, absolute tensor $g_{\alpha\beta}$ was the frame-invariance of the third principle, as expressed by the tensor equation (12.6). It is also possible to exhibit a scalar expression, the frame-invariance of which is guaranteed by the tensor property (12.2) of $g_{\alpha\beta}$. It is the quadratic form $g_{\alpha\beta}\,dx^\alpha\,dx^\beta$. But the invariance of this form, by itself, does not explain why we insist on the existence of a frame in which $g_{\alpha\beta}$ has the canonical form (12.1). The latter requirement is, of course, a consequence of the aether relations (11.2).

Quantities that transform as tensors under the restricted class of Lorentz transformations are called *world tensors*. Any general space-time tensor is a world tensor, but the converse statement is not true. The tensor $g_{\alpha\beta}$ is a *constant* world tensor, because the Lorentz transformations are, by definition, those that preserve the canonical form (12.1). If Σ is a Lorentz frame and P a point which is moving relative to Σ with a velocity $\mathbf{v}$, the *4-velocity* of P is defined as the 4-symbol which, in the frame Σ, has the components

$$(w^\alpha) = \left(\frac{\mathbf{v}}{\sqrt{1-v^2/c^2}}, \frac{1}{\sqrt{1-v^2/c^2}}\right). \tag{12.19}$$

In any other Lorentz frame Σ' the same point P will have a velocity $\mathbf{v}'$, and the components $(w^{\alpha'})$ in Σ' can be obtained, *either* from the 'primed' formula that corresponds to (12.19), *or* by applying to w^α the Lorentz transformation that leads from Σ to Σ'. Thus the 4-velocity of a point P is a world vector. Its components are not independent: they satisfy the world equation $g_{\alpha\beta}w^\alpha w^\beta = -c^2$.

Since Lorentz transformations have $J = \pm 1$, the square of the relative scalar $F \cdot$ dual F, which is equal to $(\mathbf{B}\cdot\mathbf{E})^2$, is a world scalar. Similarly, the scalar density $f \cdot F$, which in an aether frame equals $B^2/\mu_0 - \epsilon_0 E^2 = \epsilon_0(c^2 B^2 - E^2)$, is a world scalar. Lorentz transformations therefore preserve the values of

$$c^2 B^2 - E^2 \quad \text{and} \quad (\mathbf{B}\cdot\mathbf{E})^2 \tag{12.20}$$

at each event. It is not necessary to consider the product $f \cdot$ dual f, because in an aether frame it is proportional to $F \cdot$ dual F.

According to (12.20), orthogonality of **B** and **E** is a world-invariant property. Only if **B** and **E** are orthogonal can **E** or **B** be 'transformed away' (made zero) by a Lorentz transformation; and whether it is **E**, or **B**, which can be transformed away will of course depend on whether the world-invariant $c^2 B^2 - E^2$ is positive or negative.

Solutions of Maxwell's equations exist in which **B** and **E** are perpendicular, their magnitudes being related by $E = cB$. For such fields, which we shall encounter in the theory of radiation, both invariants of (12.20) vanish. They will therefore continue to vanish after a Lorentz transformation has been applied: these fields retain their properties, $\mathbf{E} \cdot \mathbf{B} = 0$ and $E = cB$, in any Lorentz frame.

Exercises

12.1. Show that the classical (non-relativistic) 4-velocity, defined as $(w^\alpha) = (\mathbf{v}, 1)$, is a contravariant vector with respect to Galilean transformations (a classical world vector).

12.2. Show that a Lorentz transformation with a velocity **u** satisfying

$$\frac{\mathbf{u}}{1 + u^2/c^2} = \frac{\mathbf{E} \times \mathbf{B}}{E^2/c^2 + B^2}$$

leads to a frame in which $\mathbf{E}'$ and $\mathbf{B}'$ are parallel.

---

13. Maxwell's equations in an aether frame

In an aether frame, Maxwell's equations can be written in the form

$$\begin{aligned}
\operatorname{div} \epsilon_0 \mathbf{E} &= q, \\
\operatorname{\mathbf{curl}} \mathbf{B}/\mu_0 - \epsilon_0 \mathbf{E}_t &= \mathbf{j}, \\
\operatorname{div} \mathbf{B} &= 0, \\
\operatorname{\mathbf{curl}} \mathbf{E} + \mathbf{B}_t &= 0.
\end{aligned} \tag{13.1}$$

These are often regarded as linear equations for the electromagnetic field, with q and **j** playing the part of sources. But this is only true in situations which are very special—as in the electrostatics of conductors—or so simple as to be degenerate—point charges, linear currents and the like—because the 'sources' q and **j** are usually not given. It is indeed normal for the charges and currents to depend in a complicated manner on the electromagnetic field itself, and perhaps even on its *history*. The equations may then become non-linear, and may even cease to be partial differential equations.

According to the aether relations, the components **H** and **D** of the charge-current potential are, respectively, proportional to **B** and **E**. There is no harm in calling **H** 'the magnetic field **H**', and **D** 'the electric field **D**', so long as we remember, *especially* in frames that are not aether frames, that **H** and **D** are really components of the charge-current potential[†].

Of the three principles of electromagnetism, the third is the easiest to forget. The two pairs of Maxwell's equations are easily memorized, but many 'paradoxes' can be manufactured by thoughtless application of (13.1) to frames that are not aether frames. We shall illustrate this by means of an example.

In an aether frame O, described by cylindrical coordinates (r, ϕ, z), let the charge-current potential have the components

$$\mathbf{D} = (0, 0, 0), \quad \mathbf{H} = \left(0, \frac{i}{2\pi r}, 0\right). \tag{13.2}$$

Obviously, the charge density vanishes everywhere. It is easy to verify that the current density also vanishes, except along the z axis; and that a linear current i flows along this axis in the positive direction (Fig. 13.1). The aether relations determine the electromagnetic field:

$$\mathbf{E} = (0, 0, 0), \quad \mathbf{B} = \left(0, \frac{\mu_0 i}{2\pi r}, 0\right). \tag{13.3}$$

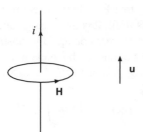

Fig. 13.1

Let O' now be an inertial frame that is moving, relative to O, with velocity u, parallel to the z axis in the direction of the current. In order to obtain the various quantities in the frame O', according to classical physics, we apply a Galilean transformation. From (11.3) we obtain

$$\mathbf{D}' = (0, 0, 0), \qquad \mathbf{H}' = \mathbf{H},$$
$$\mathbf{E}' = \left(-\frac{\mu_0 u i}{2\pi r}, 0, 0\right), \qquad \mathbf{B}' = \mathbf{B}. \tag{13.4}$$

[†]In the old literature, **H** is called the *magnetic force* or the *magnetic intensity*, and **B** the *magnetic induction*.

Since the charge-current potential is the same as in O, the charge and current distributions are also the same: no charge anywhere, and a linear current i flowing along the z axis. The magnetic field $\mathbf{B}$ is also the same as in O, but there is an electric field directed towards the z axis. If there is no charge anywhere, where does this electric field come from?

This question arises because the third principle has been forgotten. The electric field 'comes from charge' only in an aether frame, through the relation $\mathbf{E} = \mathbf{D}/\epsilon_0$. The frame O' is not an aether frame, hence $\mathbf{E}' \neq \mathbf{D}'/\epsilon_0 = 0$, and there is no contradiction between the absence of charge and a non-zero electric field.

The relativistic treatment of this particular example is trouble-free. According to the theory of relativity, inertial frames, aether frames and Lorentz frames are all synonymous, and they are connected by Lorentz transformations. The inertial frame O', now obtained from O by a Lorentz transformation, is an aether frame as well. Instead of (13.4), we have, according to (12.11)–(12.14),

$$\mathbf{D}' = \left(-\gamma \frac{ui}{2\pi r c^2}, 0, 0\right), \qquad \mathbf{H}' = \mathbf{H},$$

$$\mathbf{E}' = \left(-\gamma \frac{\mu_0 ui}{2\pi r}, 0, 0\right), \qquad \mathbf{B}' = \mathbf{B}, \tag{13.5}$$

with $\gamma = (1 - u^2/c^2)^{-1/2}$. Now, from $\int q'\, dV' = \oint D'_n\, dS'$, or from (12.18), there is also a negative line charge, of amount $\gamma ui/c^2$ per unit length, on the z axis. It would now be correct to say that $\mathbf{E}'$, which differs from the classical $\mathbf{E}'$ of (13.4)$_3$ only by the factor γ, 'comes from' this line charge, because O' is an aether frame.

Maxwell's equations (13.1) were obtained by substituting the aether relations in the first pair. The same substitution can be made in the more general integral laws (6.10)–(6.11). Equation (6.10) then becomes *Gauss's law*

$$\epsilon_0 \oint E_n\, dS = \int q\, dV. \tag{13.6}$$

The integral law (6.11) becomes

$$\oint_c \mathbf{B} \cdot \mathbf{ds} - \frac{1}{c^2} \frac{d}{dt} \int E_n\, dS = \mu_0 \int j_n\, dS. \tag{13.7}$$

Both (13.6) and (13.7) hold *only in an aether frame*. If the 'relativistic', second term on the left-hand side of (13.7) vanishes, or if it can be neglected, the result is *Ampère's law*,

$$\oint_c \mathbf{B} \cdot \mathbf{ds} = \mu_0 \int j_n\, dS. \tag{13.8}$$

6

Charged particles

14. Multipole moments

The simplest solution of Maxwell's equations is the one corresponding to a point charge e, resting at the origin of an aether frame. This elementary solution is

$$\mathbf{B} = 0, \qquad \mathbf{E} = \frac{e}{4\pi\epsilon_0}\frac{\mathbf{n}}{r^2}, \tag{14.1}$$

where $\mathbf{n} = \mathbf{r}/r$ is a unit vector in the radial direction. Correspondingly, we have the elementary potentials (cf. (10.8))

$$\mathbf{A} = 0, \qquad V = \frac{e}{4\pi\epsilon_0}\frac{1}{r}. \tag{14.2}$$

Since Maxwell's equations are linear in q and $\mathbf{j}$, solutions corresponding to several point charges, all at rest, are obtained by superposing the solutions (14.1)–(14.2). Usually, one simply adds up the V's of $(14.2)_2$, rather than the $\mathbf{E}$'s of $(14.1)_2$, because scalars are easier to add than vectors.

For a set of particles with charges e_a, resting at positions $\mathbf{r}_a$, superposition of elementary solutions like $(14.2)_2$ gives

$$V(\mathbf{r}) = \frac{1}{4\pi\epsilon_0}\sum\frac{e_a}{|\mathbf{r} - \mathbf{r}_a|}. \tag{14.3}$$

We choose the origin at some point O which is at a finite distance from the particles and consider V at great distance from the system ($r \gg r_a$). In order to do this, we expand each $|\mathbf{r} - \mathbf{r}_a|^{-1}$ according to the formula

$$\frac{1}{|\mathbf{r} - \mathbf{r}_a|} = \frac{1}{r} - x_a^i\frac{\partial}{\partial x^i}\frac{1}{r} + \frac{1}{2}x_a^i x_a^j\frac{\partial^2}{\partial x^i\partial x^j}\frac{1}{r} - \cdots. \tag{14.4}$$

Correspondingly, we write

$$V = V^{(0)} + V^{(1)} + V^{(2)} + \cdots. \tag{14.5}$$

The lowest order term of V is

$$V^{(0)} = \frac{1}{4\pi\epsilon_0} \frac{\sum e_a}{r} = \frac{1}{4\pi\epsilon_0} \frac{e}{r}. \tag{14.6}$$

It is the potential at $\mathbf{r}$ of a single charge of amount $\sum e_a$ located at the origin; $e = \sum e_a$ is called the *monopole moment* of the system. The corresponding field is

$$\mathbf{E}^{(0)} = -\mathbf{grad}\, V^{(0)} = \frac{1}{4\pi\epsilon_0} \frac{e}{r^2} \mathbf{n}, \tag{14.7}$$

where $\mathbf{n}$ is a unit vector along $\mathbf{r}$.

The first-order term is

$$\begin{aligned}
V^{(1)} &= -\frac{1}{4\pi\epsilon_0} \sum e_a x_a^i \frac{\partial}{\partial x^i} \frac{1}{r} \\
&= \frac{1}{4\pi\epsilon_0} \sum e_a x_a^i \frac{n_i}{r^2} \\
&= \frac{1}{4\pi\epsilon_0} \frac{\mathbf{d} \cdot \mathbf{n}}{r^2},
\end{aligned} \tag{14.8}$$

where

$$\mathbf{d} = \sum e_a \mathbf{r}_a \tag{14.9}$$

is the *dipole moment* of the system. The dipole field corresponding to the dipole potential (14.8) is

$$\mathbf{E}^{(1)} = -\mathbf{grad}\, V^{(1)} = \frac{1}{4\pi\epsilon_0} \frac{3(\mathbf{d} \cdot \mathbf{n})\mathbf{n} - \mathbf{d}}{r^3}. \tag{14.10}$$

The dipole moment is independent of the choice of origin if the monopole moment $e = \sum e_a$ vanishes. For let O' be such that $OO' = \mathbf{c}$. With respect to O', then,

$$\mathbf{d}' = \sum e_a \mathbf{r}_a' = \sum e_a (\mathbf{r}_a - \mathbf{c}) = \mathbf{d} - \mathbf{c} \sum e_a = \mathbf{d}.$$

In particular, for two charges with $e_1 = e$ and $e_2 = -e$,

$$\mathbf{d} = e_1 \mathbf{r}_1 + e_2 \mathbf{r}_2 = e(\mathbf{r}_1 - \mathbf{r}_2) = e\mathbf{l}, \tag{14.11}$$

where $\mathbf{l}$ is the position of $e_1 = e$ relative to e_2. We now take the limit when e is very large and l very small, in such a way that the product el has a finite value $\mathbf{d}$. Such a combination is called an *electric dipole*, and $\mathbf{d}$ is its *moment*. Dipoles serve as models for many molecules in physical chemistry. Since the foregoing limit is taken at constant $\mathbf{d}$, the potential and field of such a point dipole are still given by (14.8)–(14.10), provided that $\mathbf{r}$ is the position of the field point P relative to the dipole.

The third term in the expansion of (14.5) according to (14.4) is

$$V^{(2)} = \frac{1}{4\pi\epsilon_0} \frac{1}{2} \sum x_a^i x_a^j \frac{\partial^2}{\partial x^i \partial x^j} \frac{1}{r}, \qquad (14.12)$$

where $\sum$ denotes summation over all the charges, as before; for each charge, there is a summation over each of the repeated indices i and j.

Exercises

14.1. Show that

$$V^{(2)} = \frac{1}{4\pi\epsilon_0} \frac{D^{ij} n_i n_j}{2r^3}, \qquad (14.13)$$

where

$$D^{ij} = \sum e_a (3x_a^i x_a^j - \mathbf{x}_a^2 \delta^{ij}). \qquad (14.14)$$

14.2. Prove that D does not depend on the choice of origin if $e = \sum e$ and $\mathbf{d}$ both vanish.

14.3. Calculate the quadrupole field $\mathbf{E}^{(2)} = -\,\mathbf{grad}\, V^{(2)}$.

The symmetric tensor D of (14.14) is called the *quadrupole moment* of the system. It is evidently traceless. Hence it has five independent components.

The expansion of the potential (14.3) according to (14.4) becomes rather unwieldy beyond the quadrupole term. It is easier to proceed to higher multipole moments by use of the formula

$$\frac{1}{|\mathbf{r} - \mathbf{r}'|} = \frac{1}{\sqrt{r^2 + r'^2 - 2rr'\cos\chi}} = \sum_{n=0}^{\infty} \frac{(r')^n}{r^{n+1}} P_n(\cos\chi), \qquad (14.15)$$

where χ is the angle between $\mathbf{r}$ and $\mathbf{r}'$, and $P_n(x)$ is the Legendre polynomial of degree n. We shall return to these matters in Chapter 8.

We have based the discussion of multipole moments on the formula (14.3). For a continuous distribution of charges, all at rest, the potential is evidently

$$V(\mathbf{r}) = \frac{1}{4\pi\epsilon_0} \int \frac{q \, dV'}{|\mathbf{r} - \mathbf{r}'|}. \qquad (14.16)$$

Since the expansion (14.4) was performed on $|\mathbf{r} - \mathbf{r}_a|^{-1}$, it is equally possible to carry it out for continuously distributed charges, starting from (14.16). The only difference is that the moments become integrals over the charge density, instead of sums over the discrete charges.

Exercises

14.4. An electric dipole is at the origin, and its direction is that of the z axis, so that the potential (14.8) is proportional to $(\cos\theta)/r^2$. An element of a field line has radial and transverse projections dr and $r\,d\theta$ such that $dr : r\,d\theta = E_r : E_\theta$. Show that the field lines are given by $(\sin^2\theta)/r = \text{const}$.

14.5. Electric charge is distributed on an infinite plane surface so that the surface density is σ. P is a point at distance a from the plane, and dS is an element of surface whose distance from P is r. Prove that the electric field at P has a component away from the plane equal to $\int a\sigma\,dS/4\pi\epsilon_0 r^3$. If σ is uniform, show that this gives a value $\sigma/(2\epsilon_0)$, and deduce that in such a case one-half of the field arises from those points of the plane that are less than $2a$ from P.

14.6. A certain distribution of electric charge is spherically symmetric about the origin, and the total charge inside a sphere of radius r is $Q(r)$. Prove that the potential $V(r)$ is given by

$$V(r) = \frac{1}{4\pi\epsilon_0}\int_r^\infty \frac{Q(r)\,dr}{r^2}.$$

Show that this may be written in the alternative form

$$V(r) = \frac{1}{4\pi\epsilon_0}\frac{Q(r)}{r} + \frac{1}{\epsilon_0}\int_r^\infty rq(r)\,dr,$$

where $q(r)$ is the density of charge at distance r from the origin.

14.7. A fixed circle is drawn of radius a and a charge e is placed at a distance $3a/4$ from the centre of the circle on a line through the centre perpendicular to the plane of the circle. Show that the flux of $\mathbf{E}$ through the circle is $e/(5\epsilon_0)$. If a second charge e' is similarly placed at a distance $5a/12$ on the opposite side of the circle, and there is no net flux through the circle, prove that $e' = 13e/20$.

14.8. Electric charge is distributed at uniform density λ per unit length on an infinite straight line. (This is known as a *line charge of strength* λ.) Apply Gauss's law to a cylinder of unit length and radius r coaxial with the straight line and deduce that the field E at distance r from the line is $\lambda/(2\pi\epsilon_0 r)$. Show that the potential is

$$V = \frac{\lambda}{2\pi\epsilon_0}\ln\frac{1}{r} + \text{const.}$$

14.9. Two equal charges e are at opposite corners of a square of side a, and an electric dipole of moment $\mathbf{d}$ is at a third corner pointing towards one of the charges. If $d = 2\sqrt{2}ea$, show that the field strength at the fourth corner of the square is

$$\sqrt{\frac{17}{2}}\frac{1}{4\pi\epsilon_0}\frac{e}{a^2}.$$

14.10. Calculate the quadrupole moment of a uniformly charged ellipsoid with respect to its centre.

15. The Lorentz force

A classical charged particle is a point with which we associate a positive mass m and a charge e, positive or negative. The adjective 'classical' means 'non-quantum': in quantum mechanics a charged particle may have additional properties; for example, spin. We define the particle's *rest frame* Σ' as that inertial frame in which the particle is momentarily at rest. Of course the rest frame depends on the time, because the particle may undergo acceleration. Moreover, at a given time, the rest frame is only determined up to constant (time-independent) rotations and translations.

In this, and in the next, section we shall deal with non-relativistic dynamics. The basic assumption we shall make about a particle is that, in an electromagnetic field, it is subject to a force $e\mathbf{E}'$, where $\mathbf{E}'$ is the electric field in the particle's rest frame. Since it is defined with respect to a unique frame, this force is absolute (frame independent), in accordance with the requirements of classical mechanics.

Let Σ be any inertial frame, in which the electric and magnetic fields are $\mathbf{E}$ and $\mathbf{B}$, and the particle is moving with velocity $\dot{\mathbf{x}}$. The rest frame Σ' is then an inertial frame which is moving, relative to Σ, with velocity $\dot{\mathbf{x}}$. Hence, according to classical (non-relativistic) mechanics, $\mathbf{E}' = \mathbf{E} + \dot{\mathbf{x}} \times \mathbf{B}$ (cf. (9.8)). Thus the force is

$$\mathbf{F} = e(\mathbf{E} + \dot{\mathbf{x}} \times \mathbf{B}). \tag{15.1}$$

This relation between the force, the charge and the electromagnetic field also furnishes a connection between their dimensions. We have denoted, in Section 1, the dimension of electric charge by C; and in Section 8, the dimension of magnetic flux by Φ. The dimensions of the second term on the right-hand side of (15.1) are therefore $C\Phi/(LT)$, and this, according to our assumption, equals the dimension N of force. Thus $C\Phi$ is dimensionally the same as NLT. If we now fix a unit for C, say, this relation will enable us to determine the unit of Φ.

In order to fix the charge unit, we consider the aether constant μ_0, with the dimensions $(\Phi/C)/(L/T)$. According to the relation we have just established, the dimensions of μ_0 are $N/(C/T)^2$, that is, force divided by squared current. But μ_0 is a *universal constant*. Expressed in terms of *known* units of force and current, it will have a definite numerical value, to be determined by a suitable experiment. Conversely, an *assumed* value for μ_0 will serve to fix the unit of current in terms of the unit of force (just as the metre was fixed by (11.12) in terms of the second). This is the method by which the unit of current, the *ampère*, is fixed in SI units (Système International d'Unités). Specifically, we set

$$\mu_0 = 4\pi \times 10^{-7} \text{newton amp}^{-2} \tag{15.2}$$

(exactly) as the SI definition of the ampère. All the electromagnetic SI units follow from this unit of current. Thus the unit of charge, the *coulomb*, is defined by 1 coulomb = 1 amp s; the unit of electric potential, the *volt*, by 1 volt = 1 newton m coulomb^{-1}; the unit of $\mathbf{B}$, the *tesla*, by 1 tesla = 1 newton amp^{-1} m^{-1}; the unit of magnetic flux, the *weber*, by 1 weber = 1 tesla m^2; etc.

We can also use (15.2) to determine the numerical value of the second universal constant, ϵ_0, in these units:

$$\frac{1}{4\pi\epsilon_0} = \frac{\mu_0 c^2}{4\pi} = 10^{-7}(\text{newton amp}^{-2}) \cdot (299792458 \text{ m s}^{-1})^2$$

$$\approx 9 \times 10^9 \text{ newton m}^2 \text{ coulomb}^{-2}. \tag{15.3}$$

In an inertial frame, the equation of motion of a charged particle, under the applied force (15.1), is

$$m\ddot{\mathbf{x}} = e(\mathbf{E} + \dot{\mathbf{x}} \times \mathbf{B}). \tag{15.4}$$

If there are other applied forces, which are not due to the electromagnetic field, they should of course be added on the right-hand side of this equation. The expression on the right-hand side of (15.1) or (15.4) is called *the Lorentz force*. If we were to include in $\mathbf{F} = e\mathbf{E}'$ the field produced by the particle itself, it would become infinite, and the equation of motion would not make any sense (at least, it would not make any classical sense; in quantum electrodynamics a particle is allowed to 'act on itself', subject to definite rules which govern the resulting infinities). We therefore stipulate that in the Lorentz force the self-field is to be left out. Logically speaking, this should be regarded as part of the definition of a classical charged particle.

From (15.4) we conclude that

$$\frac{d}{dt}\tfrac{1}{2}m\dot{\mathbf{x}}^2 = e\dot{\mathbf{x}} \cdot \mathbf{E}. \tag{15.5}$$

The magnetic field can do no work on a particle because its contribution to the Lorentz force is normal to the path. If the electric field vanishes, the motion can still be very complicated, but the speed will be constant.

Exercises

15.1. Two equal charges $+e$ are fixed at the points $(\pm a, 0, 0)$. A third charge $-e$ of mass m revolves around the x axis under the influence of its attraction to the two fixed charges. Show that if it describes a circle of radius r, its velocity v is given by

$$mv^2 = \frac{1}{4\pi\epsilon_0}\frac{2e^2r^2}{(r^2 + a^2)^{3/2}}.$$

(This was an early model for the motion of electrons in simple molecules.)

15.2. Equal charges $+e$ are fixed at the four corners of a square of side $\sqrt{2}a$. A fifth charge $+e$, whose mass is m, is now placed at the centre of the square and is free to move. Show that it is in equilibrium at this point, and that the equilibrium is stable for all small displacements in the plane of the charges, the period of small oscillations in any direction in this plane being $\pi \sqrt{(4\pi\epsilon_0 2ma^3/e^2)}$, but that it is unstable with respect to motion perpendicular to the plane.

Problem 15.1 Show that an electric dipole $\mathbf{d}$ placed in an electric field $\mathbf{E}$ experiences a resultant force $(\mathbf{d} \cdot \mathbf{grad})\mathbf{E}$.

Solution We regard the dipole as the limit $el \to \mathbf{d}$ of a charge e at $\mathbf{r} + \frac{1}{2}\mathbf{l}$ and a charge $(-e)$ at $\mathbf{r} - \frac{1}{2}\mathbf{l}$. The resultant electric force on the pair is

$$\mathbf{f} = e\mathbf{E}\left(\mathbf{r} + \tfrac{1}{2}\mathbf{l}\right) - e\mathbf{E}\left(\mathbf{r} - \tfrac{1}{2}\mathbf{l}\right).$$

To the lowest order in l, the i'th component of the force is $e(\mathbf{l} \cdot \mathbf{grad})E_i(\mathbf{r})$. In the limit $el \to \mathbf{d}$, the resultant force becomes $\mathbf{f} = (\mathbf{d} \cdot \mathbf{grad})\mathbf{E}$.

Problem 15.2 Show that an electric dipole $\mathbf{d}$ placed in a uniform electric field $\mathbf{E}$ experiences a resultant couple $\mathbf{d} \times \mathbf{E}$.

Solution In the notation of the previous Problem, the torque is the limit $el \to \mathbf{d}$ of

$$\begin{aligned} \mathbf{t} &= \left(\mathbf{r} + \tfrac{1}{2}\mathbf{l}\right) \times e\mathbf{E}\left(\mathbf{r} + \tfrac{1}{2}\mathbf{l}\right) - \left(\mathbf{r} - \tfrac{1}{2}\mathbf{l}\right) \times e\mathbf{E}\left(\mathbf{r} - \tfrac{1}{2}\mathbf{l}\right) \\ &= \mathbf{r} \times e\left[\mathbf{E}\left(\mathbf{r} + \tfrac{1}{2}\mathbf{l}\right) - \mathbf{E}\left(\mathbf{r} - \tfrac{1}{2}\mathbf{l}\right)\right] + e\mathbf{l} \times \tfrac{1}{2}\left[\mathbf{E}\left(\mathbf{r} + \tfrac{1}{2}\mathbf{l}\right) + \mathbf{E}\left(\mathbf{r} - \tfrac{1}{2}\mathbf{l}\right)\right]. \end{aligned}$$

In the limit, this becomes

$$\mathbf{t} = \mathbf{r} \times (\mathbf{d} \cdot \mathbf{grad})\mathbf{E} + \mathbf{d} \times \mathbf{E} = \mathbf{r} \times \mathbf{f} + \mathbf{d} \times \mathbf{E}.$$

If $\mathbf{E}$ is uniform the resultant force $\mathbf{f}$ vanishes, and the resultant torque becomes a couple, independent of the point of reference (which in the last formula was taken as the origin, from which $\mathbf{r}$ was measured):

$$\mathbf{t} = \mathbf{d} \times \mathbf{E}.$$

Problem 15.3 Find the force that a dipole $\mathbf{d}_1$ exerts on a dipole $\mathbf{d}_2$.

Solution According to Problem 15.1, the force is $\mathbf{f} = (\mathbf{d}_2 \cdot \mathbf{grad})\mathbf{E}$, where

$$\mathbf{E} = \frac{1}{4\pi\epsilon_0}\left[\frac{3(\mathbf{d}_1 \cdot \mathbf{r})\mathbf{r}}{r^5} - \frac{\mathbf{d}_1}{r^3}\right]$$

is the field (14.10) of $\mathbf{d}_1$, and $\mathbf{r}$ is the vector from $\mathbf{d}_1$ to $\mathbf{d}_2$. By direct differentiation one arrives at the result

$$\mathbf{f} = \frac{1}{4\pi\epsilon_0}\left[3\frac{(\mathbf{d}_1 \cdot \mathbf{r})\mathbf{d}_2 + (\mathbf{d}_2 \cdot \mathbf{r})\mathbf{d}_1 + (\mathbf{d}_1 \cdot \mathbf{d}_2)\mathbf{r}}{r^5} - 15\frac{(\mathbf{d}_1 \cdot \mathbf{r})(\mathbf{d}_2 \cdot \mathbf{r})\mathbf{r}}{r^7}\right].$$

16. Non-relativistic motion of a charged particle

The simplest case of motion under the Lorentz force is when the magnetic field vanishes, and **E** is constant and uniform. Under the constant acceleration $(e/m)\mathbf{E}$, the velocity must ultimately cease to be small compared with c, and a relativistic treatment becomes necessary. We shall deal with this case in Section 18.

If, conversely, $\mathbf{E} = 0$, then we know from (15.5) that the speed must be constant. If it is initially non-relativistic, it will remain so. The equation of motion (15.4), if $\mathbf{E} = 0$, becomes

$$\dot{\mathbf{v}} = -\frac{e}{m}\mathbf{B} \times \mathbf{v}, \tag{16.1}$$

which means (compare with $\dot{\mathbf{r}} = \mathbf{\Omega} \times \mathbf{r}$) that the velocity vector rotates around an axis parallel to the magnetic field. Only if **B** is constant and uniform will the rotation be rigid. The angular frequency has the magnitude

$$\Omega = \frac{eB}{m}. \tag{16.2}$$

It is called the *gyration frequency*. Viewed along the magnetic field, the velocity vector gyrates in a counter-clockwise sense for a positive charge (clockwise for a negative charge); this is due to the minus sign in (16.1). If the direction of **B** is taken to be the z axis, the Cartesian components of (16.1) are

$$\dot{v}_x = \Omega v_y, \qquad \dot{v}_y = -\Omega v_x, \qquad \dot{v}_z = 0. \tag{16.3}$$

Thus v_z is constant. Multiplying the y component of (16.3) by $i = \sqrt{-1}$ and adding the x component gives

$$\frac{d}{dt}(v_x + iv_y) = -i\Omega(v_x + iv_y), \tag{16.4}$$

which has the solution $v_x + iv_y = ae^{-i\Omega t}$, where a is a complex constant. Writing $a = v_\perp e^{-i\alpha}$, where $v_\perp$ and α are real, we have

$$v_x + iv_y = v_\perp e^{-i(\Omega t + \alpha)}. \tag{16.5}$$

Thus $v_\perp$ is the constant magnitude of the projection of the particle's velocity in the xy plane, and α is its initial phase. Finally, we integrate (16.5), and separate real and imaginary parts. The result is

$$x = x_0 + r\sin(\Omega t + \alpha), \qquad y = y_0 + r\cos(\Omega t + \alpha), \tag{16.6}$$

where

$$r = \frac{v_\perp}{\Omega} = \frac{mv_\perp}{eB} \tag{16.7}$$

is the *radius of gyration*. Since, according to (16.3)$_3$, $v_z = v_\parallel$ is constant,

$$z = z_0 + v_\parallel t. \tag{16.8}$$

Thus the charged particle describes a helix with axis along the magnetic field. After each gyration, which the particle completes in a time $2\pi/\Omega$, it advances a distance $v_\parallel(2\pi/\Omega)$ along the axis of the helix. Thus the pitch of the helix, which is the ratio of the latter distance to the gyration radius, is $2\pi v_\parallel/v_\perp$, which is determined by the initial velocities. If $v_\parallel = 0$, the motion is along a circle in the plane perpendicular to the field. In general the average motion—the term signifies averaging over a gyration period—is that of a point moving along the axis of the helix. This point is called *the guiding centre*.

Next, we consider the case of mutually perpendicular, non-vanishing, constant and uniform **E** *and* **B**. We know, from the discussion at the end of Section 12, that one of these fields can be transformed away. If $c^2 B^2 > E^2$, this is the electric field. Indeed the transformation to a frame moving with velocity

$$\mathbf{u} = \frac{\mathbf{E} \times \mathbf{B}}{B^2} \tag{16.9}$$

will accomplish this, as is easily verified from the transformation formula (12.11) for **E**. We shall assume, more stringently, that

$$E^2 \ll c^2 B^2, \tag{16.10}$$

rather than merely $E^2 < c^2 B^2$. The **u** of (16.9) is then small compared with c, and we can replace the Lorentz by the Galilean transformation. In the new frame, we have (cf. (9.8)) $\mathbf{E}' = 0$ and $\mathbf{B}' = \mathbf{B}$; in other words, helical motion. Hence, in the original frame, we have the helical motion, superposed on the rectilinear motion at the constant *drift velocity* (16.9), which is perpendicular to both **E** and **B** (Fig. 16.1). This is one of those rare cases in which words are simpler than formulae.

If **E** and **B** are not perpendicular, **E** cannot be transformed away, and the motion—in any frame—must ultimately become relativistic.

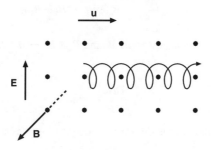

Fig. 16.1

A more complicated case arises when the magnetic field is not uniform. If the non-uniformity is slight, that is, if **B** changes only slightly over the gyration radius, a method of approximation[†] suggests itself, resulting in a motion of the guiding centre which is no longer rectilinear. We shall assume, for the time being, that $v_\parallel = 0$, so that the particle moves in a circle.

We write the trajectory of the particle in the form $\mathbf{x} = \mathbf{R}(t) + \mathbf{s}(t)$, where $\mathbf{R}(t)$ is the trajectory of the guiding centre, a slowly varying vector, while $\mathbf{s}(t)$, a rapidly varying vector, is the gyration motion around the guiding centre. We approximate **B** by

$$\mathbf{B}(\mathbf{x}) = \mathbf{B}(\mathbf{R}) + (\mathbf{s} \cdot \mathbf{grad})\mathbf{B}(\mathbf{R}), \tag{16.11}$$

and average the magnetic force $e\dot{\mathbf{x}} \times \mathbf{B}$ over a gyration period. The terms of first order in **s** will vanish on averaging. The second-order term is the average of

$$\mathbf{f} = e\dot{\mathbf{s}} \times (\mathbf{s} \cdot \mathbf{grad})\mathbf{B}. \tag{16.12}$$

According to eqns (16.6)–(16.7), the gyration vector **s** satisfies

$$\dot{\mathbf{s}} = \frac{e}{m}\mathbf{s} \times \mathbf{B}, \qquad s = \frac{mv_\perp}{eB}. \tag{16.13}$$

Hence

$$\mathbf{f} = \frac{e^2}{m}(\mathbf{s} \times \mathbf{B}) \times (\mathbf{s} \cdot \mathbf{grad})\mathbf{B}$$

$$= \frac{e^2}{m}[\mathbf{B}(\mathbf{s} \cdot (\mathbf{s} \cdot \mathbf{grad})\mathbf{B}) - \mathbf{s}(\mathbf{B} \cdot (\mathbf{s} \cdot \mathbf{grad})\mathbf{B})]. \tag{16.14}$$

In averaging the product $s_i s_j$ of two components of **s**, which is a rotating vector in the plane perpendicular to **B**, we note that the average of $s_i s_j$ must vanish when $i \neq j$, and that the averages of s_1^2 and s_2^2 must be equal. Hence the average of the product $s_i s_j$ is $\frac{1}{2}s^2\delta_{ij}$. Using this result, we find that the average of the force (16.14) is

$$f_i = \frac{\frac{1}{2}mv_\perp^2}{B^2}(B_i\partial_k B_k - B_k\partial_i B_k). \tag{16.15}$$

The sum $\partial_k B_k$ vanishes because it is div **B**, and $B_k\partial_i B_k = \partial_i \frac{1}{2}B_k B_k = \partial_i \frac{1}{2}B^2$. Thus, finally,

$$\mathbf{f} = -\frac{\frac{1}{2}mv_\perp^2}{B}\,\mathbf{grad}\, B. \tag{16.16}$$

The average force is thus in the direction of decreasing magnetic field strength.

[†]Alfvén (1940).

The foregoing derivation has assumed that $v_\parallel = 0$. If $v_\parallel$ does not vanish, the guiding centre is moving along a magnetic field line. Let ρ be the radius of curvature of the field line. We transform to a reference frame which is rotating around the centre of curvature of the field line with angular velocity $v_\parallel/\rho$. In this frame there is no longitudinal velocity, but there is a centrifugal force of magnitude $mv_\parallel^2/\rho$, directed away from the centre of curvature. This centrifugal force, and the force (16.16), are to be regarded as acting, not on the original charged particle, but on a substitute particle carried along by the guiding centre.

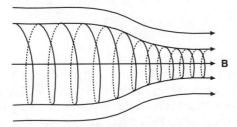

Fig. 16.2

In order to illustrate the method, we first consider the 'magnetic bottle' of Fig. 16.2, with the guiding centre moving along the central field line, which we take as the z axis. Since this field line is straight, there is no centrifugal force, and the equation of motion for the substitute particle, with the force (16.16), is

$$m\dot{v}_\parallel = -\frac{\frac{1}{2}mv_\perp^2}{B}\frac{dB}{dz}.$$

Multiplying by $v_\parallel$ and noting that $v_\parallel dB/dz = \dot{B}$, we obtain

$$\left(\tfrac{1}{2}mv_\parallel^2\right)^{\cdot} = -\frac{\frac{1}{2}mv_\perp^2}{B}\dot{B}. \tag{16.17}$$

Now $v_\parallel^2 = v^2 - v_\perp^2$, where v is the constant speed of the (real) particle. This gives $(\tfrac{1}{2}mv_\perp^2)^{\cdot} = (\tfrac{1}{2}mv_\perp^2/B)\dot{B}$ or

$$\frac{\frac{1}{2}mv_\perp^2}{B} = \mu, \tag{16.18}$$

another constant of motion. We can now write (16.17) in the form

$$\tfrac{1}{2}mv_\parallel^2 + \mu B = \text{const.},$$

where the (energy) constant is determined by the initial conditions. The guiding centre is thus slowed down by the increasing magnetic field strength in the bottle-neck, and may even be reflected. From (16.18) and (16.7) we can also infer that

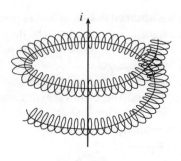

Fig. 16.3

$\pi r^2 B$ is constant. This means that the real particle encloses a constant magnetic flux throughout its helical motion: it moves on the surface of a flux tube.

A more difficult case is presented by the magnetic field of a straight, current-carrying wire, as in Fig. 16.3. In the lowest approximation, the guiding centre moves in a circle along one of the field lines. Since the magnetic field strength is, according to (13.3), inversely proportional to the distance from the wire—the cylindrical coordinate r—the force (16.16) is $\frac{1}{2}mv_\perp^2/r$, directed away from the wire. The centrifugal force, also directed away from the wire, is $mv_\parallel^2/r$ because r is the radius of curvature of the field line. The resultant of the two forces is equivalent to an electric field of strength

$$ E = \frac{m}{er}\left(v_\parallel^2 + \tfrac{1}{2}v_\perp^2\right), $$

directed away from the wire (for positive e). Since this electric field is perpendicular to **B**, it causes, according to (16.9), a drift velocity

$$ v_z = \frac{v_\parallel^2 + \tfrac{1}{2}v_\perp^2}{\Omega r}. $$

Thus the guiding centre moves in a helix which has the wire as its axis. The *real* particle moves in a helix wound around this latter helix, that is, in a double helix. Again, this motion is one which words describe more simply than formulae. Nor is it easy to see how this solution could have been obtained, let alone guessed, directly from 'first principles', that is, from the original equation of motion (15.4).

17. Lagrangian and Hamiltonian formulation

We now seek a Lagrangian formulation for (15.4). In non-relativistic mechanics forces are absolute, that is, frame-independent. The Lorentz force $e\mathbf{E}'$ satisfies this requirement, because $\mathbf{E}'$ is the electric field in a unique frame. For the Lagrangian we similarly propose the function $L = \frac{1}{2}m\dot{x}^2 - eV'$, where V' is the electric potential in the particle's rest frame Σ'. According to (10.11), $V' = V - \dot{\mathbf{x}} \cdot \mathbf{A}$, where V

and $\mathbf{A}$ are the electric and magnetic (vector) potentials in Σ. Thus our proposed Lagrangian function is

$$L(\mathbf{x}, \dot{\mathbf{x}}, t) = \tfrac{1}{2}m\dot{x}^2 + e[\dot{\mathbf{x}} \cdot \mathbf{A}(\mathbf{x}, t) - V(\mathbf{x}, t)] \tag{17.1}$$

But there is a second requirement, besides frame-independence of force, which L must satisfy: the equation of motion (15.4) is unaffected by a gauge transformation,

$$\tilde{\mathbf{A}} = \mathbf{A} + \mathbf{grad}\,\chi, \qquad \tilde{V} = V - \chi_t \tag{17.2}$$

(cf. (10.7)), because (15.4) involves the fields, not the potentials; hence the change in L, when a gauge transformation is made, must be such that Lagrange's equations will not be affected. Now, under the gauge transformation (17.2), the Lagrangian (17.1) will change to

$$\tilde{L} = L + e[\dot{\mathbf{x}} \cdot \mathbf{grad}\,\chi + \chi_t] = L + \frac{d}{dt}e\chi(\mathbf{x}(t), t). \tag{17.3}$$

This is quite satisfactory: according to Lagrangian mechanics, adding to L the time derivative of a function has no effect on Lagrange's equations.

It remains to show that (17.1) is indeed a Lagrangian for the equation of motion (15.4). The proof is straightforward. One must of course make use of the relations (10.6) between the derivatives of $\mathbf{A}$ and V and the fields:

$$\mathbf{B} = \mathbf{curl}\,\mathbf{A}, \qquad \mathbf{E} = -\mathbf{A}_t - \mathbf{grad}\,V. \tag{17.4}$$

Exercise

17.1. Let $L_{\mathbf{x}}$ be the vector with i'th component $\partial L/\partial x^i$. Show that Lagrange's equations,

$$\frac{d}{dt}L_{\dot{\mathbf{x}}} - L_{\mathbf{x}} = 0, \tag{17.5}$$

with (17.1), lead to the equation of motion (15.4).

If we define, for any path $\mathbf{x}(t)$, an *action* by

$$S = \int_{t_1}^{t_2} L(\mathbf{x}(t), \dot{\mathbf{x}}(t), t)\, dt, \tag{17.6}$$

then Lagrange's equations (17.5) are equivalent to the action principle

$$\delta S = 0, \tag{17.7}$$

where the variation is to be carried out with respect to all paths $\mathbf{x}(t)$ with given end points $\mathbf{x}_1 = \mathbf{x}(t_1)$ and $\mathbf{x}_2 = \mathbf{x}(t_2)$. For an assembly of particles, we take the action to be a sum of individual actions, one for each particle. By varying the action with respect to the path of the nth particle we are then led to the equation of motion (15.4) for that particle.

From the Lagrangian (17.1), it is a straightforward matter to pass to a Hamiltonian formulation. We define

$$\mathbf{p} = L_{\dot{\mathbf{x}}} = m\dot{\mathbf{x}} + e\mathbf{A}(\mathbf{x}, t) \tag{17.8}$$

as the generalized momentum and note that this relation may serve to express $\dot{\mathbf{x}}$ in terms of $\mathbf{x}$, $\mathbf{p}$ and t. We now form the Hamiltonian

$$H(\mathbf{x}, \mathbf{p}, t) = \mathbf{p} \cdot \dot{\mathbf{x}}(\mathbf{x}, \mathbf{p}, t) - L\big(\mathbf{x}, \dot{\mathbf{x}}(\mathbf{x}, \mathbf{p}, t), t\big). \tag{17.9}$$

Exercise

17.2. Show that

$$H(\mathbf{x}, \mathbf{p}, t) = \frac{(\mathbf{p} - e\mathbf{A})^2}{2m} + eV, \tag{17.10}$$

and that (17.8) and (15.4) follow from Hamilton's equations

$$\dot{\mathbf{x}} = H_{\mathbf{p}}, \qquad \dot{\mathbf{p}} = -H_{\mathbf{x}}. \tag{17.11}$$

The main advantage of the Lagrangian formulation is that the Cartesian coordinates $x_i(t)$ may be replaced by generalized coordinates $q_j(t)$; for example, cylindrical coordinates, or latitude and longitude for a body which is constrained to move on a sphere. We denote the collection of q's by $\mathbf{q}$, and the collection of $\dot{q}$'s by $\dot{\mathbf{q}}$. Lagrange's equations are then

$$\frac{d}{dt}L_{\dot{\mathbf{q}}} - L_{\mathbf{q}} = 0, \tag{17.12}$$

where $L_{\mathbf{q}}$ and $L_{\dot{\mathbf{q}}}$ are, respectively, the vectors with i'th component $\partial L/\partial q^i$ and $\partial L/\partial \dot{q}^i$. These equations are equivalent to the action principle

$$\delta \int_{t_1}^{t_2} L\big(\mathbf{q}(t), \dot{\mathbf{q}}(t), t\big) \, dt = 0, \qquad \delta\mathbf{q}(t_1) = \delta\mathbf{q}(t_2) = 0. \tag{17.13}$$

The generalized momenta are defined by

$$\mathbf{p} = L_{\dot{\mathbf{q}}}(\mathbf{q}, \dot{\mathbf{q}}, t). \tag{17.14}$$

If $\dot{\mathbf{q}}(\mathbf{q}, \mathbf{p}, t)$ are the solutions of these equations, the Hamiltonian is defined by

$$H(\mathbf{q}, \mathbf{p}, t) = \mathbf{p} \cdot \dot{\mathbf{q}}(\mathbf{q}, \mathbf{p}, t) - L(\mathbf{q}, \dot{\mathbf{q}}(\mathbf{q}, \mathbf{p}, t), t). \tag{17.15}$$

In terms of the q's, the p's and H, Lagrange's equations (17.12) and the equations (17.14) are equivalent to Hamilton's equations

$$\dot{\mathbf{q}} = H_{\mathbf{p}}, \qquad \dot{\mathbf{p}} = -H_{\mathbf{q}}. \tag{17.16}$$

Moreover,

$$\dot{H} = H_t(\mathbf{q}, \mathbf{p}, t) = -L_t(\mathbf{q}, \dot{\mathbf{q}}, t). \tag{17.17}$$

From the Hamiltonian formulation, it is a straightforward matter to proceed to a formulation in terms of Poisson brackets. Such a canonical formulation is a pre-requisite for setting up a quantum mechanics of charged particles. Indeed one of the objects of the classical theory of charged particles is to suggest a Hamiltonian for quantum mechanics. But the classical Lagrangian and Hamiltonian formulations, too, have some advantages.

Problem 17.1 Let (r, ϕ, z) be cylindrical coordinates. According to Problem 10.1, any poloidal magnetic field is derivable from a toroidal vector potential $\mathbf{A} = (0, A_\phi(r, z), 0)$. Prove that, for a particle moving in any such field, there exists a constant of motion γ such that

$$mr^2 \dot{\phi} + er A_\phi = m\gamma.$$

If, furthermore, the electric field vanishes, prove that the available (r, z) region for the particle is restricted by

$$\left| \gamma - \frac{e}{m} r A_\phi(r, z) \right| \leq rv,$$

where v is the constant speed of the particle.

Solution The velocity is $\mathbf{v} = (\dot{r}, r\dot{\phi}, \dot{z})$, and the Lagrangian

$$L = \frac{m}{2}(\dot{r}^2 + r^2\dot{\phi}^2 + \dot{z}^2) + e[r\dot{\phi}A_\phi(r, z) - V(r, z)]$$

is independent of ϕ. Hence

$$p_\phi = L_{\dot{\phi}} = mr^2\dot{\phi} + er A_\phi$$

is a constant $m\gamma$ (say). If the electric field vanishes, the speed v is constant. Obviously $|r\dot{\phi}| \leq v$. Thus

$$|r^2\dot{\phi}| = \left| \gamma - \frac{e}{m} r A_\phi(r, z) \right| \leq rv.$$

The last inequality can be written as

$$\gamma - rv \leq (e/m) r A_\phi(r, z) \leq \gamma + rv.$$

According to Problem 10.2, $r A_\phi(r, z) = $ const. is the equation of a field line. A charged particle with given angular momentum constant γ is therefore restricted to field lines in the vicinity of the field line $r A_\phi(r, z) = m\gamma/e$, the vicinity depending on the constant speed v and on the distance r from the magnetic axis. This result is used in explaining the Earth's radiation belts, and the *aurorae* (*aurora borealis*, or the northern lights, in the northern polar regions, and *aurora australis* in the southern polar regions).

---

18. Relativistic motion of charged particles

Since particles can be accelerated to relativistic speeds, the foregoing non-relativistic theory does not suffice. We indicate the changes required for rendering the formulation relativistic. First, we take the action to be

$$S = \int [-mc^2 \sqrt{1 - \dot{x}^2/c^2} + e(\dot{\mathbf{x}} \cdot \mathbf{A} - V)] \, dt; \tag{18.1}$$

the first term is now $-\int mc^2 \, d\tau$, where τ is the *proper time* along the path $\mathbf{x}(t)$; the second is $\int e w^\alpha A_\alpha \, d\tau$, where $(w^\alpha) = (\dot{\mathbf{x}}, 1)/\sqrt{1 - \dot{x}^2/c^2}$ is the 4-velocity (12.19) and $(A_\alpha) = (\mathbf{A}, -V)$ is the electromagnetic potential. Clearly, the action (18.1) is Lorentz–invariant. The relativistic Lagrangian

$$L = -mc^2 \sqrt{1 - \dot{x}^2/c^2} + e(\dot{\mathbf{x}} \cdot \mathbf{A} - V) \tag{18.2}$$

now leads to the generalized momentum

$$\mathbf{p} = \frac{m\dot{\mathbf{x}}}{\sqrt{1 - \dot{x}^2/c^2}} + e\mathbf{A} = \mathbf{p}_k + e\mathbf{A}, \tag{18.3}$$

where the kinetic momentum

$$\mathbf{p}_k = \frac{m\dot{\mathbf{x}}}{\sqrt{1 - \dot{x}^2/c^2}} \tag{18.4}$$

replaces the former $m\dot{\mathbf{x}}$. Lagrange's equations with the Lagrangian (18.2) lead to the relativistic equation of motion

$$\dot{\mathbf{p}}_k = e(\mathbf{E} + \dot{\mathbf{x}} \times \mathbf{B}). \tag{18.5}$$

We define

$$\mathcal{K} = \frac{mc^2}{\sqrt{1 - \dot{x}^2/c^2}} \tag{18.6}$$

and call this *the relativistic kinetic energy*. For small speeds it reduces to $\frac{1}{2}m\dot{x}^2 + mc^2$. Thus $\mathcal{K}$ includes the constant *rest-energy* mc^2.

Exercise

18.1. Prove the relations

$$\mathbf{p}_k = \mathcal{K}\dot{\mathbf{x}}/c^2, \qquad \dot{\mathcal{K}} = \dot{\mathbf{x}} \cdot \dot{\mathbf{p}}_k. \tag{18.7}$$

From (18.5) and (18.7)$_2$ we obtain the relativistic analogue of (15.5):

$$\dot{\mathcal{K}} = e\dot{\mathbf{x}} \cdot \mathbf{E}. \tag{18.8}$$

It is still true that, if $\mathbf{E} = 0$, the speed is constant. Substitution of (18.7)$_1$ in (18.5), and the use of (18.8), yield

$$m\ddot{\mathbf{x}} = e\sqrt{1 - \dot{x}^2/c^2}\left(\mathbf{E} + \dot{\mathbf{x}} \times \mathbf{B} - \frac{1}{c^2}(\dot{\mathbf{x}} \cdot \mathbf{E})\dot{\mathbf{x}}\right), \tag{18.9}$$

which is the relativistic generalization of (15.4).

The relativistic Hamiltonian is obtained by substituting from (18.2) and (18.3) in the definition (17.9). A straightforward calculation leads to

$$H = \mathcal{K} + eV = \sqrt{m^2c^4 + (\mathbf{p} - e\mathbf{A})^2c^2} + eV. \tag{18.10}$$

For an assembly of charged particles, the relativistic action is taken to be the sum of relativistic actions, one for each particle. Variation of the action then delivers equations of motion for each one of the particles.

An alternative to Hamilton's equations is provided by the Hamilton–Jacobi partial differential equation

$$\frac{\partial S}{\partial t} + H\left(\mathbf{q}, \frac{\partial S}{\partial \mathbf{q}}, t\right) = 0. \tag{18.11}$$

For the Hamiltonian (17.10), this equation becomes

$$\frac{\partial S}{\partial t} + \frac{(\mathrm{grad}\, S - e\mathbf{A})^2}{2m} + eV = 0. \tag{18.12}$$

For the relativistic Hamiltonian (18.10), it is

$$\frac{\partial S}{\partial t} + c\sqrt{m^2c^2 + (\mathrm{grad}\, S - e\mathbf{A})^2} + eV = 0. \tag{18.13}$$

Since only the derivatives of S appear in (18.11), its solution S will contain an arbitrary additive constant. Any solution

$$S = S(\mathbf{q}, \mathbf{a}, t) \tag{18.14}$$

of the Hamilton–Jacobi equation (18.11) which depends, *apart* from the additive constant, on n constants $\mathbf{a}$, n being the number of q's, and satisfies the condition

$$\det\left(\frac{\partial^2 S}{\partial q_r \partial a_s}\right) \neq 0, \tag{18.15}$$

is called a *complete integral*. There may be many of them.

The Hamilton–Jacobi theorem states: if S is *any* complete integral of the form (18.14), the solutions $q_s(\mathbf{a}, \mathbf{b}, t)$ of the equations

$$\frac{\partial S}{\partial a_r} = -b_r, \tag{18.16}$$

the b_r being n arbitrary constants, satisfy Lagrange's equations (17.12)[†]; and these solutions $q_s(\mathbf{a}, \mathbf{b}, t)$ of (18.16), together with the momenta defined by

$$p_r = \frac{\partial S}{\partial q_r}, \tag{18.17}$$

provide the solutions of Hamilton's equations (17.16).

We shall use the Hamilton-Jacobi theorem to solve two problems. The first one is the motion of a charged particle in a constant electric field, of magnitude E in the x direction, say. This is obviously described by the potentials

$$\mathbf{A} = 0, \qquad V = -Ex. \tag{18.18}$$

We shall seek a (one-dimensional) complete integral $S(x, \alpha, t)$ of the relativistic Hamilton–Jacobi equation (cf. (18.13))

$$\frac{\partial S}{\partial t} + c\sqrt{m^2c^2 + \left(\frac{\partial S}{\partial x}\right)^2} - eEx = 0 \tag{18.19}$$

in the form

$$S = -\alpha t + \phi(x), \tag{18.20}$$

where α is a constant. Substituting (18.20) in (18.19), we obtain for $\phi(x)$ the differential equation

$$-\alpha + c\sqrt{m^2c^2 + (\phi')^2} - eEx = 0. \tag{18.21}$$

[†]It is in order to ensure the solvability of (18.16) that we require the property (18.15) of a complete integral. In practice one guesses a solution $S(\mathbf{q}, \mathbf{a}, t)$ and proceeds to solve (18.16), without bothering about (18.15). If (18.16) can be solved for the $q_s(\mathbf{a}, \mathbf{b}, t)$, then $S(\mathbf{q}, \mathbf{a}, t)$ must be a complete integral.

Thus

$$S = -\alpha t + \frac{1}{c} \int^x \sqrt{f(\xi)} \, d\xi, \qquad f(x) = (\alpha + eEx)^2 - m^2 c^4, \qquad (18.22)$$

is a complete integral, provided we can solve (18.16), which in the present case reduces to a single equation. The indefinite integral in (18.22) contains an arbitrary additive constant, but we have agreed to ignore this. In order to find the solution of Lagrange's problem, we use (18.16), with the name t_0 for the arbitrary b_1:

$$-t_0 = \frac{\partial S}{\partial \alpha} = -t + \frac{1}{c} \int^x \frac{\alpha + eE\xi}{\sqrt{f(\xi)}} \, d\xi. \qquad (18.23)$$

The last integral is easy, and we obtain the solution

$$\sqrt{f(x)} = eEc(t - t_0) + \text{const.}, \qquad (18.24)$$

which shows that the position x increases monotonically with t. The velocity $\dot{x}$, which can be computed from (18.24), is always less than c, but approaches the latter as t, and x, go to infinity.

19. The relativistic Kepler problem

The second problem to which we apply the Hamilton–Jacobi method is the orbit of an electron (charge $-e$) in the inverse square field of a nucleus (charge Ze); the latter is assumed to remain at rest at the origin of an aether frame. The Hamiltonian for this two-dimensional problem is

$$H = c\sqrt{m^2 c^2 + p_r^2 + r^{-2} p_\phi^2} - mc^2 + U,$$

$$U = -\frac{k}{r}, \quad k = \frac{Ze^2}{4\pi\epsilon_0}. \qquad (19.1)$$

We have subtracted mc^2 from the kinetic energy, in order to facilitate the comparison with the non-relativistic limit at each stage of the calculation. According to (17.17), since H has no explicit time dependence, there is a constant of energy $E = H$. We set

$$S = -Et + W(r, \phi), \qquad (19.2)$$

and obtain for W the modified Hamilton–Jacobi equation

$$c\sqrt{m^2 c^2 + \left(\frac{\partial W}{\partial r}\right)^2 + \frac{1}{r^2}\left(\frac{\partial W}{\partial \phi}\right)^2} = E - U + mc^2. \qquad (19.3)$$

The fact that ϕ is cyclic in the Hamiltonian suggests that we seek a W of the form

$$W = \alpha\phi + R(r), \tag{19.4}$$

where α is the second (angular momentum) constant. Substitution in (19.3) gives the differential equation

$$(R')^2 = 2m(E - U) + \frac{(E - U)^2}{c^2} - \frac{\alpha^2}{r^2} \equiv f(r). \tag{19.5}$$

Thus

$$W = \alpha\phi + \int^r \sqrt{f(s)}\, ds. \tag{19.6}$$

The first of the eqns (18.16), now with the name t_0 for the arbitrary b_1, is

$$t - t_0 = \frac{\partial W}{\partial E} = \frac{1}{c^2} \int^r \frac{mc^2 + E - U}{\sqrt{f(s)}}\, ds, \tag{19.7}$$

and the second equation, with the name β for the arbitrary b_2, is

$$-\beta = \frac{\partial W}{\partial \alpha} = \phi - \int^r \frac{\alpha/s^2}{\sqrt{f(s)}}\, ds. \tag{19.8}$$

The last equation, which does not contain the time, determines the orbit $r(\phi)$. Equation (19.7) describes the motion along the orbit. For the U of (19.1)$_2$, the f of (19.5) is

$$f(r) = E\left(2m + \frac{E}{c^2}\right) + 2k\left(m + \frac{E}{c^2}\right)\frac{1}{r} - \alpha^2\left(1 - \frac{k^2}{c^2\alpha^2}\right)\frac{1}{r^2}, \tag{19.9}$$

which has two positive real roots if $-2mc^2 < E < 0$ and $k < c\alpha$; these conditions are satisfied under normal circumstances, on account of the largeness of c. Denoting the smaller of these roots by r_1, and the larger by r_2, we can write

$$f(r) = \alpha^2\lambda^2\left(\frac{1}{r_1} - \frac{1}{r}\right)\left(\frac{1}{r} - \frac{1}{r_2}\right), \quad \lambda^2 = 1 - \frac{k^2}{c^2\alpha^2}. \tag{19.10}$$

In the orbit equation (19.8) we can redefine the arbitrary constant β by choosing r_1 as the lower limit of the integral. In order to evaluate the integral, we write

$$\frac{1}{r} = \frac{1}{2}\left(\frac{1}{r_1} + \frac{1}{r_2}\right) + \frac{1}{2}\left(\frac{1}{r_1} - \frac{1}{r_2}\right)\cos\psi, \tag{19.11}$$

and make a change of variable from r to ψ. This gives

$$\lambda(\phi + \beta) = \psi. \tag{19.12}$$

Thus the equation of the orbit is

$$\frac{1}{r} = \frac{1}{2}\left(\frac{1}{r_1} + \frac{1}{r_2}\right) + \frac{1}{2}\left(\frac{1}{r_1} - \frac{1}{r_2}\right)\cos\lambda(\phi + \beta). \qquad (19.13)$$

For $\lambda = 1$ this would be an ellipse. According to $(19.10)_2$, however, $\lambda < 1$. Thus the orbit is not closed: to recover the same distance r, ϕ must change by $2\pi/\lambda > 2\pi$. If the difference $2\pi/\lambda - 2\pi$ is small, the orbit may be regarded as an ellipse with slowly rotating, or *precessing*, axes.

If we replace k by GMm, where G is the Newtonian gravitational constant and M is the mass of the sun, (19.13) gives the relativistic orbit of a planet of mass m. (That is why (19.1) is called the Kepler problem.) Einstein realized that, in the case of Mercury, the precession $2\pi(1/\lambda - 1)$ of the orbit gave only one sixth of the observed value. The correct amount of precession was obtained when he replaced Newtonian gravitation by his own gravitational theory, the general theory of relativity.

7

Polarization and magnetization

20. Material response functions

Experience shows that many materials *respond*, as it were, to an electromagnetic field by setting up charge and current distributions. This takes place in a myriad of forms, depending on the kind of material, on its state, and even on its history. Indeed, the resulting charge and current distributions are so diverse that, in constructing a theory of these phenomena, it is wise to keep an open mind. In particular, we should not impose on them any restrictions beyond those that we regard as absolutely necessary. Such a restriction is the principle of charge conservation, which we are certainly not about to renounce. A simple way of imposing it is to use what we have already learned and characterize the charge and current distributions of material response by a charge-current potential,

$$\mathbf{D}_R = -\mathbf{P}, \qquad \mathbf{H}_R = \mathbf{M}, \qquad (20.1)$$

where $\mathbf{P}$ (the sign is conventional) and $\mathbf{M}$ are two piecewise smooth vector fields which depend in the most general way on the material, on its state—such as its density, temperature, elastic strain—and on the electromagnetic field, that is, $\mathbf{E}$ and $\mathbf{B}$. Indeed, $\mathbf{P}$ and $\mathbf{M}$ may at any time depend, not only on the values of these variables at that time, but even on their histories. The field $\mathbf{M}$ is called the (*Minkowski*) *magnetization*; the field $\mathbf{P}$, the *dielectric polarization*, or simply the *polarization*.

Of course a constant $\mathbf{P}$, or a constant $\mathbf{M}$, are examples of a particularly simple response, just as $f(x) = c$ is a particularly simple function of x. Such constant response potentials describe *permanent polarization*, or *permanent magnetization*.

The charge and current distributions that correspond to $\mathbf{P}$ and $\mathbf{M}$—often called *bound* charges and currents—can now be obtained from the relations of Section 6. Wherever $\mathbf{P}$ and $\mathbf{M}$ are smooth,

$$q_R = -\operatorname{div}\mathbf{P}, \qquad \mathbf{j}_R = \operatorname{curl}\mathbf{M} + \mathbf{P}_t \qquad (20.2)$$

(cf. (6.5)). These q_R and $\mathbf{j}_R$ obey the law of charge conservation (1.14), as they must. On a surface across which $\mathbf{M}$ or $\mathbf{P}$ suffer discontinuities—we expect this to happen, for example, on the boundary between two materials—there will be surface charges and currents (cf. (7.8)):

$$\sigma_R = -\mathbf{n} \cdot [\![\mathbf{P}]\!], \qquad \mathbf{K}_R = \mathbf{n} \times [\![\mathbf{M}]\!] - v_n [\![\mathbf{P}]\!]. \tag{20.3}$$

Under Galilean transformations, $\mathbf{P}$ and $\mathbf{M}$ will transform as follows:

$$\begin{aligned} \mathbf{P}' &= \mathbf{P}, \\ \mathbf{M}' &= \mathbf{M} + \mathbf{u} \times \mathbf{P} \end{aligned} \tag{20.4}$$

(cf. (6.8)). Thus $\mathbf{P}$ and q_R are Galilean invariants, but $\mathbf{M}$ and $\mathbf{j}_R$ are not.

As in (5.9) and (6.9), we can construct two Galilean invariants:

$$\begin{aligned} \mathcal{M} &= \mathbf{M} + \mathbf{v} \times \mathbf{P}, \\ \mathcal{J}_R &= \mathbf{j}_R - q_R \mathbf{v} = \mathbf{curl}\, \mathbf{M} + \mathbf{P}_t + \mathbf{v}\, \mathrm{div}\, \mathbf{P}. \end{aligned} \tag{20.5}$$

The first invariant, $\mathcal{M}$, is called the *Lorentz magnetization*.

Under a Lorentz transformation, $-\mathbf{P}$ and $\mathbf{M}$ will transform like the $\mathbf{D}$ and $\mathbf{H}$ of (12.13)–(12.14):

$$\begin{aligned} \mathbf{P}' &= \gamma(\mathbf{P} - \mathbf{u} \times \mathbf{M}/c^2) - \frac{\gamma^2}{\gamma + 1} \mathbf{u}(\mathbf{u} \cdot \mathbf{P})/c^2, \\ \mathbf{M}' &= \gamma(\mathbf{M} + \mathbf{u} \times \mathbf{P}) - \frac{\gamma^2}{\gamma + 1} \mathbf{u}(\mathbf{u} \cdot \mathbf{M})/c^2. \end{aligned} \tag{20.6}$$

We shall now discuss some simple properties of the foregoing charge-current distributions. Consider, first, an isolated body (in empty space) which is polarized, but not magnetized, i.e. $\mathbf{M} = 0$. It then has a volume charge density $- \mathrm{div}\, \mathbf{P}$ and (if we choose $\mathbf{n}$ to point outward) a surface charge density $-\mathbf{n} \cdot [\![\mathbf{P}]\!] = \mathbf{n} \cdot \mathbf{P} = P_n$, since $\mathbf{P} = 0$ in the empty space outside the body. Taken together, the total bound charge on the body is $\int - \mathrm{div}\, \mathbf{P}\, dV + \oint P_n\, dS = 0$, by Gauss's theorem.

Problem 20.1   Prove that, for an isolated, polarized body,

$$\int (\mathbf{x} - \mathbf{x}_O) q_R\, dV + \oint (\mathbf{x} - \mathbf{x}_O) \sigma_R\, dS = \int \mathbf{P}\, dV, \tag{20.7}$$

where $\mathbf{x}_O$ is the position of a fixed point O and $\mathbf{x}$ is the position at which the integrand is evaluated.

Solution   It is required to show that

$$\int (\mathbf{x} - \mathbf{x}_O)_i (- \mathrm{div}\, \mathbf{P})\, dV + \oint (\mathbf{x} - \mathbf{x}_O)_i P_n\, dS = \int P_i\, dV.$$

By Gauss's theorem

$$\oint dS(\mathbf{x} - \mathbf{x}_O)_i P_n = \int dV \operatorname{div}[(\mathbf{x} - \mathbf{x}_O)_i \mathbf{P}]$$

$$= \int dV[(\mathbf{x} - \mathbf{x}_O)_i \operatorname{div} \mathbf{P} + \operatorname{\mathbf{grad}}(\mathbf{x} - \mathbf{x}_O)_i \cdot \mathbf{P}]$$

$$= \int dV(\mathbf{x} - \mathbf{x}_O)_i \operatorname{div} \mathbf{P} + \int P_i \, dV.$$

The expression on the left-hand side of (20.7) is the *dipole moment* with respect to O of the charge distribution. Equation (20.7) shows that the dipole moment does not depend on O (this is because the total charge vanishes), and that $\mathbf{P}$ is the dipole moment per unit volume.

We can now similarly discuss some analogous properties of isolated, magnetized bodies: $\mathbf{M} \neq 0$, but $\mathbf{P} = 0$. In such a body there is a current density $\mathbf{j}_R = \operatorname{\mathbf{curl}} \mathbf{M}$ and (again choosing $\mathbf{n}$ to point outward) a surface current density $\mathbf{K}_R = \mathbf{n} \times [\![\mathbf{M}]\!] = \mathbf{M} \times \mathbf{n}$.

Problem 20.2   An isolated, magnetized body is bisected (Fig. 20.1) by a surface S. Use Stokes's theorem to prove that the net total magnetization current through S vanishes.

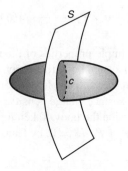

Fig. 20.1

Solution   The total magnetization current consists of two parts. The first, due to the magnetization current density $\mathbf{j}_R = \operatorname{\mathbf{curl}} \mathbf{M}$, is

$$\int \mathbf{j}_R \cdot \mathbf{dS} = \int \operatorname{\mathbf{curl}} \mathbf{M} \cdot \mathbf{dS} = \oint \mathbf{M} \cdot \mathbf{ds},$$

where, by Stokes's theorem, the last line integral is taken around the intersection curve c of S with the surface of the body. The second part is due to the surface current $\mathbf{K}_R = \mathbf{M} \times \mathbf{n}$ flowing through c along the surface of the body. Taking the scalar product of $\mathbf{K}_R$ with $\mathbf{n} \times \mathbf{ds}$, which has length ds and the correct direction of flow through c, the surface current is

$$\oint (\mathbf{M} \times \mathbf{n}) \cdot (\mathbf{n} \times \mathbf{ds}) = \oint [(\mathbf{n} \cdot \mathbf{M})(\mathbf{ds} \cdot \mathbf{n}) - (\mathbf{ds} \cdot \mathbf{M})] = -\oint \mathbf{M} \cdot \mathbf{ds},$$

because $\mathbf{n}$ is normal to $\mathbf{ds}$. The two parts of the magnetization current are therefore equal and opposite.

Problem 20.3 Prove that, in an isolated, magnetized body (with the notation of Problem 20.1),

$$\tfrac{1}{2} \int (\mathbf{x} - \mathbf{x}_O) \times \mathbf{j}_R \, dV + \tfrac{1}{2} \oint (\mathbf{x} - \mathbf{x}_O) \times \mathbf{K}_R \, dS = \int \mathbf{M} \, dV. \qquad (20.8)$$

Solution The integrand of the surface integral is

$$(\mathbf{x} - \mathbf{x}_O) \times \mathbf{K}_R = (\mathbf{x} - \mathbf{x}_O) \times (\mathbf{M} \times \mathbf{n})$$
$$= [(\mathbf{x} - \mathbf{x}_O) \cdot \mathbf{n}] \mathbf{M} - [(\mathbf{x} - \mathbf{x}_O) \cdot \mathbf{M}] \mathbf{n}.$$

We use the general form of Gauss's formula,

$$\int dV \partial_j = \oint dS n_j,$$

and obtain

$$\tfrac{1}{2} \oint (\mathbf{x} - \mathbf{x}_O) \times \mathbf{K}_R \, dS = \int \mathbf{M} \, dV - \tfrac{1}{2} \int (\mathbf{x} - \mathbf{x}_O) \times \operatorname{curl} \mathbf{M} \, dV$$
$$= \int \mathbf{M} \, dV - \tfrac{1}{2} \int (\mathbf{x} - \mathbf{x}_O) \times \mathbf{j}_R \, dV.$$

The expression on the left-hand side of (20.8) is called the *magnetic moment* with respect to O of the current distribution. Equation (20.8) shows that the magnetic moment is independent of O, and that $\mathbf{M}$ is the magnetic moment per unit volume.

The foregoing properties of polarization and magnetization in isolated bodies are often used to *define* $\mathbf{P}$ and $\mathbf{M}$ on the basis of the atomic constitution of matter. This requires a rather involved process of averaging over microscopic charges and currents, with one or two infinite series of 'higher-order multipole moments' cast away as one moves away from the isolated body, presumably in the direction of a measuring apparatus.

Quite apart from the fact that polarization and magnetization emerge as quantities that are only approximations, one is left wondering whether any quantity can ever be averaged in the absence of complete information regarding its distribution, whether electrons and nuclei can be assumed to have definite (even if unknown) positions *and* momenta, in defiance of the uncertainty principle, whether $(20.3)_2$ can be correctly obtained in this way, and so on.

There are also the usual danger signs that one has come to expect in theories of this kind—philosophical reflections on the imperfection inherent in all physical quantities, the finite accuracy of measurements, etc. In any case, we know that all this is unnecessary, because the definition of **P** and **M** follows quite naturally from the principle of charge conservation. The atomic constitution of matter is as irrelevant here as the observation, in mechanics, that a rigid body, like the physical pendulum, 'really' consists of many atoms. How many atoms are there in a *mathematical* pendulum, and what difference does it make?

21. The partial potentials: Maxwell's equations in media

The bound charges and currents that arise from the response of a material to an electromagnetic field do not necessarily constitute the total charge residing in the material or the total current passing through it. Indeed, as we have seen, the total polarization charge (volume and surface) in an isolated body always vanishes: the net charge of such a body, if it is not zero, cannot therefore be accounted for by the bound charges alone. Similarly, the total magnetization current (volume and surface) passing through an isolated body always vanishes, and cannot therefore account for a non-zero net current.

Since bodies may in general be charged or carry currents, allowance must be made for other charges and currents, beside the bound ones. We call these charges and currents *free*, and denote them by q_F and $\mathbf{j}_F$. Thus, if q and $\mathbf{j}$ stand for the total charge density and total current density, then

$$q_F = q - q_R, \qquad \mathbf{j}_F = \mathbf{j} - \mathbf{j}_R. \tag{21.1}$$

Correspondingly, we define the *partial (free) charge-current potentials* as

$$\mathbf{D}_F = \mathbf{D} - \mathbf{D}_R = \mathbf{D} + \mathbf{P}, \qquad \mathbf{H}_F = \mathbf{H} - \mathbf{H}_R = \mathbf{H} - \mathbf{M}. \tag{21.2}$$

In terms of these, the first pair of Maxwell's equations becomes

$$\operatorname{div} \mathbf{D}_F = q_F,$$
$$\operatorname{curl} \mathbf{H}_F - \frac{\partial \mathbf{D}_F}{\partial t} = \mathbf{j}_F. \tag{21.3}$$

The jump conditions, too, can be obtained by this process of subtraction:

$$\sigma_F = \mathbf{n} \cdot [\![\mathbf{D}_F]\!],$$
$$\mathbf{K}_F = \mathbf{n} \times [\![\mathbf{H}_F]\!] + v_n [\![\mathbf{D}_F]\!]. \tag{21.4}$$

To these we must now add the second pair of Maxwell's equations *and* the aether relations. If we substitute the latter in (21.2), we obtain $\mathbf{D}_F = \mathbf{D} + \mathbf{P} = \epsilon_0 \mathbf{E} + \mathbf{P}$, $\mathbf{H}_F = \mathbf{H} - \mathbf{M} = \mathbf{B}/\mu_0 - \mathbf{M}$. We recall that the aether relations refer to the *total* potentials $\mathbf{H}$ and $\mathbf{D}$, not to the *partial* potentials $\mathbf{H}_F$ and $\mathbf{D}_F$. Thus, in an aether frame,

$$\mathbf{D}_F = \epsilon_0 \mathbf{E} + \mathbf{P},$$
$$\mathbf{H}_F = \mathbf{B}/\mu_0 - \mathbf{M}. \tag{21.5}$$

A glance at (21.3)–(21.5) shows that the total charge-current $(\mathbf{j}, q)$ and the total charge-current potential $\{\mathbf{H}, \mathbf{D}\}$ have disappeared, and the equations contain only the partial, or free, quantities. We may therefore leave out the subscript $_F$ and rewrite all the equations:

$$\text{div } \mathbf{D} = q,$$
$$\text{curl } \mathbf{H} = \mathbf{j} + \mathbf{D}_t,$$
$$\text{div } \mathbf{B} = 0,$$
$$\text{curl } \mathbf{E} = -\mathbf{B}_t, \tag{21.6}$$
$$\mathbf{D} = \epsilon_0 \mathbf{E} + \mathbf{P},$$
$$\mathbf{H} = \mathbf{B}/\mu_0 - \mathbf{M}.$$

The first two equations $(21.6)_{1,2}$ *look* like (7.7), but they are not the same, because in (21.6) q and $\mathbf{j}$ are the free charge and current densities, and $\mathbf{D}$ and $\mathbf{H}$ the corresponding partial potentials. That is, of course, why $\mathbf{P}$ and $\mathbf{M}$ now appear in the aether relations $(21.6)_{5,6}$. The partial potential $\mathbf{D}$ is called the *electric displacement*, or the *displacement*.

It should perhaps be emphasized that there is nothing arbitrary in the division of the charge-current, or its potential, into bound and free parts: it is the material that decides on $\mathbf{P}$ and $\mathbf{M}$—in response to $\mathbf{E}$ and $\mathbf{B}$—just as a gas decides on its volume when it is held under a given pressure, and at a given temperature.

From (21.6) one gets, by applying Gauss's formula to the first equation and Stokes's formula to the second one, the basic integral laws

$$\oint D_n \, dS = \int q \, dV,$$
$$\oint_c \mathbf{H} \cdot \mathbf{ds} = \int j_n \, dS + \frac{d}{dt} \int D_n \, dS. \tag{21.7}$$

which now involve the free charge and current densities, and the partial potentials. Of course the surface in the second of these equations must be fixed, as it was in (6.11) or (8.1). Otherwise the time differentiation of $(21.6)_2$ cannot be taken outside the surface integral. In Section 25 we shall derive an alternative to $(21.7)_2$, which applies to a moving surface.

We can also leave out the subscript in (21.4). The complete list of jump conditions becomes

$$\mathbf{n} \cdot [\![\mathbf{D}]\!] = \sigma,$$
$$\mathbf{n} \times [\![\mathbf{H}]\!] + v_n [\![\mathbf{D}]\!] = \mathbf{K},$$
$$\mathbf{n} \cdot [\![\mathbf{B}]\!] = 0,$$
$$\mathbf{n} \times [\![\mathbf{E}]\!] - v_n [\![\mathbf{B}]\!] = 0. \tag{21.8}$$

Again, σ and $\mathbf{K}$ are now the free surface charge and surface current densities, and $\mathbf{D}$ and $\mathbf{H}$ are related to $\mathbf{E}$ and $\mathbf{B}$ through $(21.6)_{5,6}$.

Finally, we can form the Galilean invariants

$$\mathcal{J} = \mathbf{j} - q\mathbf{v}, \qquad \mathcal{H} = \mathbf{H} - \mathbf{v} \times \mathbf{D}, \tag{21.9}$$

in which the subscripts $_F$ have again been omitted on the right-hand sides. The $\mathcal{J}$ defined by $(21.9)_1$ is called the (free) *conduction current density*. The $\mathcal{H}$ defined by $(21.9)_2$ is called the *magnetomotive intensity*, in analogy with the electromotive intensity $\mathcal{E} = \mathbf{E} + \mathbf{v} \times \mathbf{B}$.

Exercise

21.1. Prove that, in an aether frame, the magnetomotive intensity is

$$\mathcal{H} = \mathbf{B}/\mu_0 - \epsilon_0 \mathbf{v} \times \mathbf{E} - \mathcal{M}, \tag{21.10}$$

where $\mathcal{M}$ is the Lorentz magnetization.

---

22. The dragging of light by a dielectric

As an example, we consider the dragging of light by a dielectric[†]. For any given point P of the material, which is moving with a velocity $\mathbf{v}$ in the aether frame, we define the *rest frame* $\Sigma'(P)$ as that inertial frame in which P is momentarily at rest. The rest frame depends on P, because different points in the material may have different velocities, and on the time, because a given point may undergo acceleration. Moreover, the rest frame of a given material point, at a given time, is only determined up to constant (time-independent) rotations and translations.

[†]Faraday introduced the term *dielectric* as synonymous with 'insulator': a material that does not allow the passage of *free* electric current. In such a material, only a *polarization current* $\mathbf{P}_t$ or a *magnetization current* **curl** M are possible $((20.2)_2)$. But engineers commonly speak of 'lossy dielectrics', which would strictly mean 'conducting insulators'. We shall use the term dielectric to describe any material capable of polarization.

We now define a linear dielectric as a material that has the response functions

$$\mathbf{P}' = \epsilon_0 \chi \mathbf{E}', \qquad \mathbf{M}' = 0 \tag{22.1}$$

at each point P; the primed vectors refer to the rest frame $\Sigma'(P)$, and χ, the *dielectric susceptibility*, is a positive constant. These equations are supposed to hold for each point of the material at all times; and since they are three-dimensional vector equations (and χ is a constant), the arbitrary constant rotations and translations, to which the rest frame may still be subject, do not matter.

The reason for defining response functions in this way is due to the fact that the relations (22.1), between quantities that are components of tensors of different types, cannot hold in all frames. We have already encountered a similar situation with the aether relations in Section 11.

Let Σ be an aether frame, in which the electromagnetic field is $\{\mathbf{B}, \mathbf{E}\}$ and the material point P is moving with velocity $\mathbf{v}$. According to classical (i.e. non-relativistic) mechanics the relations between the quantities in Σ and in the rest frame $\Sigma'(P)$ are given by $(9.8)_2$ and (20.4):

$$\mathbf{P}' = \mathbf{P}, \qquad \mathbf{E}' = \mathbf{E} + \mathbf{v} \times \mathbf{B}, \qquad \mathbf{M}' = \mathbf{M} + \mathbf{v} \times \mathbf{P}.$$

Substitution of these in (22.1) gives

$$\mathbf{P} = \epsilon_0 \chi (\mathbf{E} + \mathbf{v} \times \mathbf{B}), \qquad \mathbf{M} + \mathbf{v} \times \mathbf{P} = 0, \tag{22.2}$$

These relations, which hold at every material point P (with velocity $\mathbf{v}$) are the response functions, or constitutive relations, of a classical (i.e. non-relativistic), linear dielectric. They can also be written in the form

$$\mathbf{P} = \epsilon_0 \chi \mathcal{E}, \qquad \mathcal{M} = 0, \tag{22.3}$$

where $\mathcal{E}$ is the electromotive intensity (9.9) and $\mathcal{M}$ is the Lorentz magnetization $(20.5)_1$. The aether relations for the dielectric are obtained by substituting (22.2) in $\mathbf{D} = \epsilon_0 \mathbf{E} + \mathbf{P}$ and $\mathbf{H} = \mathbf{B}/\mu_0 - \mathbf{M}$:

$$\mathbf{D} = \epsilon_0 \big[\mathbf{E} + \chi (\mathbf{E} + \mathbf{v} \times \mathbf{B}) \big],$$
$$\mathbf{H} = \mathbf{B}/\mu_0 + \mathbf{v} \times \mathbf{P} = \frac{1}{\mu_0} \Big[\mathbf{B} + \frac{\chi}{c^2} \mathbf{v} \times (\mathbf{E} + \mathbf{v} \times \mathbf{B}) \Big]. \tag{22.4}$$

Consider now, in an aether frame, an electromagnetic shock propagating through the dielectric in the absence of any *free* charges or currents; the *bound* charges and currents are of course a matter to be decided by the response functions (22.1). As in Section 11, it is easy to see that the shock must be moving; otherwise all jumps must vanish. Again, it is sufficient to consider the two jump conditions

$$\mathbf{n} \times [\![\mathbf{H}]\!] + u_n [\![\mathbf{D}]\!] = 0,$$
$$\mathbf{n} \times [\![\mathbf{E}]\!] - u_n [\![\mathbf{B}]\!] = 0, \tag{22.5}$$

where we have denoted the normal shock velocity by u_n in order to avoid confusing it with the velocity of the dielectric. For simplicity, we take the latter to be normal to the shock. If we now substitute (22.4) in (22.5)$_1$ we arrive, after a simple calculation, at the following condition for non-zero jumps:

$$(1 + \chi)u_n^2 - 2\chi u_n v + \chi v^2 - c^2 = 0. \tag{22.6}$$

This is a quadratic equation for the shock speed u_n. If $\chi = 0$, it gives $u_n = \pm c$; the motion of the material through the aether has no effect at all in this case. For a dielectric at rest (in an aether frame) $v = 0$, and the solutions of (22.6) become

$$u_n = \pm\frac{c}{n}, \qquad n = \sqrt{1 + \chi}. \tag{22.7}$$

The two possible velocities are smaller than the speed of light c in vacuum by the factor n, called the *index of refraction* of the dielectric[†]. If the dielectric is moving, the solution (to the first order in v/c) is

$$u_n = \pm\frac{c}{n} + \left(1 - \frac{1}{n^2}\right)v. \tag{22.8}$$

This dragging of light by a moving dielectric was predicted by Fresnel and confirmed by Fizeau in an experiment that used flowing water. Equation (22.8) shows that the dragging is imperfect, because only a fraction of v is added to $\pm c/n$.

The foregoing derivation, and the definition of a classical linear dielectric on which it was based, were clearly non-relativistic. They must be replaced by a relativistic treatment if the dielectric is moving, relative to the aether frame, at a speed that is not small compared with c. In order to do this, we must first determine the *relativistic* response functions of a linear dielectric.

23. Relativistic response functions

According to the theory of relativity, all inertial frames are Lorentz frames, connected with each other by Lorentz transformations. The rest frame $\Sigma'(P)$, in particular, will be obtained from an aether frame—with respect to which the material point P is moving with velocity $\mathbf{v}$—by applying a Lorentz transformation with $\mathbf{u} = \mathbf{v}$. According to (12.11) and (20.6), the response functions (22.1) will then take the form

$$\mathbf{P} - \mathbf{v} \times \mathbf{M}/c^2 - \frac{\gamma}{\gamma + 1}\mathbf{v}(\mathbf{v} \cdot \mathbf{P})/c^2$$

$$= \epsilon_0 \chi \left[\mathbf{E} + \mathbf{v} \times \mathbf{B} - \frac{\gamma}{\gamma + 1}\mathbf{v}(\mathbf{v} \cdot \mathbf{E})/c^2\right], \tag{23.1}$$

$$\mathbf{M} + \mathbf{v} \times \mathbf{P} - \frac{\gamma}{\gamma + 1}\mathbf{v}(\mathbf{v} \cdot \mathbf{M})/c^2 = 0,$$

[†]The formula $n^2 = 1 + \chi$ is called *Maxwell's relation*.

where γ now stands for $(1 - v^2/c^2)^{-1/2}$. In order to obtain the relativistic analogues of (22.4), we must first solve for $\mathbf{P}$ and $\mathbf{M}$, since in (23.1) they appear implicitly. Taking the scalar product of $(23.1)_{1,2}$ with $\mathbf{v}$, we obtain

$$\mathbf{v} \cdot \mathbf{P} = \epsilon_0 \chi (\mathbf{v} \cdot \mathbf{E}), \qquad \mathbf{v} \cdot \mathbf{M} = 0. \tag{23.2}$$

If we use (23.2), we can solve (23.1) for $\mathbf{P}$ and $\mathbf{M}$:

$$\mathbf{P} = \frac{\epsilon_0 \chi}{1 - v^2/c^2} \left[\mathbf{E} + \mathbf{v} \times \mathbf{B} - \frac{1}{c^2} \mathbf{v}(\mathbf{v} \cdot \mathbf{E}) \right], \qquad \mathbf{M} = -\mathbf{v} \times \mathbf{P}. \tag{23.3}$$

We can now substitute these in $\mathbf{D} = \epsilon_0 \mathbf{E} + \mathbf{P}$ and $\mathbf{H} = \mathbf{B}/\mu_0 - \mathbf{M}$ and obtain

$$\mathbf{D} = \epsilon_0 \left[\mathbf{E} + \frac{\chi}{1 - v^2/c^2} (\mathbf{E} + \mathbf{v} \times \mathbf{B}) - \frac{\chi}{c^2(1 - v^2/c^2)} \mathbf{v}(\mathbf{v} \cdot \mathbf{E}) \right],$$

$$\mathbf{H} = \frac{1}{\mu_0} \left[\mathbf{B} + \frac{\chi}{c^2(1 - v^2/c^2)} \mathbf{v} \times (\mathbf{E} + \mathbf{v} \times \mathbf{B}) \right]. \tag{23.4}$$

Clearly, the relativistic response functions (23.3) and the aether relations (23.4) reduce to the corresponding classical expressions (22.2) and (22.4) if one neglects all terms of order v^2/c^2. If one substitutes (23.4) in the jump conditions, the resulting equation for the shock speeds is

$$(1 + \chi) u_n^2 - \frac{2 \chi u_n v}{1 - v^2/c^2} + \frac{\chi v^2}{1 - v^2/c^2} - c^2 = 0, \tag{23.5}$$

which should be compared with the classical equation (22.6). If $\chi = 0$ one gets $u_n = \pm c$, and if $v = 0$, which means that the dielectric is at rest, the shock speeds are again given by $\pm c/n$. This is as it should be, because under these circumstances it does not matter whether the treatment is classical or relativistic. In general, when χ and v do not vanish, the two relativistic shock speeds u_n will differ from the classical ones by terms of order v^2/c^2. But if the desired end-result is a formula for the dragging of light which is correct to order v/c, it will obviously be the same as Fresnel's formula (22.8).

Finally, it is perhaps worth noting that the assumption of a strictly constant dielectric susceptibility can be somewhat relaxed. There is nothing to prevent us from assuming, in the classical case, that χ depends on Galilean invariants, such as the density or the temperature. Such a χ will behave like a constant when we carry out the Galilean transformation from the rest frame to the aether frame. Similarly, in the relativistic case, we may assume that χ depends on Lorentz-invariant (or world scalar) arguments.

Exercises

23.1. A *linearly magnetizable* material is defined by the rest frame response functions

$$\mathbf{M}' = \frac{\chi_B}{\mu_0} \mathbf{B}', \qquad \mathbf{P}' = 0, \tag{23.6}$$

where χ_B is the *magnetic susceptibility*. Determine the classical and relativistic response functions and aether relations for such a material in an aether frame, relative to which the material is moving with velocity $\mathbf{v}$.

23.2. A *linearly conducting* or *Ohmic* material is defined by the rest frame relation, *Ohm's law*,

$$\mathbf{j}' = C\mathbf{E}', \qquad (23.7)$$

where C is the *conductivity*[†]. Show that, in terms of the conduction current density $\mathcal{J} = \mathbf{j} - q\mathbf{v}$ (cf. (21.9)), the classical and relativistic forms of Ohm's law are, respectively,

$$\mathcal{J} = C(\mathbf{E} + \mathbf{v} \times \mathbf{B}),$$
$$\mathcal{J} = \gamma C[\mathbf{E} + \mathbf{v} \times \mathbf{B} - \mathbf{v}(\mathbf{v} \cdot \mathbf{E})/c^2]. \qquad (23.8)$$

It is important to note that we consider the relativistic treatment to be superior to the classical one—to which it reduces in the limit of small velocities—because of its evident successes when applied to experiments. But it should be remembered that those experiments involve motion at relativistic speeds. In a laboratory such speeds can only be attained by accelerated particles, or by light itself. Macroscopic bodies are known to attain relativistic speeds in some astronomical situations, but not in terrestrial laboratories. Relativistic response functions, or constitutive relations, for *materials* are therefore appropriate in discussions of the universe. For terrestrial experiments they are not worth the trouble, because the terms of order v^2/c^2 are negligible.

24. The initial value problem for Maxwell's equations

In the system of eqns (21.6) there are two time derivatives:

$$\mathbf{D}_t = \operatorname{curl} \mathbf{H} - \mathbf{j},$$
$$\mathbf{B}_t = -\operatorname{curl} \mathbf{E}. \qquad (24.1)$$

Does this mean that these equations describe the temporal evolution of $\mathbf{D}$ and $\mathbf{B}$? If so, what about the evolution of all the other vector fields? We shall discuss this initial value problem in a mathematically naïve manner. Let $\mathbf{D}(\mathbf{x}, 0)$ and $\mathbf{B}(\mathbf{x}, 0)$ be the initial distributions of the displacement and magnetic field $\mathbf{D}(\mathbf{x}, t)$ and $\mathbf{B}(\mathbf{x}, t)$. Since, according to (21.6)$_3$, $\operatorname{div} \mathbf{B} = 0$ holds at all times, we must assume that $\mathbf{B}(\mathbf{x}, 0)$ satisfies $\operatorname{div} \mathbf{B}(\mathbf{x}, 0) = 0$.

[†]The standard notation is σ, but we use C in order to avoid confusing the conductivity with the surface charge density. Of course Ohm's law is a constitutive relation, not a law.

The equations (24.1) ought to allow us, by a process of integration over time, to calculate $\mathbf{D}(\mathbf{x}, t)$ and $\mathbf{B}(\mathbf{x}, t)$. But, as they stand, these equations are underdetermined: in order to use them, we must know $\mathbf{H}$, $\mathbf{j}$ and $\mathbf{E}$. Now the response functions for $\mathbf{P}$ and $\mathbf{M}$, which express the dependence of these fields on $\mathbf{E}$ and $\mathbf{B}$, together with the aether relations $(21.6)_{5,6}$, provide constitutive relations $\mathbf{D}(\mathbf{E}, \mathbf{B})$ and $\mathbf{H}(\mathbf{E}, \mathbf{B})$. We shall assume that in a conductor the current density, too, is given by a constitutive relation $\mathbf{j}(\mathbf{E}, \mathbf{B})$ (Ohms's law is an example of such a relation). If we now assume, in addition, that the aether relation $\mathbf{D}(\mathbf{E}, \mathbf{B})$ can be solved for $\mathbf{E}$ in terms of $\mathbf{D}$ and $\mathbf{B}$ (this is a condition that must be met by acceptable constitutive relations for $\mathbf{P}$), the system (24.1) becomes determined, and we may assume that it can be integrated.

What about the equation $\operatorname{div} \mathbf{B} = 0$, which we have only satisfied at $t = 0$, and the equation $\operatorname{div} \mathbf{D} = q$ (cf. $(21.6)_1$), which we have not mentioned at all? The first question is easy: according to $(24.1)_2$, the solution will satisfy $(\operatorname{div} \mathbf{B})_t = 0$. Now the initial $\mathbf{B}$ was solenoidal; hence $\operatorname{div} \mathbf{B} = 0$ is guaranteed. As for the second question, since $\mathbf{D}(\mathbf{x}, t)$ has already been determined by the foregoing integration, the only way to read the equation $\operatorname{div} \mathbf{D} = q$ is from right to left:

$$q(\mathbf{x}, t) = \operatorname{div} \mathbf{D}(\mathbf{x}, t). \tag{24.2}$$

It delivers the charge density. Of course $q(\mathbf{x}, 0) = \operatorname{div} \mathbf{D}(\mathbf{x}, 0)$, and if $q(\mathbf{x}, 0)$ is prescribed, the initial distribution $\mathbf{D}(\mathbf{x}, 0)$ must satisfy this equation. Will the q of (24.2) satisfy the principle of electric charge conservation, especially in view of the fact that we have, perhaps recklessly, agreed to accept an arbitrary constitutive relation for $\mathbf{j}$? The answer is yes, because if we take the divergence of $(24.1)_1$ and use (24.2), we obtain $q_t + \operatorname{div} \mathbf{j} = 0$.

In an insulator $\mathbf{j}$ vanishes by definition, and if there is a q, it must be permanent. The equation $q = \operatorname{div} \mathbf{D}$, which must be satisfied initially, will continue to be satisfied, because there is no $\mathbf{j}$, so that $(\operatorname{div} \mathbf{D} - q)_t = \operatorname{div} \mathbf{curl} \mathbf{H} - 0 = 0$. A vacuum is of course an insulator with $q = 0$.

On the boundary between a conductor and an insulator, or between any two conductors with different constitutive relations, discontinuities may appear. These have to satisfy the jump conditions (21.8). We have seen that the charge density is given by $q = \operatorname{div} \mathbf{D}$. The jump conditions $\mathbf{n} \cdot [\![\mathbf{D}]\!] = \sigma$ and $\mathbf{n} \times [\![\mathbf{H}]\!] + v_n [\![\mathbf{D}]\!] = \mathbf{K}$ are similarly read from right to left: they deliver the surface charge and current densities.

Is the equation $q = \operatorname{div} \mathbf{D}$ ever needed after the initial instant? Yes, for we shall show, in Chapter 15, that the force density (the force per unit volume) acting on a body contains a term $q\mathcal{E}$, where $\mathcal{E}$ is the electromotive intensity (9.9). The force, in turn, determines the motion, on which the constitutive relations usually depend (cf., for example, (22.4), (23.4) or (23.8)). Similarly, the surface charge density $\sigma = \mathbf{n} \cdot [\![\mathbf{D}]\!]$ and the surface current density $\mathbf{K} = \mathbf{n} \times [\![\mathbf{H}]\!] + v_n [\![\mathbf{D}]\!]$ determine the stress, that is, the force per unit surface area of a body.

As an example, we consider a sphere of radius a, made of a material which is both a linear dielectric and a linear conductor. The sphere is at rest in an aether frame. Thus the aether relations and the response function for the current density are (cf. (22.4) and (23.8))

$$\mathbf{D} = \epsilon \mathbf{E}, \qquad \mathbf{B} = \mu_0 \mathbf{H}, \qquad \mathbf{j} = C \mathbf{E}, \tag{24.3}$$

where $\epsilon = (1 + \chi)\epsilon_0$. We shall assume that the initial $\mathbf{D}$ has only a radial component $D_0(r)$, a function of the distance from the centre of the sphere; that it is continuous at $r = a$; that it has a finite divergence inside the sphere; and that the initial $\mathbf{B}$ vanishes everywhere.

According to these assumptions, the initial charge distribution is

$$q_0(r) = \text{div } \mathbf{D}_0 = \frac{1}{r^2} \frac{\partial}{\partial r} [r^2 D_0(r)]. \tag{24.4}$$

The surface charge density vanishes initially (since the radial component $D_0(r)$ is continuous). Outside, in the vacuum, $D_0(r)$ must satisfy (24.4) with $q_0 = 0$. Thus, for $r > a$, $r^2 D_0(r) = $ const., the constant being $a^2 D_0(a)$. The initial electric field is $D_0(r)/\epsilon$ inside the sphere, and $(a/r)^2 D_0(a)/\epsilon_0$ outside. Since $\mathbf{E}$ is radial and a function of r only, its curl vanishes. Thus the magnetic field will continue to vanish, and so will $\mathbf{H} = \mathbf{B}/\mu_0$.

According to (24.1)$_1$ and (24.3)$_{1,3}$ we shall have, inside the sphere,

$$D_t + \frac{C}{\epsilon} D = 0. \tag{24.5}$$

This equation has the solution

$$D(r, t) = D_0(r) e^{-Ct/\epsilon}. \tag{24.6}$$

The charge density $q = \text{div } \mathbf{D}$ will therefore decay exponentially:

$$q(r, t) = q_0(r) e^{-Ct/\epsilon}. \tag{24.7}$$

Outside, we have $D_t = 0$. Thus D outside remains equal to $(a/r)^2 D_0(a)$. Hence the surface charge density $\sigma = \mathbf{n} \cdot [\![\mathbf{D}]\!]$ is given by

$$\sigma(t) = D_0(a) - D(a, t) = D_0(a)(1 - e^{-Ct/\epsilon}). \tag{24.8}$$

It is readily verified that $\dot\sigma = j(a, t)$: the surface charge increases at precisely the rate at which electric charge is being deposited there by the current. Ultimately, the whole initial charge will be uniformly distributed over the surface of the sphere.

What we have discussed in this section was the initial value problem for Maxwell's equations in matter, where we assumed response functions, or constitutive equations, for $\mathbf{P}$, $\mathbf{M}$ and $\mathbf{j}$. An entirely different situation presents itself when we have, instead, a number of charged particles, which are moving along trajectories in vacuum. The charge density q is then a sum of terms of the form $e_i \delta(\mathbf{x} - \mathbf{R}_i)$, where e_i and $\mathbf{R}_i(t)$ are the charge and trajectory of the i-th particle, regarded as a point charge; and $\mathbf{j}$ is a sum of terms like $e_i \dot{\mathbf{R}}_i \delta(\mathbf{x} - \mathbf{R}_i)$. The solution of Maxwell's equations with such q and $\mathbf{j}$ is the subject matter of the theory of electromagnetic radiation from charged particles. It will be dealt with in Chapter 12. Although the charged particles 'respond' to the electromagnetic field—it exerts forces that influence the trajectories, and may even lead to break-up or coalescence of particles—there are no $\mathbf{P}$ or $\mathbf{M}$, because the response creates no charge or current distributions beside those that are already accounted for by the sums of $e_i \delta(\mathbf{x} - \mathbf{R}_i)$ and $e_i \dot{\mathbf{R}}_i \delta(\mathbf{x} - \mathbf{R}_i)$.

25. The flux derivative

The second of the integral laws (21.7),

$$\oint_c \mathbf{H} \cdot \mathbf{ds} = \int j_n \, dS + \frac{d}{dt} \int D_n \, dS, \tag{25.1}$$

applies to a fixed, open surface S. There is a more general law which applies to a moving (and deforming) surface. Let $\mathbf{v}(\mathbf{x}, t)$ be a differentiable velocity field (not necessarily the velocity of any actual material). We now consider an open surface S, with boundary c, which is swept along by the velocity field (Fig. 25.1). We wish to calculate the rate of change of the instantaneous flux $\int \mathbf{A} \cdot \mathbf{n} \, dS$ of a vector field $\mathbf{A}(\mathbf{x}, t)$ through S.

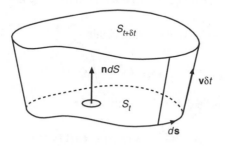

Fig. 25.1

If δ denotes a change throughout the time interval δt then, by the rule of differentiating a product,

$$\delta \int \mathbf{A} \cdot \mathbf{n} \, dS = \int \delta \mathbf{A} \cdot \mathbf{n} \, dS + \int \mathbf{A} \cdot \delta(\mathbf{n} \, dS). \tag{25.2}$$

Since we are going to take the limit of a difference quotient, we need only calculate to the first order in δt. In the first term on the right-hand side, $\delta \mathbf{A} = (\partial \mathbf{A}/\partial t)\delta t$ and

$\mathbf{n} \, dS$ is the vector surface element at t. In the second term, $\mathbf{A}$ is at t and $\delta(\mathbf{n} \, dS) = (\mathbf{n} \, dS)_{t+\delta t} - (\mathbf{n} \, dS)_t$. Thus

$$\delta \int \mathbf{A} \cdot \mathbf{n} \, dS = \delta t \int \frac{\partial \mathbf{A}}{\partial t} \cdot \mathbf{n} \, dS + \int \mathbf{A} \cdot (\mathbf{n} \, dS)_{t+\delta t} - \int \mathbf{A} \cdot (\mathbf{n} \, dS)_t. \quad (25.3)$$

In Fig. 25.1, S_t and $S_{t+\delta t}$ are at the bottom and top, and $\mathbf{n}$ is pointing to the same side (upward) on each surface. We now apply Gauss's divergence theorem to the volume shown in the figure:

$$\int \text{div} \, \mathbf{A} \, dV = \int \mathbf{A} \cdot (\mathbf{n} \, dS)_{t+\delta t} - \int \mathbf{A} \cdot (\mathbf{n} \, dS)_t + \oint_c \mathbf{A} \cdot (\mathbf{ds} \times \mathbf{v}\delta t). \quad (25.4)$$

The minus sign before the second term on the right-hand side is because $\mathbf{n}$ at the bottom points *into* the volume, whereas Gauss's theorem concerns the *outward* flux. The last term is the outward flux through the sides: the vector product $\mathbf{ds} \times \mathbf{v}\delta t$ has the magnitude of an area strip with base $\mathbf{ds}$ and length $\mathbf{v}\delta t$, and points outward. Since the element of volume is $dV = \mathbf{n} \, dS \cdot \mathbf{v}\delta t$ (a prism with base $\mathbf{n} \, dS$ and length $\mathbf{v}\delta t$), (25.4) becomes

$$\delta t \int (\mathbf{v} \, \text{div} \, \mathbf{A}) \cdot \mathbf{n} \, dS = \int \mathbf{A} \cdot \delta(\mathbf{n} \, dS) + \delta t \oint_c (\mathbf{v} \times \mathbf{A}) \cdot \mathbf{ds}.$$

In accordance with the remarks following (25.2), $\mathbf{A}$ is at time t in all the terms of this equation. Applying Stokes's theorem to the line integral, we have

$$\int \mathbf{A} \cdot \delta(\mathbf{n} \, dS) = \delta t \int [\mathbf{v} \, \text{div} \, \mathbf{A} - \text{curl} \, (\mathbf{v} \times \mathbf{A})] \cdot \mathbf{n} \, dS. \quad (25.5)$$

Substituting (25.5) in (25.3), we finally obtain

$$\frac{d}{dt} \int A_n \, dS = \int \overset{*}{\mathbf{A}} \cdot \mathbf{n} \, dS, \quad (25.6)$$

where, for any vector field $\mathbf{A}$ and velocity field $\mathbf{v}$,

$$\overset{*}{\mathbf{A}} = \frac{\partial \mathbf{A}}{\partial t} + \mathbf{v} \, \text{div} \, \mathbf{A} - \text{curl} \, (\mathbf{v} \times \mathbf{A}) \quad (25.7)$$

is called the *flux derivative* of $\mathbf{A}$. Now, according to the first pair of Maxwell's equations $(21.6)_{1,2}$,

$$\text{curl} \, (\mathbf{H} - \mathbf{v} \times \mathbf{D}) = \mathbf{j} + \frac{\partial \mathbf{D}}{\partial t} - \text{curl} \, (\mathbf{v} \times \mathbf{D})$$

$$= \mathbf{j} - \mathbf{v} \, \text{div} \, \mathbf{D} + \frac{\partial \mathbf{D}}{\partial t} + \mathbf{v} \, \text{div} \, \mathbf{D} - \text{curl} \, (\mathbf{v} \times \mathbf{D})$$

$$= \mathbf{j} - q\mathbf{v} + \overset{*}{\mathbf{D}}. \quad (25.8)$$

In terms of the Galilean invariants (21.9), this equation can be written in the form

$$\text{curl } \mathcal{H} = \mathcal{J} + \overset{*}{\mathbf{D}}. \tag{25.9}$$

From (25.6) we now obtain, for a moving surface,

$$\oint_c \mathcal{H} \cdot \mathbf{ds} = \int \mathcal{J}_n \, dS + \frac{d}{dt} \int D_n \, dS, \tag{25.10}$$

Finally, we can manipulate the second pair of Maxwell's equations in the same way. This leads to the local equation

$$\text{curl } \mathcal{E} = -\overset{*}{\mathbf{B}}, \tag{25.11}$$

where $\mathcal{E} = \mathbf{E} + \mathbf{v} \times \mathbf{B}$ is the electromotive intensity (9.9), and to the integral law

$$\oint_c \mathcal{E} \cdot \mathbf{ds} = -\frac{d}{dt} \int B_n \, dS, \tag{25.12}$$

which is Faraday's law of induction for a moving surface.

Although the local equations (25.9) and (25.11) are written in terms of Galilean invariants, they are merely alternative forms of the corresponding Maxwell's equations. Similarly, the integral laws (25.10) and (25.12) are exact (not non-relativistic approximations).

If the conductivity C of any linearly conducting material becomes infinite, then, according to Ohm's law—in *either* of the two forms of (23.8)—the electromotive intensity $\mathcal{E}$ must vanish. According to (25.12) we shall then have

$$\frac{d}{dt} \int B_n \, dS = 0 \tag{25.13}$$

for any surface which is moving along with the matter. This is called *magnetic flux freezing*. It occurs in astronomical plasmas, in which the electric conductivities and the length and time scales are often such that the left-hand side of (25.12) is extremely small compared with the various terms that make up the right-hand side (Chapter 16).

8

Electrostatics

In the absence of any motion (through the aether) and any changes in time, Maxwell's equations (21.6) and the electromagnetic jump conditions (21.8) are

$$\operatorname{div} \mathbf{D} = q, \qquad \mathbf{n} \cdot [\![\mathbf{D}]\!] = \sigma,$$
$$\operatorname{curl} \mathbf{E} = 0, \qquad \mathbf{n} \times [\![\mathbf{E}]\!] = 0,$$
$$\mathbf{D} = \epsilon_0 \mathbf{E} + \mathbf{P},$$
$$\operatorname{div} \mathbf{B} = 0, \qquad \mathbf{n} \cdot [\![\mathbf{B}]\!] = 0,$$
$$\operatorname{curl} \mathbf{H} = \mathbf{j}, \qquad \mathbf{n} \times [\![\mathbf{H}]\!] = \mathbf{K},$$
$$\mathbf{H} = \mathbf{B}/\mu_0 - \mathbf{M}.$$

It is obvious that any connection between the first and second halves of this system of equations can only be provided by $\mathbf{P}, \mathbf{j}$ or $\mathbf{M}$ (for a material at rest the Minkowski $\mathbf{M}$ and the Lorentz $\mathcal{M}$ are the same; cf. $(20.5)_1$). Further discussion therefore depends on material properties.

If $\mathbf{j} = \mathbf{M} = 0$ and $\mathbf{P}$ is independent of $\mathbf{B}$, the first half of the equations is independent of the second; its study is the subject of *electrostatics*. Similarly, if $\mathbf{j} = \mathbf{P} = 0$ and $\mathbf{M}$ is independent of $\mathbf{E}$, the second half is independent of the first; its study is the subject of *magnetostatics*. In spite of the similarity in the differential equations and aether relations, it is not true that every problem (with $q = 0$) in electrostatics translates into one in magnetostatics, and vice versa; for the jump conditions are different. But the mathematical techniques for solving problems in electrostatics and magnetostatics are quite similar. That is why only one of them, usually electrostatics, is studied in detail.

Finally, if the currents do not vanish, the whole set of equations forms the subject of *steady currents*.

26. Electrostatics of conductors

In electrostatics a conductor is a material that has the following two constitutive properties: any non-zero rest frame electric field $\mathbf{E}'$ is associated with a non-zero current density $\mathbf{j}'$; and if the material is polarizable, $\mathbf{P}'$ vanishes when $\mathbf{E}'$ does. We do not insist that the material be a linear conductor, or a linear dielectric. The requirements on the functions $\mathbf{E}'(\mathbf{j}')$ and $\mathbf{P}'(\mathbf{E}')$ are just that $\mathbf{E}'(0) = 0$ and $\mathbf{P}'(0) = 0$.

In the absence of motion all fields are already rest frame fields, and we conclude that in electrostatics, that is, when $\mathbf{j} = 0$, the electric field $\mathbf{E}$ and the polarization $\mathbf{P}$ must both vanish in the interior of any conductor. Furthermore, since $q = \operatorname{div}\mathbf{D}$ and $\mathbf{D} = \epsilon_0\mathbf{E} + \mathbf{P} = 0$, there can be no charge density inside a conductor: if the conductor is charged, the charge must reside wholly on its surface.

Outside, where there is (by definition) an *insulator*, perhaps an insulating vacuum, $\mathbf{E}$ need not vanish, but the jump condition $\mathbf{n} \times [\![\mathbf{E}]\!] = 0$ must hold at the interface. Hence $\mathbf{E}$ on the outer side must be normal to the boundary. If we take the unit normal $\mathbf{n}$ to point out of the conductor, the jump conditions are

$$\mathbf{n} \times \mathbf{E} = 0,$$
$$D_n = \epsilon_0 E_n + P_n = \sigma, \tag{26.1}$$

where σ is the surface density of (free) charge on the conductor and all fields refer to the outer side; inside, they all vanish. If there are several conductors— separated by insulators, of course—the conditions (26.1) hold on the surface of each conductor.

The conditions (26.1) are just boundary conditions on the field in the insulating medium outside the conductors, and everything else depends on the properties of this medium. If it is a dielectric, we shall need information regarding $\mathbf{P}$—in fact, a constitutive relation. We shall deal with dielectrics in Chapter 9. If, as we shall assume throughout this chapter, the insulator is non-polarizable, it is governed by the equations

$$\operatorname{div}\epsilon_0\mathbf{E} = q,$$
$$\mathbf{curl}\,\mathbf{E} = 0. \tag{26.2}$$

From $(26.2)_2$ we have $\mathbf{E} = -\mathbf{grad}\,V$, where the electric potential V includes an arbitrary additive constant (this is what the gauge transformation reduces to in electrostatics). In terms of the potential, eqns (26.2) become

$$\Delta V = -q/\epsilon_0. \tag{26.3}$$

This is Poisson's equation. It is to be solved subject to the condition that V be constant on the surface of each conductor, which follows from $(26.1)_1$; each conductor is therefore an *equipotential*. Further conditions may be required, and we shall address them later on.

Having found any particular solution, corresponding to a given distribution of q throughout the insulating medium outside the conductors, we can obtain other solutions by adding solutions of the *homogeneous* equation $\Delta V = 0$—Laplace's equation—which also satisfy conditions of constancy of V on each conductor. For every solution, we obtain the surface charge density on any conductor from the boundary condition $(26.1)_2$ (with $\mathbf{P} = 0$)

$$\sigma = -\epsilon_0 \frac{\partial V}{\partial n}. \tag{26.4}$$

The total charge on a conductor is

$$Q = -\epsilon_0 \oint \frac{\partial V}{\partial n} \, dS, \tag{26.5}$$

the integral being taken over the complete surface of the conductor.

The electric field exerts a stress—a force per unit area—on the conductor. We cannot determine this stress from the principles we have laid down so far, because a conductor, like any macroscopic body, is subject to the principles of electromagnetism, mechanics *and* thermodynamics. Formulae for the stress on various kinds of bodies will be derived in Chapter 15. For the moment, we shall be content with quoting the result for a conductor: at any point on its surface, the electric stress, or force per unit area, is

$$\mathbf{t} = \tfrac{1}{2} D_n \mathbf{E} = \tfrac{1}{2} \sigma \mathbf{E}, \tag{26.6}$$

where, in the second equation, we have used the relation $D_n = \sigma$. This force is (like $\mathbf{E}$) normal to the conductor and directed outwards—a tension, or a negative pressure. It tends to inflate the conductor at the expense of the surrounding insulator; this effect is called *electrostriction*. The electric tension is counteracted by the elastic stresses in the conductor and in the adjacent insulator (unless it is a vacuum). The equilibrium state cannot, therefore, be determined without information regarding the elastic stresses.

As a simple example, consider a spherical conductor of radius a that carries a total charge Q. If the charge density outside the conductor vanishes,

$$V = \frac{Q}{4\pi\epsilon_0 r}, \tag{26.7}$$

where r is the distance from the centre of the sphere, is a solution of the electrostatic problem, for the potential (26.7) is constant on the surface $r = a$, and the surface charge density $\sigma = -\epsilon_0 \partial V/\partial n$ has the uniform value $Q/(4\pi a^2)$, which adds up to Q over the whole surface. It remains to verify that $\Delta V = 0$ outside the sphere. If $\mathbf{r}$ denotes the vector (x, y, z), with magnitude $r = (x^2 + y^2 + z^2)^{1/2}$, then, for $r > 0$,

$$\mathbf{grad}\, \frac{1}{r} = -\frac{1}{r^2}\, \mathbf{grad}\, r = -\frac{\mathbf{r}}{r^3}. \tag{26.8}$$

Furthermore (for $r > 0$),

$$\text{div } \mathbf{grad} \, \frac{1}{r} = - \text{div } \frac{\mathbf{r}}{r^3}$$

$$= -\frac{1}{r^3} \text{div } \mathbf{r} - \mathbf{r} \cdot \mathbf{grad} \, \frac{1}{r^3}$$

$$= -\frac{3}{r^3} + \mathbf{r} \cdot \frac{3}{r^4} \, \mathbf{grad} \, r$$

$$= 0. \tag{26.9}$$

Any function U is called *harmonic* in a region if $\Delta U = 0$ in that region. Equation (26.9) shows that $1/r$ is harmonic in any region that does not include the point $r = 0$. We have thus demonstrated that (26.7) solves the electrostatic problem for a conductor whose outer boundary is a sphere. Since the force per unit area (26.6) is constant over the sphere for this solution, its resultant vanishes by symmetry. It does not matter whether the sphere is hollow or not. Equation (26.7) provides a solution even if the sphere contains an irregular worm-hole in its interior. Whether it is the *only* solution is another matter. The following problem shows that it is not.

Problem 26.1 If $\mathbf{E}_0$ is a constant vector, prove that

$$\mathbf{grad}(\mathbf{E}_0 \cdot \mathbf{r}) = \mathbf{E}_0. \tag{26.10}$$

Hence prove that

$$V = \frac{Q}{4\pi \epsilon_0 r} + \left(\frac{a^3}{r^3} - 1\right)(\mathbf{E}_0 \cdot \mathbf{r}) \tag{26.11}$$

is another solution for a spherical conductor of radius a with a total charge Q. This solution corresponds to a uniform electric field $\mathbf{E}_0$ at infinity. Prove that the resultant force $\oint \mathbf{t} \, dS$ is $Q\mathbf{E}_0$, equal to the force that the field $\mathbf{E}_0$ would exert on a particle carrying a charge Q[†]. This is *Coulomb's law*.

Solution If $\mathbf{a} = (a_x, a_y, a_z)$ is constant

$$\mathbf{grad}(\mathbf{a} \cdot \mathbf{r}) = \mathbf{grad}(a_x x + a_y y + a_z z) = (a_x, a_y, a_z) = \mathbf{a}.$$

The V of (26.11) is manifestly constant on $r = a$. In the form

$$V = -\mathbf{E}_0 \cdot \mathbf{r} + \frac{Q}{4\pi \epsilon_0 r} + a^3 \frac{\mathbf{E}_0 \cdot \mathbf{r}}{r^3}$$

it is a superposition of a constant field ($\mathbf{E}_0 = \mathbf{grad}(\mathbf{E}_0 \cdot \mathbf{r})$) potential, a monopole potential and a dipole potential (cf. (14.6) and (14.8)). Each of these is by itself a solution of Laplace's

---

[†]Because of the factor $\frac{1}{2}$ in (26.6), this result should not be regarded as obvious.

equation (for $r > 0$). The electric field is

$$\mathbf{E} = -\operatorname{grad} V = \mathbf{E}_0 + \frac{Q\mathbf{r}}{4\pi\epsilon_0 r^3} + \frac{a^3}{r^5}[3(\mathbf{E}_0 \cdot \mathbf{r})\mathbf{r} - r^2\mathbf{E}_0].$$

This corresponds to a charge density

$$\sigma = \epsilon_0 E_n = \frac{Q}{4\pi a^2} + 3\epsilon_0(\mathbf{E}_0 \cdot \mathbf{n}).$$

The force is

$$\oint \tfrac{1}{2}\sigma \mathbf{E}\, dS = \oint \frac{1}{2\epsilon_0}\left[\frac{Q}{4\pi a^2} + 3\epsilon_0(\mathbf{E}_0 \cdot \mathbf{n})\right]^2 \mathbf{n} a^2\, d\Omega = Q\mathbf{E}_0.$$

The potential (26.11) differs from (26.7) by a solution V_0 of Laplace's equation $\Delta V = 0$ that does not disturb the constancy of the potential over the conductor, or its charge (26.5). In order to eliminate such V_0's, that is, force them to vanish, we must impose further boundary conditions, in addition to the condition (26.5) on the surface of the conductor. For example, $V = 0$ at infinity.

As a second example, consider a pair of infinite, parallel, conducting planes in vacuum. Let $y = \pm\tfrac{1}{2}d$ be the equations of the two planes. The linear function $V = (v/d)y$, with constant v (the potential difference between the plates), satisfies $\Delta V = 0$ and assumes the constant values $\pm\tfrac{1}{2}v$ on the conductors. Hence it is a solution. The surface charge densities on the conductors are $\pm\epsilon_0 v/d$, since $\partial V/\partial n$ is $\partial V/\partial y$ on one and $-\partial V/\partial y$ on the other. The total charge on each infinite plane is of course infinite. The electric field is uniform and normal to the planes; its direction is from the positively charged conductor to the negatively charged one. Finally, each conductor is attracted towards the other one by a force $\tfrac{1}{2}\epsilon_0 v^2/d^2 = \tfrac{1}{2}\epsilon_0 E^2$ per unit area.

Exercises

26.1. A spherical soap bubble carries a charge Q. Its surface tension provides an inward force of $2\alpha/r$ per unit area. Determine its equilibrium radius.

26.2. A conducting spherical shell of inner radius a and outer radius b is charged with a charge Q. There is a point charge e at the centre $r = 0$. Determine the electric field in the inner region $0 < r < a$ and outside the shell.

---

27. Green's identities

In order to discuss the properties of solutions of Poisson's equation (26.3) we shall make use of two identities, which follow from Green's divergence theorem (Gauss's formula). Let U and V be any two sufficiently smooth functions. If we substitute the vector $\mathbf{a} = U \, \mathbf{grad} \, V$ in the formula $\int \text{div} \, \mathbf{a} \, d^3x = \int a_n \, dS$, we obtain

$$\int (U \Delta V + \mathbf{grad} \, U \cdot \mathbf{grad} \, V) \, d^3x = \int U \frac{\partial V}{\partial n} \, dS. \qquad (27.1)$$

This is Green's first identity. By interchanging U and V and subtracting we obtain his second identity,

$$\int (V \Delta U - U \Delta V) \, d^3x = \int \left(V \frac{\partial U}{\partial n} - U \frac{\partial V}{\partial n} \right) dS. \qquad (27.2)$$

For any static distribution of volume and surface charges, the potential can always be written as a superposition of the elementary solutions $V = e/(4\pi \epsilon_0 r)$ (cf. (14.2) and (14.16)),

$$V = \frac{1}{4\pi \epsilon_0} \left(\int \frac{q \, d^3x}{r} + \int \frac{\sigma \, dS}{r} \right), \qquad (27.3)$$

provided of course that the integrals converge, as they will for any finite system of conductors and charges. The first integral extends over all space and the second over all the conductors. If V is to have the value V_0, rather than zero, at infinity, we just add V_0 on the right-hand side of (27.3). If its gradient is to have the value $-\mathbf{E}_0$ at infinity, we add the term $-\mathbf{E}_0 \cdot \mathbf{r}$. Thus

$$V = V_0 - \mathbf{E}_0 \cdot \mathbf{r} + \frac{1}{4\pi \epsilon_0} \left(\int \frac{q \, d^3x}{r} + \int \frac{\sigma \, dS}{r} \right). \qquad (27.4)$$

Equation (27.4) is sometimes called the solution of Poisson's equation (26.3). It is no such thing, because σ stands for $-\epsilon_0 \partial V/\partial n$ and is in general unknown. The electrostatic problem is to find a solution of (26.3) which is constant over each conductor, and which yields the correct total charge (26.5) on each conductor with prescribed charge. When this problem has been solved, (27.4) is still true, and of as little use as it was at the outset.

Unfortunately, there is no general method that leads with certainty to a solution of any given electrostatic problem. One is therefore forced to resort to guesses and tricks. But, having succeeded in obtaining a solution in one way or another, one may wonder whether there are others. Suppose, then, that two solutions of Poisson's equation exist which satisfy the same conditions: both are constant over each conductor; both yield the same total charges on those conductors whose charges are given, and the same potentials on those whose potentials are prescribed; finally, both have the same required value, or the same required gradient (electric field) at infinity. Now apply Green's first identity (27.1), with $V = U$, to the *difference* U of the two solutions. Since $\Delta U = 0$,

$$\int (\mathbf{grad} \, U)^2 \, d^3x = \int U \frac{\partial U}{\partial n} \, dS. \qquad (27.5)$$

On the left-hand side, the integral extends over all space outside the conductors and inside a very large sphere S, which we let recede to infinity. The right-hand side is a sum of integrals over the conductors and the sphere at infinity. There is no contribution from the sphere at infinity, because either U or **grad** U vanishes there. In each of the integrals over the conductors, U is the difference of two constants and is therefore itself a constant. Each of these integrals is therefore U/ϵ_0 times the difference between the total charges corresponding to the two solutions. On a conductor with prescribed potential, the first factor vanishes; on a conductor with prescribed charge, the second factor does. Hence the left-hand side of (27.5) vanishes, and U is a constant everywhere. If the conditions of the problem prescribe the potential *somewhere*, the difference U will vanish there, and therefore everywhere. Otherwise any solution is determined to within an additive constant, and we conclude, again, that any two solutions are essentially the same.

Equation (27.5) can also be used to prove that a function which is harmonic in a region is uniquely determined by its values (constant or not) on the complete boundary of the region (Dirichlet's problem); alternatively, it is determined up to an additive constant by the values of its normal derivative on the complete boundary (Neumann's problem).

The potential problem in its various forms does not have more than one solution. But does it have one? This question of *existence* is far more difficult than that of uniqueness, and has occupied the best mathematicians for a century. It has been proved that there is a solution if the boundary is sufficiently regular and the prescribed values on it sufficiently smooth. A precise statement of these conditions lies outside the scope of this book. We shall simply assume that they are satisfied in every problem we shall discuss. After all, the mathematical examples we usually treat are stated in terms of analytic functions (the smoothest one can think of) for the equations of the surfaces and for the prescribed values of V or $\partial V/\partial n$. The experimenter, on the other hand, either models his arrangement in mathematical terms in order to apply the theory, and then solves a mathematical example; or he just measures capacities and potential differences, in which case he is not concerned with the question of their existence.

Exercises

27.1. Show that, if a function $f(\mathbf{x})$ has a minimum at P, its Laplacian at P is positive. Deduce that, in a system of charged conductors with no volume charges between them, the potential attains its minimum or its maximum on one of the conductors or at infinity.

27.2. Show that, in a system of charged conductors with zero total net charge, and with no volume charges, at least one conductor is everywhere charged positively, and one everywhere negatively. A capacitor (see the next section) is a specially simple case to which this theorem applies.

27.3. A series of conductors carrying charges $Q_1, Q_2, \ldots$ have potentials $V_1, V_2, \ldots$. Let $Q_1', Q_2', \ldots$ be a second set of charges on the same conductors, corresponding to potentials

V_1', V_2', ... Prove Green's reciprocal theorem:

$$\sum Q_a V_a' = \sum Q_a' V_a.$$

Show that the result still holds if some, or all, of the conductors are replaced by point charges, and the potential V_i at a point charge Q_i is defined as the potential due to all charges except Q_i.

27.4. Apply Green's reciprocal theorem to the case when all the charges are zero, except Q_1 and Q_2', the latter being a point charge. Deduce that the potential of an uncharged conductor under the influence of a unit charge at a point P is the same as the potential at P due to a unit charge placed on the conductor. Thus a unit charge at a distance d from the centre of an uncharged conducting sphere raises the latter to potential $(4\pi\epsilon_0 d)^{-1}$.

Problem 27.1 Let V be harmonic in a region R. Prove the mean value theorem: the value of V at the centre P of any sphere lying wholly in R is equal to the mean value of V on the surface of the sphere, that is,

$$V(P) = \frac{1}{4\pi r^2} \oint V \, dS,$$

where r is the radius of the sphere. Deduce that V cannot have a minimum, nor a maximum, except on the boundary of R.

Solution By Gauss's theorem we have, for any sphere of radius r lying wholly in R,

$$0 = \int \Delta V \, d^3x = \oint \frac{\partial V}{\partial r} \, dS = r^2 \int \frac{\partial V}{\partial r} \, d\Omega.$$

Thus the last integral must vanish for $r > 0$. Hence

$$\int_0^r dr \int \frac{\partial V}{\partial r} \, d\Omega = \int [V(r, \theta, \phi) - V(P)] \, d\Omega = 0.$$

From this follows the mean value theorem

$$V(P) = \frac{1}{4\pi} \int V \, d\Omega = \frac{1}{4\pi r^2} \oint V \, dS.$$

If V had a maximum (minimum) at a point P that does not lie on the boundary of R, a sphere with centre at P, lying wholly within R, would exist on which V would be everywhere less (greater) than $V(P)$. According to the mean value theorem this is impossible.

According to the last problem, V cannot attain a minimum at any point P in an uncharged region (where V must be harmonic). If a movable point charge e be placed at P, it cannot be in stable equilibrium because its potential energy eV cannot be at a minimum. This is *Earnshaw's theorem*. It tells us that no stationary system of charges can be in equilibrium under their own attractions or repulsions. For example, the electrons in an atom or molecule must be *moving*.

28. Capacitors

For a single conductor in empty space, let V be the regular solution that corresponds to a constant potential V_1 and a charge Q_1 on the conductor. If λ is any constant then, by the uniqueness theorem we have just proved, λV is *the* regular solution that corresponds to the constant potential λV_1 and the charge λQ_1 on the conductor. The ratio

$$C = \frac{Q}{V} = -\frac{\epsilon_0}{V} \int \frac{\partial V}{\partial n} \, dS \qquad (28.1)$$

is therefore a property of the conductor which is purely geometric. It is called the *capacity* of the conductor. Its SI unit is the *farad*, defined by $1 \text{farad} = 1 \text{coulomb volt}^{-1}$. For a conductor of given shape, it is proportional to the size, as is evident from (28.1). It is easy to prove that $C > 0$. According to (26.7), $C = 4\pi\epsilon_0 a$ for a spherical conductor of radius a.

A *capacitor*, or *condenser*, is an instrument for storing charge. It consists of two conductors of arbitrary shapes called the *plates* of the capacitor. The positive plate carries a positive charge Q_+ and the negative plate a charge $Q_- = -Q_+$, so that the net charge on the capacitor as a whole vanishes. The capacity of a capacitor is defined by

$$C = \frac{Q_+}{V_+ - V_-}. \qquad (28.2)$$

Again, this is a purely geometrical constant depending on the positions and shapes of the two plates.

The arrangement we have considered in Section 26 of two infinite, parallel, plane conductors constitutes a capacitor. Since the charges on the plates are infinite, we define a capacity per unit area, equal to the charge per unit area divided by the potential difference, or *voltage*. We have seen that the charge per unit area is $\pm\epsilon_0 v/d$, where v is the voltage. Hence the capacity per unit area is ϵ_0/d. If the parallel plates are finite rather than infinite, there will be edge effects, but we expect these to be small so long as the plate diameter is very large compared with the distance between them. For a parallel-plate capacitor in which the area of a plate is S ($S \gg d^2$), the capacity is

$$C = \frac{\epsilon_0 S}{d}. \qquad (28.3)$$

Problem 28.1   A capacitor consists of two coaxial cylinders of radii a and b, each of length L. Neglecting edge effects, determine its capacity.

Solution   Neglecting edge effects, $\mathbf{D}$ in this problem has only a cylindrical radial component D, and $\operatorname{div} \mathbf{D} = r^{-1} d(rD)/dr = 0$, where r now denotes the cylindrical radial coordinate. Hence rD is constant, except for jumps at $r = a$ and b. The constant rD is zero in the central region $r < a$, for otherwise D would be singular on the axis. At $r = a$, D must jump by an amount equal to the surface charge density σ_a. Hence $D = a\sigma_a/r$ for $a < r < b$. Thus

$$V_a - V_b = \int_a^b E \, dr = \int_a^b \frac{D}{\epsilon_0} \, dr = \int_a^b \frac{a\sigma_a}{\epsilon_0 r} \, dr = \frac{a\sigma_a}{\epsilon_0} \ln(b/a).$$

The capacity is

$$C = \frac{Q_a}{V_a - V_b} = \frac{2\pi a L \sigma_a}{V_b - V_a} = \frac{2\pi \epsilon_0 L}{\ln(b/a)}. \tag{28.4}$$

Exercise

28.1. Three concentric hollow conducting spheres have radii a, b and c. The inner and outer are connected by a fine wire and form one plate of a capacitor. The middle sphere is the other plate. Show that the capacity is $4\pi \epsilon_0 b^2 (c - a)/[(b - a)(c - b)]$.

---

29. Expansions in series of harmonic functions

We have already noted that there exists no general method for solving every given electrostatic problem. There are, however, several methods which have been found successful in dealing with certain classes of problems. A new problem may, with luck, turn out to belong to one of these classes; or it may yield to a combination of these methods.

The first method is to expand the potential V as a series of harmonic functions with coefficients that must be determined from the boundary conditions—the prescribed potentials or charges on conductors and the behaviour at infinity.

In spherical coordinates (r, θ, ϕ) the simplest harmonic functions are

$$r^n P_n(\cos \theta) \qquad \text{and} \qquad r^{-(n+1)} P_n(\cos \theta),$$

where n is an integer. For $n \leq 1$ these are

$$1, \quad r \cos \theta, \qquad \text{and} \qquad 1/r, \quad \cos \theta/r^2.$$

The functions $r^n P_n(\cos \theta)$ are finite at $r = 0$, but not at infinity; the opposite is true of the functions $r^{-(n+1)} P_n(\cos \theta)$. Both types have rotational symmetry around the axis $\theta = 0$, and their use is therefore confined to problems that have this symmetry. Otherwise, each Legendre polynomial $P_n(\cos \theta)$ must be replaced by a linear combination of the $2n + 1$ spherical harmonics $P_n^m(\cos \theta) \cos m\phi$ or $P_n^m(\cos \theta) \sin m\phi$, with $-n \leq m \leq n$. Among the functions in this extended set we find the simple harmonics $x = r \sin \theta \cos \phi$ and $y = r \sin \theta \sin \phi$. The simple harmonic $z = r \cos \theta$ is the $n = 1$ member of the axisymmetric set $r^n P_n(\cos \theta)$.

In cylindrical coordinates (r, ϕ, z) the standard harmonic functions are

$$J_m(nr) \cos m\phi e^{\pm nz} \quad \text{and} \quad J_m(nr) \sin m\phi e^{\pm nz},$$

where $J_m(nr)$ is the Bessel function of order m and argument nr.

In two-dimensional polar coordinates (r, θ) the corresponding functions are

$$\ln r, \quad r^{\pm n} \cos n\theta \quad \text{and} \quad r^{\pm n} \sin n\theta.$$

With the latter it may be necessary to restrict n to be an integer if we want the potential to be single valued over $0 \leq \theta \leq 2\pi$.

There are many other harmonic functions besides those mentioned above, but we shall not require them in this book. Indeed we shall not even use tesseral functions $P_n^m(\cos \theta)$ or Bessel functions.

As an example, consider the problem of a spherical conductor in a uniform field $\mathbf{E}_0$ parallel to the z axis. We already know that the (unique) solution is given by (26.11), but now we wish to derive it. At infinity, V must tend to $-E_0 z = -E_0 r \cos \theta$; hence the difference $V - (-E_0 r \cos \theta)$ must be finite. We therefore try

$$V = -E_0 r \cos \theta + \sum_{n=0}^{\infty} \frac{A_n}{r^{n+1}} P_n(\cos \theta). \tag{29.1}$$

This satisfies the equation $\Delta V = 0$ and the boundary condition $V \to -\mathbf{E}_0 \cdot \mathbf{r}$. It remains to impose the conditions that V be a constant V_0 on the sphere $r = a$, and that the total charge on the conductor be a given Q. The first gives

$$0 = \left(\frac{A_0}{a} - V_0 \right) P_0(\cos \theta) + \left(\frac{A_1}{a^2} - E_0 a \right) P_1(\cos \theta)$$

$$+ \sum_{n=2}^{\infty} \frac{A_n}{a^{n+1}} P_n(\cos \theta)$$

for all θ. Now the Legendre polynomials are linearly independent. Hence

$$A_0 = a V_0, \qquad A_1 = a^3 E_0,$$

and $A_2 = A_3 = \cdots = 0$. The charge density on $r = a$ is

$$\sigma = -\epsilon_0 \frac{\partial V}{\partial r} = \epsilon_0 E_0 \cos \theta + \sum_{n=0}^{\infty} (n+1) \frac{\epsilon_0 A_n}{a^{n+2}} P_n(\cos \theta).$$

The total charge is $Q = \int \sigma \, dS = a^2 \int \sigma \, d\omega$, but only the term with $P_0(\cos \theta)$ can survive this integration. Hence $Q = 4\pi \epsilon_0 A_0$. The solution is therefore

$$V = -E_0 r \cos \theta + \frac{Q}{4\pi \epsilon_0 r} + \frac{E_0 a^3}{r^2} \cos \theta$$

$$= -\mathbf{E}_0 \cdot \mathbf{r} + \frac{Q}{4\pi \epsilon_0 r} + a^3 \frac{\mathbf{E}_0 \cdot \mathbf{r}}{r^3}.$$

This is (26.11): a superposition of a uniform field term, a point charge term and a dipole potential, the direction of the dipole being that of $\mathbf{E}_0$.

Problem 29.1 A conducting cylinder of radius a and infinite length is placed with its axis perpendicular to a uniform field E_0. Calculate the field at all points and deduce that the greatest surface density of induced charge is $2\epsilon_0 E_0$.

Solution This problem can be easily solved by writing

$$V = -E_0 r \cos\theta + \sum \frac{A_n \cos n\theta + B_n \sin n\theta}{r^n}$$

and then determining the coefficients from the conditions of the problem. It is easier, and more instructive, to attempt a superposition of a uniform field $\mathbf{E}_0$ and a two-dimensional dipole field.

The potential of a two-dimensional dipole in the direction $\mathbf{p}$ is obtained, like (14.8), by differentiating the potential $V = \text{const} \ln r$ of a line charge in the z direction:

$$V = \mathbf{p} \cdot \mathbf{grad} \ln r = \frac{\mathbf{p} \cdot \mathbf{n}}{r}.$$

In this equation $\mathbf{n}$ is the unit vector in the direction of $\mathbf{r} = (x, y)$. The dipole field is

$$\mathbf{E} = -\,\mathbf{grad}\, V = -\,\mathbf{grad}\, \frac{\mathbf{p} \cdot \mathbf{r}}{r^2} = \frac{2(\mathbf{p} \cdot \mathbf{n})\mathbf{n} - \mathbf{p}}{r^2}.$$

The last two formulae are the two-dimensional analogues of (14.8) and (14.10).

In order to find the field $\mathbf{E} = (E_x, E_y)$ around a cylinder in a uniform field, we try the superposition

$$\mathbf{E} = \mathbf{E}_0 + \alpha \left(\frac{a}{r}\right)^2 [2(\mathbf{E}_0 \cdot \mathbf{n})\mathbf{n} - \mathbf{E}_0],$$

where the non-dimensional coefficient α determines the strength of the dipole, and we require that $\mathbf{n} \times \mathbf{E} = 0$ on $r = a$. This gives $\alpha = 1$. Hence

$$\sigma = \epsilon_0 E_n = 2\epsilon_0 (\mathbf{E}_0 \cdot \mathbf{n}),$$

which has the maximal value $2\epsilon_0 E_0$.

Exercises

29.1. Show that the least charge Q that can be given to a conducting sphere of radius a so that, when the system is placed in a uniform field E_0 no part of the sphere is negatively charged, is $Q = 12\pi\epsilon_0 a^2 E_0$.

29.2. A charge e is placed at a point P outside an uncharged conducting sphere of radius a. The distance of P from the centre O of the sphere is x. Show that the potential at any point whose distance from P is R is given by

$$4\pi\epsilon_0 V = \frac{e}{R} - e \sum_{n=1}^{\infty} \frac{a^{2n+1}}{x^{n+1} r^{n+1}} P_n(\cos\theta).$$

Show that this may be written as

$$4\pi\epsilon_0 V = \frac{e}{R} - \frac{ea/x}{R_1} + \frac{ea/x}{r},$$

where R_1 is the distance from the inverse point of P with respect to the sphere—the point along OP at a distance $x' = a^2/x$ from O. (Charges ea/x at O and $-ea/x$ at the inverse point are called images of e at P.)

29.3. The charge e in the last exercise is replaced by a dipole of moment d pointing away from O. Find the potential at all points by differentiation of the V of the last exercise along the direction of the dipole.

29.4. A charge e is placed at a point C between two concentric spheres, of radii a and b, at a distance c from the common centre O of the spheres. The two spheres are put to earth (zero potential). Show that between the spheres the potential is given by

$$4\pi\epsilon_0 V = \frac{e}{R} + \sum(A_n r^n + B_n r^{-(n+1)})P_n(\cos\theta),$$

where R is the distance from C, r and θ are measured from O and the line OC respectively, and

$$A_n = -\frac{e}{c^{n+1}}\left(\frac{c^{2n+1} - a^{2n+1}}{b^{2n+1} - a^{2n+1}}\right),$$

$$B_n = -\frac{ea^{2n+1}}{c^{n+1}}\left(\frac{b^{2n+1} - c^{2n+1}}{b^{2n+1} - a^{2n+1}}\right).$$

29.5. Calculate the potential of a uniformly charged circular wire of radius a. Use the expansion $V = \sum(A_n r^n + B_n/r^{n+1})P_n(\cos\theta)$ and determine the coefficients A_n and B_n from the potential on the axis, where $r = z$ and $P_n(\cos\theta) = 1$. The latter can be obtained from (27.3) and expanded in powers of z.

---

30. The method of images

The second method is the method of images[†]. We present it by an example. If an electron has been ejected from a metal (perhaps as a result of 'thermal agitation'), with what force will it be pulled back? Consider, then, a particle with charge $-e$ at a distance x from a conducting half-space. The conductor is left with the charge $+e$, which must be distributed over its plane surface. We wish to determine the field outside the conductor.

 If we replace the charge on the conductor by a particle of charge e placed at the *image* point—with respect to the surface of the conductor—of the electron, the

[†]It is due to Newton and Kelvin.

potential becomes

$$V = \frac{1}{4\pi\epsilon_0}\left(\frac{e}{r'} - \frac{e}{r}\right); \tag{30.1}$$

here r is the distance from the electron and r' is the distance from the image point. Outside the conductor this potential satisfies Laplace's equation. It is also constant on the surface of the conductor. Hence it is *the* solution. The field of the image has the magnitude $e/(16\pi\epsilon_0 x^2)$ at the position of the electron. The force is therefore $e^2/(16\pi\epsilon_0 x^2)$.

Exercise

30.1. Calculate the surface charge density σ on the conductor of the foregoing example. Prove that $\int \sigma \, dS = e$.

The method of images is evidently capable of generalization. The problem of several particles in front of a plane is as easy as the one we have considered: for each particle, we introduce an image. For a finite body in front of a plane, we introduce its finite image, and so on. But the method is not confined to images in a plane.

Let e be a point charge outside a spherical conductor of radius a, at a distance x ($> a$) from the centre. We place an image charge $-e'$ inside the sphere, at a distance x' from the centre along the line joining the centre to e, and seek to determine e' and x' in such a way that

$$V = \frac{1}{4\pi\epsilon_0}\left(\frac{e}{r} - \frac{e'}{r'}\right) \tag{30.2}$$

will vanish on the sphere (r and r' are the distances from e and e').

Exercise

30.2. Prove that e' and x' in the foregoing discussion are given by

$$e' = \frac{a}{x}e, \qquad x' = \frac{a^2}{x}. \tag{30.3}$$

With e' and x' given by (30.3), the potential (30.2) vanishes on the sphere. By Gauss's theorem it is the solution corresponding to the case in which the spherical conductor carries a charge $-e'$. If the conductor is to have a total charge Q, we add a point charge $Q + e'$ at the centre; this will not disturb the constancy of V on the sphere.

Determining the images, and their positions, obviously requires some ingenuity. But the method is always subject to the basic rule that images are never to be placed in the region for which one evaluates the potential; they can only be in the *other* regions.

Exercises

30.3. A line charge of strength λ is placed at a distance a from an infinite conducting plane at zero potential. Show that at a point P on the plane the surface density of induced charge is $-\lambda a/(\pi r^2)$, where r is the shortest distance from P to the line charge. Deduce that one-half of the charge induced on the plane lies within $a\sqrt{2}$ from the line charge.

30.4. Two equal charges e are placed a distance a apart, and each one of them is $a/2$ from an infinite conducting plane at zero potential. The charges are both on the same side of the plane. Prove that the force on each one of the charges is

$$\frac{1}{4\pi\epsilon_0}\frac{3e^2}{2a^2}.$$

If the sign of one of the charges is reversed, why are the forces not simply reversed?

30.5. A point charge e is placed inside a spherical cavity of radius a cut out of a conducting block of metal at zero potential. If the distance of the charge from the centre of the cavity is x, show that the force on it is

$$\frac{1}{4\pi\epsilon_0}\frac{e^2 ax}{(a^2 - x^2)^2}.$$

30.6. An electric dipole of moment $\mathbf{d}$ is at a distance x from the centre of a conducting sphere of radius a kept at zero potential. The dipole points away from the sphere. Prove that the image is a dipole of moment $a^3 d/x^3$ and a charge ad/x^2 at the inverse point.

30.7. An electric dipole of moment d is held at a distance a from an infinite conducting plane at zero potential. Show that the image is an equal dipole. Deduce that if it is free to turn about its centre, it will take up a position perpendicular to the plane, and the period of small oscillations about this direction is $4\pi\sqrt{8\pi\epsilon_0 a^3 I/d}$, where I is the moment of inertia.

30.8. A charge e is placed a distance x from the centre of a conducting sphere of radius a. Show that the least positive charge, which must be given to the sphere so that the surface charge density is everywhere positive, is $ea[(x + a)/(x - a)^2 - 1/x]$.

---

31. The complex potential method

The third method uses functions of a complex variable for potential problems in two dimensions. Laplace's equation is then

$$\Delta V = V_{xx} + V_{yy} = 0. \tag{31.1}$$

We set $z = x + iy$, where $i = \sqrt{-1}$, and consider the analytic function

$$w = f(z) = V(x, y) + iU(x, y); \tag{31.2}$$

we have denoted the real part of w by V and the imaginary part by U. Differentiation of $w = f(x + iy)$ gives $w_{xx} + w_{yy} = 0$. Separating real and imaginary parts, we have

$$\Delta V = 0, \qquad \Delta U = 0. \tag{31.3}$$

Hence V and U are harmonic. Furthermore,

$$w_y = if'(z) = iw_x. \tag{31.4}$$

Equating the real and imaginary parts, we obtain the Cauchy–Riemann relations

$$V_x = U_y, \qquad U_x = -V_y. \tag{31.5}$$

Exercise

31.1. Prove that $V(x, y) = $ const. and $U(x, y) = $ const. are two families of curves which cut orthogonally.

If we regard the real part V of w as the electric potential, the lines $V = $ const. become the equipotentials; therefore the lines $U = $ const. are the electric field lines. With this interpretation, $w(z) = V + iU$ is called the complex potential[†].

Exercises

31.2. Prove that the magnitude of the electric field is

$$E = |w'(z)|. \tag{31.6}$$

and that $-\arg w'(z)$ is the angle that $\mathbf{E}$ makes with the x axis. These results suggest that it is often easier to work directly with w than with U or V.

31.3. Let c be a closed curve in the xy plane, and let $\delta(U)$ be the change in U upon traversing c in a counterclockwise direction. Prove that $-\epsilon_0 \delta(U)$ is the charge (per unit length in the z direction) inside c. This provides a ready means for calculating capacities.

[†]Some authors identify the potential with the imaginary part of w; others write $w = V - iU$.

As a simple example, consider the complex potential

$$w = -\frac{\lambda}{2\pi\epsilon_0} \ln z. \tag{31.7}$$

If we use polar coordinates (r, θ), then $z = re^{i\theta}$ and $\ln z = \ln r + i\theta$. The real part of (31.7) becomes

$$V = -\frac{\lambda}{2\pi\epsilon_0} \ln r, \tag{31.8}$$

which is the potential of a line, perpendicular to the xy plane through the origin, that carries a charge λ per unit length (cf. Exercise 14.8). The imaginary part of w is $-\lambda\theta/(2\pi\epsilon_0)$. Hence $\theta = $ const. gives the electric field lines. If we substitute $z - z_0$ for z in (31.7) we get the complex potential for a line through $z_0 = x_0 + iy_0$ instead of through the origin:

$$w = -\frac{\lambda}{2\pi\epsilon_0} \ln(z - z_0). \tag{31.9}$$

The 'reason' for the minus signs in (31.7)–(31.9) is $-\ln z = \ln(1/z)$.

A uniform field E_0 in the x direction is given by $V = -E_0 x$, $U = -E_0 y$, hence $w = -E_0 z$. If it is in the y direction, then $V = -E_0 y$, $U = +E_0 x$, hence $w = iE_0 z$, and if the field makes an angle α with the x axis,

$$w = -E_0 e^{-i\alpha} z. \tag{31.10}$$

A *line doublet* is the two-dimensional analogue of a dipole. Consider a pair of line charges, one of charge λ per unit length at $z = \frac{1}{2}l$ and the other of charge $-\lambda$ at $z = -\frac{1}{2}l$. The separation l is a complex number. In the limit $\lambda l \to d$ and $l \to 0$ we obtain

$$w = -\frac{\lambda}{2\pi\epsilon_0} \ln\left(z + \tfrac{1}{2}l\right) + \frac{\lambda}{2\pi\epsilon_0} \ln\left(z - \tfrac{1}{2}l\right) \to -\frac{d}{2\pi\epsilon_0 z}.$$

If d is real, the doublet is in the x direction; if it is imaginary, the doublet is in the y direction. Generally, the direction is arg d. If the doublet is at z_0 the complex potential is

$$w = -\frac{d}{2\pi\epsilon_0(z - z_0)}. \tag{31.11}$$

Obviously, the method of complex potentials (also called the method of conformal mapping) can be combined with the method of images to solve potential problems in two dimensions. As an example, we consider a line charge of strength λ at the point z_0 above the infinite plane $y = 0$, which is kept at zero potential. By the method of images the potential is the same as that due to a line charge λ at z_0 and a line charge $-\lambda$ at the image point $\bar{z}_0$. Hence

$$
\begin{aligned}
w &= -\frac{\lambda}{2\pi\epsilon_0} \ln(z - z_0) + \frac{\lambda}{2\pi\epsilon_0} \ln(z - \bar{z}_0) \\
&= -\frac{\lambda}{2\pi\epsilon_0} \ln \frac{z - z_0}{z - \bar{z}_0}.
\end{aligned}
\tag{31.12}
$$

From this it is easy to calculate the field and the charge distribution.

As a further example, we consider the circle theorem[†], which combines the complex potential with the method of images. Let $f(z)$ be a complex potential with sources (which may be line charges, line doublets or a uniform field) lying outside the circle $|z| = a$. If we now introduce a cylindrical conductor $|z| = a$ at zero potential with its axis through the origin, what is the new complex potential? Consider the complex function

$$
w = f(z) - \bar{f}(a^2/z),
\tag{31.13}
$$

where, for any $f = V + iU$, $\bar{f} = V - iU$. Note that $\bar{f}(z)$ is *not* the same as $\overline{f(z)} = \bar{f}(\bar{z})$. Since $f(z)$ has no singularities inside the circle, $\bar{f}(a^2/z)$ has none outside, and its subtraction will not alter the distribution of sources outside the circle. When $|z| = a$ we have $a^2/z = \bar{z}$ so that for all points on the circle $|z| = a$ we obtain $\bar{f}(a^2/z) = \bar{f}(\bar{z}) = \overline{f(z)}$. Thus w is purely imaginary on the circle, and its real part V vanishes there. Hence (31.13) provides the solution. The singularities of $\bar{f}(a^2/z)$, which lie inside the circle, are the images required for securing $V = 0$ on the cylindrical conductor.

Finally, we consider an example that really demonstrates the power of the method of complex potentials. Let $w(z)$ be the complex potential defined (implicitly) through

$$
z = 1 + iw + e^{iw}.
\tag{31.14}
$$

Separating real and imaginary parts, we have

$$
\begin{aligned}
x &= 1 - U + e^{-U} \cos V, \\
y &= V + e^{-U} \sin V.
\end{aligned}
\tag{31.15}
$$

Consider now the two equipotentials $V = \pm\pi$. Along these,

$$
x = 1 - U - e^{-U}, \qquad y = \pm\pi.
$$

Since $e^{-U} > 1 - U$ for any U, these equipotentials are the straight half-lines $x < 0$, $y = \pm\pi$. We identify them with the two plates of a capacitor. If U is eliminated between the two members of (31.15), the result,

$$
x = 1 + \ln(y - V) - \ln \sin V + (y - V) \cot V,
\tag{31.16}
$$

[†]Milne–Thomson (1940).

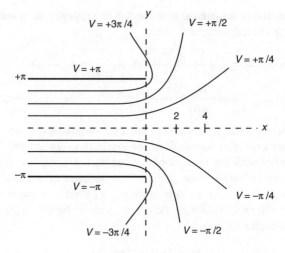

Fig. 31.1

gives an equipotential $x = x(y)$ for each value of V between $-\pi$ and π. Some of these are shown in Fig. 31.1. We see how the equipotentials fan out near the edge of a parallel-plate capacitor. If the plates are at the potentials $\pm v/2$, instead of $\pm \pi$, we must replace w by $wv/(2\pi)$ in (31.14); if they are at $y = \pm d/2$, instead of $\pm \pi$, we must replace z by $zd/(2\pi)$.

The methods we have introduced in the previous two sections, and in this one, are mathematical tools for solving Laplace's equation with appropriate boundary conditions. Their applications reach far beyond the electrostatics of conductors in vacuum. As we shall see, problems involving linear dielectrics, linear magnets and steady currents in linear conductors (those that obey Ohm's law) can all be mathematically formulated in terms of Laplace's equation, and can therefore be solved by the same methods. Indeed, almost every solution of an electrostatic problem provides a solution of corresponding problems in the other contexts. For example, every formula for the capacity of a condenser is also a formula for the resistance of a corresponding Ohmic conductor.

Exercises

31.4. Show that $w = -E_0(z - a^2/z)$ gives the potential for a conducting cylinder of radius a at zero potential in a uniform field E_0.

31.5. Find the potential of a line charge lying outside and parallel to an uncharged cylinder.

31.6. Show that the capacity per unit length of two parallel cylinders of radius R a distance $2D$ apart is given by

$$C = \frac{\pi \epsilon_0}{\cosh^{-1}(D/R)}.$$

9

Linear Dielectrics

32. Electrostatics of dielectrics

If the insulating medium in an electrostatic problem is a linear dielectric, we must consider its constitutive equations

$$\mathbf{P}' = \epsilon_0 \chi \mathbf{E}', \qquad \mathbf{M}' = 0$$

(cf. (22.1)), where χ is the dielectric susceptibility and the primed quantities refer to the material rest frame. In the electrostatic case we are, by definition, in the rest frame. Thus

$$\mathbf{D} = \epsilon_0 \mathbf{E} + \mathbf{P} = \epsilon_0 (1 + \chi) \mathbf{E} = \epsilon \mathbf{E}, \tag{32.1}$$

where

$$\epsilon = (1 + \chi) \epsilon_0 \tag{32.2}$$

is called the *permittivity*. The dielectric susceptibility χ is always positive. It ranges from less than 10^{-3} (in gases at room temperature and atmospheric pressure) to more than 2000. Water has a χ of about 80. The permittivity is of course positive, too; indeed $\epsilon > \epsilon_0$. The ratio $K = \epsilon/\epsilon_0$ is called the *dielectric constant*.

In the presence of linear dielectrics, electrostatics is governed by the equations

$$\operatorname{curl} \mathbf{E} = 0,$$
$$\operatorname{div} \mathbf{D} = \operatorname{div} \epsilon \mathbf{E} = q, \tag{32.3}$$
$$\mathbf{n} \times [\![\mathbf{E}]\!] = 0, \qquad \mathbf{n} \cdot [\![\mathbf{D}]\!] = \sigma.$$

Wherever there is a vacuum, we simply set $\epsilon = \epsilon_0$. If ϵ is uniform, these equations lead to Poisson's equation $\Delta V = -q/\epsilon$; in the absence of free charges (which would have no means of getting into an insulating dielectric) they lead to Laplace's equation $\Delta V = 0$. At a boundary between two insulating dielectrics, both $\mathbf{n} \times \mathbf{E}$ and D_n are continuous.

At a stationary surface of discontinuity the jump condition $(10.17)_2$ reduces to $[\![V]\!] = 0$. Since V always contains an arbitrary constant, the continuity of V fixes the arbitrary constant on one side of the surface in terms of the constant on the other side. The latter remains arbitrary, and is fixed by some convention; for example, $V = 0$ at infinity, or on all conductors that are 'put to earth'. In terms of the potential, the jump conditions are then

$$V \quad \text{and} \quad \epsilon \frac{\partial V}{\partial n} \quad \text{are continuous.} \tag{32.4}$$

As an example, we consider a dielectric sphere of radius a in an external uniform field $\mathbf{E}_0$ which is parallel to the z direction. Let ϵ_i be the permittivity inside the sphere, and ϵ_e the permittivity of the medium outside it. For the potential V_e outside the sphere we assume the form (29.1), which satisfies $V_e = 0$ at infinity. Inside the sphere we set $V_i = \sum B_n r^n P_n(\cos\theta)$, which is finite at $r = 0$ and includes an undetermined constant B_0. On applying the conditions (32.4), one finds that only the terms with $n \leq 1$ survive. For the fields one obtains

$$\mathbf{E}_e = \mathbf{E}_0 + \frac{a^3}{r^3} \frac{\epsilon_i - \epsilon_e}{2\epsilon_e + \epsilon_i} \left[3(\mathbf{n} \cdot \mathbf{E}_0)\mathbf{n} - \mathbf{E}_0 \right],$$

$$\mathbf{E}_i = \frac{3\epsilon_e}{2\epsilon_e + \epsilon_i} \mathbf{E}_0. \tag{32.5}$$

Somewhat unexpectedly, the field inside the sphere turns out to be uniform, and independent of the size of the sphere. Its strength depends on the permittivities; more exactly, on their ratio. Outside there is a dipole field, superimposed on the uniform $\mathbf{E}_0$.

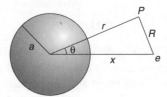

Fig. 32.1

The next example is one in which all the P_n's are needed. Consider a charge e placed at a distance x from the centre of an uncharged dielectric sphere of radius a, as in Fig 32.1. For the potentials inside and outside the sphere we write

$$4\pi\epsilon_0 V = \sum A_n r_n P_n(\cos\theta),$$

$$4\pi\epsilon_0 V_e = \frac{e}{R} + \sum \frac{B_n}{r^{n+1}} P_n(\cos\theta), \tag{32.6}$$

where the first term on the right-hand side of $(32.6)_2$ is the potential due to the point charge alone; R is of course the distance from e. In order to apply the conditions

(32.4) at $r = a$ we use (for $r < x$) the expansion

$$\frac{1}{R} = \sum \frac{r^n}{x^{n+1}} P_n(\cos\theta).$$ (32.7)

On equating the coefficients of the linearly independent P_n's, we obtain

$$A_n = \frac{e}{x^{n+1}} + \frac{B_n}{a^{2n+1}},$$

$$\epsilon n A_n = \epsilon_0 \left(\frac{ne}{x^{n+1}} - \frac{(n+1)B_n}{a^{2n+1}} \right),$$ (32.8)

from which all the A's and B's can be found. In order to obtain the force experienced by the charge, we calculate the radial field

$$-\frac{\partial}{\partial r}\left(V_e - \frac{e}{4\pi\epsilon_0 R} \right),$$

(leaving out the self-field of the charge) and substitute $\theta = 0$, $r = x$. Since, for $\theta = 0$, each $P_n(\cos\theta) = 1$, the force (e times the field) is

$$\frac{e}{4\pi\epsilon_0} \sum \frac{(n+1)B_n}{x^{n+2}},$$ (32.9)

in the direction OP.

At a conductor–dielectric boundary there may still be a free surface charge density, and the relation (26.4) is replaced by

$$\sigma = D_n = -\epsilon \frac{\partial V}{\partial n}.$$ (32.10)

The boundary condition $\mathbf{n} \times \mathbf{E} = 0$ (which means that V is constant) still holds on the surface of each conductor.

As a simple example, we consider a spherical conductor of radius a that carries a charge Q and is embedded in a dielectric medium with permittivity ϵ. The potential is then given by (26.7), with ϵ_0 replaced by ϵ.

Similarly, if we substitute ϵ for ϵ_0 in the solution for the infinite, parallel-plate capacitor, we obtain the solution for the case in which the medium between the plates is a dielectric with permittivity ϵ. The capacity per unit area is ϵ/d, larger than the ϵ_0/d for a vacuum. Filling the capacitor with a dielectric therefore enables it to store a larger charge $Q_+ = C(V_+ - V_-)$ for a given voltage.

Finally, we consider the cylindrical capacitor: a hollow cylinder of length L, with inner radius a and outer radius b, consisting of dielectric material with permittivity ϵ. The inside and outside surfaces are coated with conducting material, and provide the plates of the capacitor. Neglecting edge effects, the capacity is given by (28.4), with ϵ_0 replaced by ϵ:

$$C = \frac{2\pi\epsilon L}{\ln(b/a)}.$$ (32.11)

Problem 32.1   A parallel-plate capacitor is filled with two dielectrics, arranged in alternating layers that are all parallel to the plates. The layers of the first dielectric, with permittivity ϵ_a, occupy a fraction a (< 1) of the volume between the plates; the layers of the second dielectric, with permittivity ϵ_b, a fraction $b = 1 - a$. Prove that the capacity is $\epsilon S/d$, where $\epsilon^{-1} = a\epsilon_a^{-1} + b\epsilon_b^{-1}$.

Solution   The displacement $D = \sigma$ is continuous across the layers, but the field $E = D/\epsilon$ is not. The voltage is

$$V = \int E\, dz = \int \frac{\sigma}{\epsilon}\, dz = \sigma\left(\frac{a}{\epsilon_a} + \frac{b}{\epsilon_b}\right)d = \frac{\sigma d}{\epsilon},$$

and $C = Q/V = \sigma S/V = \epsilon S/d$.

Problem 32.2   The same as Problem 32.1, but with the two dielectrics arranged in prisms (or columns) perpendicular to the plates. Prove that the capacity is $\epsilon S/d$, with $\epsilon = a\epsilon_a + b\epsilon_b$.

Solution   The field $E = V/d$ is continuous because it is parallel to the columns, but the displacement $D = \epsilon E$ and surface charge density $\sigma = D$ are not. The charge is

$$Q = \int \sigma\, dS = E \int \epsilon\, dS = E(a\epsilon_a + b\epsilon_b)S = E\epsilon S.$$

and $C = Q/V = E\epsilon S/(Ed) = \epsilon S/d$.

Problem 32.3   A cylindrical capacitor with inside and outside surfaces $r = a$ and $r = b$, and with vertical axis, is filled up to a fraction λ of its height with a liquid of dielectric constant K, the remaining part being empty. If the arrangement is to provide a liquid level gauge, find its sensitivity $d \ln C/d \ln \lambda$.

Solution   From div $\mathbf{D} = 0$ the radial component D of the displacement is $D = a\sigma_a/r$. The horizontal field $E = D/\epsilon$ must be continuous at the liquid–air interface. Hence the surface charge is proportional to ϵ, and the capacity is

$$C = \frac{Q}{V} = \frac{2\pi\epsilon_0 L}{\ln(b/a)}(1 - \lambda + K\lambda).$$

Thus

$$\frac{d \ln C}{d \ln \lambda} = \frac{\lambda(K-1)}{1 + \lambda(K-1)}.$$

Exercises

32.1.   Extend the statements in Exercises 27.1, 27.2 and 27.3 to charged conductors in a linear dielectric medium.

32.2. Show from a consideration of the jump conditions that at a change of dielectric medium the electric field lines are refracted according to the law $\epsilon_1 \cot \theta_1 = \epsilon_2 \cot \theta_2$, where the angles are those between the field lines and the common normal to the boundary, as in Fig. 32.2.

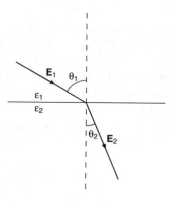

Fig. 32.2

32.3. A spherical uncharged conductor of radius a is surrounded by a dielectric whose outer boundary is a concentric sphere of radius b. It is placed in a uniform field E_0. Show that the total positive and negative charges induced on the conductor are

$$\pm \frac{9\pi \epsilon \epsilon_0 a^2 b^3 E_0}{(\epsilon + 2\epsilon_0)b^3 + 2(\epsilon - \epsilon_0)a^3}.$$

32.4. A electric point dipole $\mathbf{p}$ is embedded at the centre of a sphere of radius a, made of linear dielectric material with dielectric constant K. Determine the electric potential.

32.5. The plates of a parallel-plate capacitor are squares of side a, and the distance between them is d. Between the plates is a dielectric slab of area $a \times a$, thickness d and permittivity ϵ. The plates are connected by wires to the electrodes of a battery, which maintains a constant voltage V. What are the charges on the plates if the dielectric slab has been pushed out a distance $x < a$ in a direction parallel to one of its sides? What current will flow in the wires if it is pushed out with velocity v?

---

33. Thomson's theorem

Let us now consider an uncharged dielectric medium of permittivity ϵ (a positive, but otherwise arbitrary, function of position) in which conductors carrying given charges have been placed. According to the first principle of electromagnetism (the first pair of Maxwell's equations), the displacement $\mathbf{D}$ must then be such that div $\mathbf{D} = 0$ in the medium and the integrals $Q_a = \oint D_n \, dS_a$ over the conductors $a = 1, 2, \ldots$ have fixed values. We shall *not* assume that **curl E** = **curl D**$/\epsilon$ = 0, i.e. $\mathbf{D} = -\epsilon$ **grad** V, or that the charge density D_n on the conductors is distributed in such a way that V is constant over each conductor. Rather, we shall consider the *functional*

$$W[\mathbf{D}] = \int \frac{D^2}{2\epsilon} d^3x, \qquad (33.1)$$

the integral being taken throughout the medium; it will exist if the system of conductors is finite in the sense of Section 27. For any vector field $\mathbf{D}$ such that $\operatorname{div} \mathbf{D} = 0$ and $\oint D_n dS_a = Q_a$, $W[\mathbf{D}]$ is positive (since $\epsilon > 0$). It can be made arbitrarily large by choosing a sufficiently tangled $\mathbf{D}$ (we recall that **curl D** is now arbitrary). But W must have a minimum; indeed a *positive* minimum (unless the charges Q_a on the conductors all vanish).

In order to find the displacement that renders $W[\mathbf{D}]$ a minimum, we introduce Lagrangian multipliers $V(\mathbf{x})$ and λ_a and take the *unrestricted* variation of

$$W'[\mathbf{D}] = W[\mathbf{D}] - \int V \operatorname{div} \mathbf{D} \, d^3x - \sum \lambda_a \oint D_n \, dS_a \qquad (33.2)$$

with respect to $\mathbf{D}$. The result, after an integration by parts (remembering that on each conductor δD_n is the normal component of $\mathbf{D}$ in the direction pointing outward, and hence *into* the dielectric medium), is

$$\delta W' = \int \delta \mathbf{D} \cdot \left(\frac{\mathbf{D}}{\epsilon} + \operatorname{grad} V \right) d^3x$$
$$- \sum \oint \delta D_n (\lambda_a - V) \, dS_a. \qquad (33.3)$$

If $\delta W'$ is to vanish for all $\delta \mathbf{D}$ throughout the medium, and for all δD_n on the conductors, then $\mathbf{D} = -\epsilon \operatorname{grad} V$, where V has the constant value λ_a over the conductor a. These are precisely the requirements we have left out. This proves *Thomson's theorem*: the charges on the conductors distribute themselves in such a way as to render $W[\mathbf{D}]$ a minimum.

Thomson's theorem is a variational formulation of the electrostatic problem for a linear dielectric medium. This may have interesting implications. It tells us, for example, that the displacement field will tend to be larger in absolute value (the lines of force of the vector field $\mathbf{D}$ will be more concentrated) in regions of high permittivity, because this minimizes the integral of D^2/ϵ.

Problem 33.1 Prove that the solution of the electrostatic problem of conductors in the presence of linear dielectrics is unique. Show, in particular, that the uniqueness does not depend on ϵ being uniform, and that it is unaffected by sudden discontinuities in the permittivity.

Solution Let $V = V_2 - V_1$, $\mathbf{D} = \mathbf{D}_2 - \mathbf{D}_1$ and $\mathbf{E} = \mathbf{E}_2 - \mathbf{E}_1$ denote the differences between the quantities corresponding to any two solutions. For any region between the conductors, in which ϵ is smooth,

$$\int \epsilon(\mathbf{grad}\, V)^2\, d^3x = \int \mathbf{D} \cdot \mathbf{E}\, d^3x$$

$$= -\int \mathbf{D} \cdot \mathbf{grad}\, V\, d^3x$$

$$= -\int [\mathrm{div}\, V\mathbf{D} - V\, \mathrm{div}\, \mathbf{D}]\, d^3x$$

$$= -\int V D_n\, dS + \int V\, \mathrm{div}\, \mathbf{D}\, d^3x.$$

The last integral vanishes because $\mathrm{div}\, \mathbf{D} = \mathrm{div}\, \mathbf{D}_2 - \mathrm{div}\, \mathbf{D}_1 = 0$. Over any surface across which ϵ is discontinuous, V and the normal component of $\mathbf{D}$ are both continuous. Since $\mathbf{n}$ points out of each region, the integrals $\int V D_n\, dS$ over such surfaces of discontinuity will have opposite signs for each pair of adjoining regions. We can therefore extend the result,

$$\int \epsilon(\mathbf{grad}\, V)^2\, d^3x = -\int V D_n\, dS,$$

to the whole of the region outside the conductors. The last surface integral is now over the conductors and the sphere at infinity. Over each conductor $V = V_2 - V_1$ is constant and $\int D_n\, dS = -(Q_2 - Q_1)$, since $\mathbf{n}$ is now pointing *out* of the region, hence *into* the conductor. For a conductor with given potential, $V_2 = V_1$. If the charge is given, $Q_2 = Q_1$. Finally at infinity either the potential is given, in which case $V = 0$, or the electric field is given, in which case $\mathbf{D} = \epsilon(\mathbf{E}_2 - \mathbf{E}_1) = 0$. Thus $\mathbf{grad}(V_2 - V_1) = 0$ everywhere, and the electric field in the two solutions is the same everywhere.

Problem 33.2   Prove that an increase in the permittivity without alteration of the charges on the conductors decreases $W = \int \frac{1}{2}\mathbf{D} \cdot \mathbf{E}\, d^3x$. This is another theorem of Thomson.

Solution   For a small, generally non-uniform, change $\delta\epsilon \geq 0$ in the permittivity,

$$\delta W = \int \frac{1}{2}\delta\left(\frac{D^2}{\epsilon}\right) d^3x$$

$$= \int \left(\frac{\mathbf{D} \cdot \delta\mathbf{D}}{\epsilon} + \frac{1}{2}D^2\delta\epsilon^{-1}\right) d^3x$$

$$= \int \left(\mathbf{E} \cdot \delta\mathbf{D} + \frac{1}{2}D^2\delta\epsilon^{-1}\right) d^3x.$$

Since $\mathrm{div}\, \mathbf{D} = 0$ the first term is

$$-\int \delta\mathbf{D} \cdot \mathbf{grad}\, V\, d^3x = -\sum \oint V\delta D_n\, dS,$$

the sum being over all the conductors. Since V is constant on each conductor, and $\oint \delta D_n\, dS = 0$ (the charges being unaltered), we have

$$\delta W = \int \frac{1}{2}D^2\delta\epsilon^{-1}\, d^3x \leq 0,$$

since $\delta\epsilon \geq 0$.

Problem 33.3   Show that the introduction of a new uncharged or earthed conductor diminishes $W = \int \frac{1}{2} \mathbf{D} \cdot \mathbf{E} \, d^3x$.

Solution   We denote the quantities resulting from the introduction of the new conductor by primes. The region R' outside the new set of conductors is of course smaller than the region R outside the original set.

$$
\begin{aligned}
W' - W &= \int_{R'} \tfrac{1}{2}\epsilon E'^2 \, d^3x - \int_R \tfrac{1}{2}\epsilon E^2 \, d^3x \\
&= \int_{R'} \tfrac{1}{2}\epsilon E'^2 \, d^3x - \int_{R'} \tfrac{1}{2}\epsilon E^2 \, d^3x - \int_{R-R'} \tfrac{1}{2}\epsilon E^2 \, d^3x \\
&= \int_{R'} \tfrac{1}{2}\epsilon (E'^2 - E^2) \, d^3x - \int_{R-R'} \tfrac{1}{2}\epsilon E^2 \, d^3x \\
&= \int_{R'} \tfrac{1}{2}\epsilon \big[-(\mathbf{E} - \mathbf{E}')^2 + 2E'^2 - 2\mathbf{E} \cdot \mathbf{E}' \big] \, d^3x - \int_{R-R'} \tfrac{1}{2}\epsilon E^2 \, d^3x \\
&= \int_{R'} \epsilon \mathbf{E}' \cdot (\mathbf{E}' - \mathbf{E}) \, d^3x - \int_{R'} \tfrac{1}{2}\epsilon (\mathbf{E} - \mathbf{E}')^2 \, d^3x - \int_{R-R'} \tfrac{1}{2}\epsilon E^2 \, d^3x.
\end{aligned}
$$

Now $\operatorname{div}(\mathbf{D}' - \mathbf{D}) = 0$ throughout R'. Thus

$$
\begin{aligned}
\int_{R'} \epsilon \mathbf{E}' \cdot (\mathbf{E}' - \mathbf{E}) \, d^3x &= -\int_{R'} \mathbf{grad}\, V' \cdot (\mathbf{D}' - \mathbf{D}) \, d^3x \\
&= -\int_{R'} \operatorname{div}[V'(\mathbf{D}' - \mathbf{D})] \, d^3x \\
&= -\sum \oint V'(D'_n - D_n) \, dS.
\end{aligned}
$$

The last sum extends over all the conductors. On each one V' is constant and can be taken out of the integral. The integrals $\oint (D'_n - D_n) \, dS$ over the original conductors vanish, because their charges are unaltered. The new conductor contributes $-V' \oint (D'_n - D_n) \, dS$, but this vanishes because $\oint D_n \, dS = 0$ (the original 'charge' of the region occupied by the new conductor vanishes) and the new conductor is either uncharged ($\oint D'_n \, dS = 0$) or earthed ($V' = 0$). Thus

$$
W' - W = -\int_{R'} \tfrac{1}{2}\epsilon (\mathbf{E} - \mathbf{E}')^2 \, d^3x - \int_{R-R'} \tfrac{1}{2}\epsilon E^2 \, d^3x < 0.
$$

Exercises

33.1.   A spherical capacitor consists of two concentric spheres of radii a and d. Concentric with these and lying between them is a spherical shell of permittivity ϵ bounded by the spheres $r = b$ and $r = c$. Show that if $a < b < c < d$, the capacity C is given by

$$
\frac{4\pi \epsilon_0}{C} = \frac{1}{a} - \frac{1}{d} + \frac{\epsilon_0 - \epsilon}{\epsilon}\left(\frac{1}{b} - \frac{1}{c}\right).
$$

33.2. A capacitor is formed of the two spheres $r = a$ and $r = b$ ($b > a$) with a dielectric of uniform ϵ between them. The dielectric strength of the material (i.e. the greatest permitted electric field strength before it conducts) is E_0. Show that the greatest voltage between the two spheres, so that the field nowhere exceeds the critical value, is $E_0 a(b - a)/b$.

33.3. If in the last exercise the voltage is gradually increased beyond the critical value so that charge can flow into part of the dielectric, show that the capacitor does not break down completely until the voltage is increased to $E_0(b - a)$.

33.4. The surfaces of an air capacitor are concentric spheres. If half the space between the spheres is filled with material of dielectric constant K, the dividing surface between the material and the air being a plane through the centre of the spheres, show that the capacity will be the same as though the whole space between the spheres were of uniform dielectric constant $\frac{1}{2}(1 + K)$.

33.5. A point charge e is placed at a point P outside a semi-infinite medium of uniform permittivity ϵ. Show that in the vacuum the potential is the same as that due to a charge e at P and e' at the image point P', and in the dielectric it is the same as that due to a charge e'' at P, where

$$e' = -\frac{\epsilon - \epsilon_0}{\epsilon + \epsilon_0} e, \qquad e'' = \frac{2\epsilon}{\epsilon + \epsilon_0} e.$$

33.6. In a medium in which the permittivity depends solely on the radial distance r, the potential is governed by the equation

$$\frac{1}{r^2} \frac{d}{dr}\left(r^2 \epsilon \frac{dV}{dr} \right) = -q.$$

A charge is placed at the origin in a medium in which $\epsilon = \epsilon_0(1 + a/r)$. Show that the potential is

$$V = \frac{e}{4\pi \epsilon_0 a} \ln \frac{r + a}{r}.$$

33.7. A point charge e is placed a small distance x from the centre O of a spherical cavity of radius a in an infinite dielectric. Show that the charge experiences a force approximately equal to

$$\frac{2(\epsilon - \epsilon_0)}{2\epsilon + \epsilon_0} \frac{e^2 x}{4\pi \epsilon_0 a^3}$$

away from O.

---

34. Wilson's experiment

Electrostatics does not make full use of the material properties of dielectrics, because the part $\dot{\mathbf{x}} \times \mathbf{B}$ of the electromotive intensity is ignored. If we wish to exhibit the influence of this term, we must examine phenomena in which a dielectric is set in motion (with respect to the aether) and made to cross magnetic field lines. We shall therefore discuss a famous experiment that was designed and conducted by Wilson in 1904 with the object of investigating such an effect.

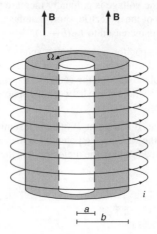

Fig. 34.1

The inner and outer surfaces, $r = a$ and $r = b$, of a hollow dielectric cylinder of length L (Fig. 34.1) are coated with conducting material and connected through sliding contacts by a wire. The cylinder is placed inside a solenoid, which (as we shall show in Section 41) provides a uniform magnetic field $\mathbf{B}_0$, parallel to the axis. When the cylinder is rotated around its axis the inner and outer conducting surfaces become oppositely charged (since the system was neutral to begin with). We wish to calculate this charge in the steady state that is established after the angular velocity Ω (parallel to $\mathbf{B}_0$) has been maintained constant for some time[†].

The equations of the problem are

$$\operatorname{div}\mathbf{D} = 0, \qquad\qquad \mathbf{n}\cdot[\![\mathbf{D}]\!] = \sigma,$$
$$\operatorname{\mathbf{curl}}\mathbf{E} = 0, \qquad\qquad \mathbf{n}\times[\![\mathbf{E}]\!] = 0,$$
$$\mathbf{D} = \epsilon_0\mathbf{E} + \mathbf{P}, \qquad\qquad \mathbf{P} = \epsilon_0\chi\,(\mathbf{E} + \dot{\mathbf{x}}\times\mathbf{B}), \qquad (34.1)$$
$$\operatorname{div}\mathbf{B} = 0, \qquad\qquad \mathbf{n}\cdot[\![\mathbf{B}]\!] = 0,$$
$$\operatorname{\mathbf{curl}}\mathbf{H} = 0, \qquad\qquad \mathbf{n}\times[\![\mathbf{H}]\!] = \mathbf{K},$$

$$\mathbf{H} = \mathbf{B}/\mu_0 + \dot{\mathbf{x}}\times\mathbf{P}.$$

Of course, $\dot{\mathbf{x}} = \Omega\times\mathbf{r}$. In the last line of (34.1) we have used the relation $\mathbf{M} = -\mathbf{v}\times\mathbf{P}$ (cf. (22.2)) for a linear, non-relativistic dielectric. Using cylindrical coordinates (r, ϕ, z), we seek a solution with $\mathbf{E} = (E, 0, 0)$ and $\mathbf{B} = (0, 0, B)$. As a result of

[†]In Section 35 we shall see that this time is practically zero.

the equations, we shall then have $\mathbf{P} = (P, 0, 0)$, $\mathbf{D} = (D, 0, 0)$ and $\mathbf{H} = (0, 0, H)$. Since the net charge on the inner and outer surfaces vanishes, $\mathbf{D}$ will vanish outside the cylinder. Similarly, the convection currents due to the rotating free and bound charges will have no magnetic effect outside: even near the cylinder we shall then have the undistorted magnetic field $\mathbf{B_0}$. Inside, however, $\mathbf{B}$ need not equal $\mathbf{B_0}$.

The equation $\text{div } \mathbf{D} = 0$ gives $rD = \alpha$, a constant. Applying the condition $\mathbf{n} \cdot [\![\mathbf{D}]\!] = \sigma$ at the surfaces, we obtain $\alpha = a\sigma_a = -b\sigma_b$, where σ_a and σ_b are the surface charge densities. Thus, for $a < r < b$,

$$D = \epsilon_0 E + P = \epsilon_0 E + \epsilon_0 \chi (E + \Omega r B) = \frac{a}{r}\sigma_a, \tag{34.2}$$

and everywhere else $D = 0$.

Turning now to the magnetic field, **curl H** $= 0$ states that

$$H = B/\mu_0 - \Omega r P = B/\mu_0 - \Omega r \epsilon_0 \chi (E + \Omega r B)$$

is constant. Applying the condition $\mathbf{n} \times [\![\mathbf{H}]\!] = \mathbf{K}$ to the surfaces, we obtain $H_i - H = \Omega a \sigma_a$ and $H - B_0/\mu_0 = \Omega b \sigma_b$, where H_i is the current potential in the inner region $r < a$. Thus $H_i = B_0/\mu_0$, and

$$H = B/\mu_0 - \Omega r \epsilon_0 \chi (E + \Omega r B) = B_0/\mu_0 - \Omega a \sigma_a. \tag{34.3}$$

The last equalities in (34.2) and (34.3) provide a pair of linear equations for B and $a\sigma_a$. The solutions are

$$B = B_0 - \Omega r E/c^2,$$

$$\frac{a}{r}\sigma_a = \left(1 - \frac{\epsilon - \epsilon_0}{\epsilon}\frac{\Omega^2 r^2}{c^2}\right)\epsilon E + (\epsilon - \epsilon_0)\Omega B_0 r, \tag{34.4}$$

where, as usual, $\epsilon = (1 + \chi)\epsilon_0$. Neglecting the fraction $(\epsilon - \epsilon_0)/\epsilon$ of the relativistically small ratio $\Omega^2 r^2/c^2$, we have

$$\frac{a}{r}\sigma_a = \epsilon E + (\epsilon - \epsilon_0)\Omega B_0 r. \tag{34.5}$$

Equation (34.5) is the same as the last equality of (34.2), with B replaced by B_0: the magnetic field B inside the cylinder differs from the field B_0 outside by a relativistically small term.

Since $E = -dV/dr$ (that is where **curl E** $= 0$ is used), and the voltage $v = V_b - V_a$ between the inner and outer surfaces of the cylinder, which are connected by a wire, is zero, (34.5) integrates to

$$a\sigma_a \ln (b/a) = \tfrac{1}{2}(\epsilon - \epsilon_0)\Omega B_0(b^2 - a^2).$$

In terms of the capacity $C = 2\pi \epsilon L/\ln (b/a)$ of the cylinder (cf. (32.11)), the last equation becomes

$$Q = 2\pi a L\sigma_a = \tfrac{1}{2}C\frac{\epsilon - \epsilon_0}{\epsilon}\Omega B_0(b^2 - a^2). \tag{34.6}$$

If the wire is now cut and the two ends connected to an electrometer[†] of capacity C_e, a voltage v is set up such that $(C + C_e)v = Q$ (capacitors connected in parallel), or

$$v = \frac{\epsilon - \epsilon_0}{\epsilon} \frac{\Omega}{2\pi} \frac{\pi(b^2 - a^2)B_0}{1 + C_e/C}. \tag{34.7}$$

We note that $\Omega/(2\pi)$ is the rate of rotation in revolutions per unit time and that $\pi(b^2 - a^2)$ is the cross-section of the hollow cylinder; the product $\pi(b^2 - a^2)B_0$ is therefore the magnetic flux through the cross-section of the cylinder. If the current in the solenoid is reversed, B_0 becomes antiparallel to Ω: Q and v then change their signs.

Wilson's experiment confirmed the linear dependence of the voltage on Ω or B_0, as well as its change of sign (and no more than that) upon reversal of the current through the solenoid. But its historical importance was in confirming the way in which the v of (34.7) depends on the dielectric constant $K = \epsilon/\epsilon_0$. At the turn of the twentieth century there was a great controversy regarding the motion of bodies through the aether. According to the principles we have adopted, $\mathbf{P} = (\epsilon - \epsilon_0)\mathcal{E}$ and $\mathbf{D} = \epsilon_0\mathbf{E} + \mathbf{P} = \epsilon_0\mathbf{E} + (\epsilon - \epsilon_0)\mathcal{E}$. This was the view of Lorentz[‡].

Hertz, on the other hand, advocated a different view, based on the belief that bodies with electromagnetic properties carried the aether along with them. According to Hertz, the relation between the electric displacement and the electromotive intensity in a moving linear dielectric was $\mathbf{D} = \epsilon\mathcal{E}$ (that is, $\mathbf{D}' = \epsilon\mathbf{E}'$ in the rest frame). Hertz's relation would have led, in Wilson's experiment, to a voltage given by (34.7), but with $(\epsilon - \epsilon_0)/\epsilon$ replaced by 1. Thus—so long as $C_e \ll C$—the voltage in Hertz's theory ought to have been independent of the dielectric permittivity. Instead, the experiment confirmed the prediction of Lorentz, that v should vary as $(\epsilon - \epsilon_0)/\epsilon = 1 - K^{-1}$.

As a matter of fact, there had been a previous experiment, by Blondlot (1901), which was also devised in order to decide between Hertz's and Lorentz's aether theories. Blondlot drove a current of air between the plates of a capacitor, which were connected by a wire, so as to be at the same potential. There was a magnetic field $\mathbf{B}$ at right angles to the air current. According to Hertz's theory, the aether relation was

$$\mathbf{D} = \epsilon(\mathbf{E} + \mathbf{v} \times \mathbf{B}),$$

and the capacitor should have become charged, since $\mathbf{E} = 0$. According to Lorentz, on the other hand, the aether relation was

$$\mathbf{D} = \epsilon\mathbf{E} + (\epsilon - \epsilon_0)\mathbf{v} \times \mathbf{B},$$

[†] An electrometer is a capacitor with a measuring device (a spring) for the force between its plates; it is used for measuring a potential difference.

[‡] H. A. Lorentz, *Versuch einer Theorie der elektrischen und optischen Erscheinungen in bewegten Körpern*, 1895.

and the charges on the plates should have been only $(\epsilon - \epsilon_0)/\epsilon = 1 - K^{-1}$ of the charges on Hertz's theory; which in the case of air would have been practically zero. The result of Blondlot's experiment was in favour of Lorentz's aether principle.

Problem 34.1 A dielectric sphere of radius a and uniform permittivity ϵ is placed in a uniform magnetic field **B** and made to rotate at constant angular velocity Ω around an axis parallel to **B**. Find the electric potential inside and outside the sphere.

Solution Since the dielectric is an insulator, there is no charge or current anywhere. As in Wilson's experiment, the magnetic field **B** inside the sphere differs from the field outside by relativistic terms, which we neglect. Inside the sphere

$$\text{div } \mathbf{D} = \text{div}[\epsilon \mathbf{E} + (\epsilon - \epsilon_0)\mathbf{v} \times \mathbf{B}] = \epsilon \, \text{div } \mathbf{E} + 2(\epsilon - \epsilon_0)(\Omega \cdot \mathbf{B}),$$

where we have used the identity $\text{div } \mathbf{v} \times \mathbf{B} = \mathbf{B} \cdot \text{curl } \mathbf{v} - \mathbf{v} \cdot \text{curl } \mathbf{B}$, together with $\text{curl } \Omega \times \mathbf{r} = 2\Omega$ and $\text{curl } \mathbf{B} = 0$. Thus

$$-\epsilon \Delta V + 2(\epsilon - \epsilon_0)(\Omega \cdot \mathbf{B}) = 0.$$

We seek a particular solution $V_0(r)$ which depends only on r:

$$\frac{1}{r^2}\frac{d}{dr}r^2\frac{dV_0}{dr} = 2\frac{\epsilon - \epsilon_0}{\epsilon}(\Omega \cdot \mathbf{B}).$$

The integrations are elementary, and give

$$V_0(r) = \tfrac{1}{3}\frac{\epsilon - \epsilon_0}{\epsilon}(\Omega \cdot \mathbf{B})(r^2 - a^2),$$

where an integration constant has been chosen so that $V_0(a) = 0$. Since V inside the sphere can only differ from the particular solution by an arbitrary harmonic function, it must be of the form

$$V = \tfrac{1}{3}\frac{\epsilon - \epsilon_0}{\epsilon}(\Omega \cdot \mathbf{B})(r^2 - a^2) + \sum A_n r^n P_n(\cos\theta),$$

which includes an arbitrary constant A_0. The polar angle θ is of course measured from the rotation axis. Outside, we have

$$V_e = \sum \frac{B_n}{r^{n+1}} P_n(\cos\theta),$$

which corresponds to $V = 0$ at infinity.

In order to satisfy the jump condition $\mathbf{n} \times [\![\mathbf{E}]\!] = 0$ we require that V be continuous at $r = a$. Thus

$$A_n a^n = \frac{B_n}{a^{n+1}}. \tag{1}$$

The second jump condition is $\mathbf{n} \cdot [\![\mathbf{D}]\!] = 0$. Inside the sphere we have $\mathbf{D} = \epsilon \mathbf{E} + (\epsilon - \epsilon_0)\mathbf{v} \times \mathbf{B}$. The velocity at the surface is $|\boldsymbol{\Omega} \times \mathbf{r}| = \Omega a \sin\theta$, and the radial component of $\mathbf{v} \times \mathbf{B}$ involves a further $\sin\theta$. Thus $D_n = \epsilon E_n + (\epsilon - \epsilon_0)(\boldsymbol{\Omega} \cdot \mathbf{B})a \sin^2\theta$, and the jump condition requires

$$-\epsilon \frac{\partial V}{\partial r} + (\epsilon - \epsilon_0)(\boldsymbol{\Omega} \cdot \mathbf{B})a \sin^2\theta = -\epsilon_0 \frac{\partial V_e}{\partial r}$$

at $r = a$. It should, perhaps, be noted that we must not blindly impose the continuity of $\epsilon \partial V / \partial r$ in this case, for $\mathbf{D}$ inside the sphere is not $\epsilon \mathbf{E}$. Since $P_2(\cos\theta) = \frac{1}{2}(3\cos^2\theta - 1)$,

$$\sin^2\theta = 1 - \cos^2\theta = 1 - \frac{2P_2 + 1}{3} = \tfrac{2}{3}(P_0 - P_2).$$

Thus

$$-\tfrac{2}{3}(\epsilon - \epsilon_0)(\boldsymbol{\Omega} \cdot \mathbf{B})a P_0 - \epsilon \sum n A_n a^{n-1} P_n + \tfrac{2}{3}(\epsilon - \epsilon_0)(\boldsymbol{\Omega} \cdot \mathbf{B})a(P_0 - P_2)$$
$$= \epsilon_0 \sum (n+1)\frac{B_n}{a^{n+2}} P_n. \tag{2}$$

Equations (1) and (2) give

$$A_2 = -\tfrac{2}{3}\frac{\epsilon - \epsilon_0}{2\epsilon + 3\epsilon_0}(\boldsymbol{\Omega} \cdot \mathbf{B}), \qquad B_2 = A_2 a^5;$$

the remaining A's and B's all vanish. Thus, finally, the potential inside the sphere is

$$V = \tfrac{1}{3}\frac{\epsilon - \epsilon_0}{\epsilon}(\boldsymbol{\Omega} \cdot \mathbf{B})(r^2 - a^2) - \tfrac{2}{3}\frac{\epsilon - \epsilon_0}{2\epsilon + 3\epsilon_0}(\boldsymbol{\Omega} \cdot \mathbf{B})r^2 P_2(\cos\theta).$$

Outside, there is a quadrupole potential

$$V_e = -\tfrac{2}{3}\frac{\epsilon - \epsilon_0}{2\epsilon + 3\epsilon_0}(\boldsymbol{\Omega} \cdot \mathbf{B})a^2 \left(\frac{a}{r}\right)^3 P_2(\cos\theta).$$

10

Steady currents in linearly conducting materials

35. Linearly conducting materials

In most common conductors of electricity the electric current is found to be associated with an electric field on the one hand, and with inhomogeneities of material properties, such as temperature ϑ or concentration c, on the other. Thus there is a constitutive relation, in the rest frame, between $\mathbf{j}'$, $\mathbf{E}'$ and the gradients of material properties. Furthermore, so long as none of these quantities is too large, the relation between them is *linear*. In a conductor at rest (in an aether frame), we need not bother with the primes. Under these circumstances, the constitutive relation in an isotropic conductor of uniform composition is of the form

$$\mathbf{E} = \mathbf{j}/\sigma + \alpha \, \mathbf{grad} \, \vartheta. \tag{35.1}$$

The coefficients σ and α, which are characteristic of the material and depend on its state (temperature, density), are respectively called the *electric conductivity* and the *Thomson coefficient*[†].

The electric conductivity has the dimensions of current density divided by electric field, or amp/(volt m). The ratio volt /amp is called the *ohm* and is the SI unit of resistance (which we shall define in the next section). The reciprocal of the ohm is called the *siemen*[‡]. Thus σ has the SI units of $(\text{ohm m})^{-1}$ or siemen m^{-1}. Metallic conductors—like copper, silver or zinc—have conductivities of a few times 10^7 siemen m^{-1}. The Thomson coefficient α has, according to (35.1), the dimensions of volts, divided by the temperature unit. Typically, α is of the order of microvolts per kelvin degree.

If the linear conductor is not isotropic, the coefficients σ^{-1} and α must be replaced by matrices. Although we shall confine ourselves to the simple relation

---

[†]The use of σ to denote the conductivity is standard. It may lead to confusion with the equally standard use of σ to denote surface charge density. In such cases we replace one of the σ's by another letter (e.g. C for conductivity).

[‡]In the older literature it was called *mho* (the word *ohm* read backwards).

(35.1), many of the methods devised for dealing with the isotropic case can be readily extended to the more general case.

If we take the line integral of (35.1)—we have a conducting wire in mind—we obtain

$$\int E_s \, ds = \int \frac{j_s}{\sigma} \, ds + \int \alpha \, d\vartheta. \tag{35.2}$$

Obviously, then, the Thomson term can be neglected whenever the voltage on the left-hand side exceeds a few millivolts, and the temperature difference is less than a hundred kelvin degrees. The constitutive relation then reduces to *Ohm's law*

$$\mathbf{j} = \sigma \mathbf{E}. \tag{35.3}$$

Substituting this in the law of charge conservation, we obtain

$$
\begin{aligned}
q_t + \operatorname{div} \mathbf{j} &= q_t + \operatorname{div} \sigma \mathbf{E} \\
&= q_t + \sigma \operatorname{div} \mathbf{E} + \mathbf{E} \cdot \mathbf{grad}\, \sigma \\
&= 0.
\end{aligned} \tag{35.4}
$$

If the conductivity and the permittivity are uniform, $q = \operatorname{div} \mathbf{D} = \epsilon \operatorname{div} \mathbf{E}$ gives

$$q_t + \frac{\sigma}{\epsilon} q = 0. \tag{35.5}$$

We conclude that the charge density q decays as $e^{-(\sigma/\epsilon)t}$ wherever (35.5) holds. The same is true of $\operatorname{div} \mathbf{j} = -q_t$. For metallic conductors, with σ of the order of several times 10^7 siemen m^{-1}, ϵ/σ is of the order of 10^{-18} seconds, and any q will disappear practically instantaneously. Since charge is conserved, it cannot really disappear, and we may rightly ask: where does it go? The answer is that it accumulates at those places where (35.5) does *not* hold; in other words, on the surfaces along which σ or ϵ are discontinuous. Electrostatics is therefore only a special example of the rule that the electric charge of a conductor must reside wholly on its surface. Except for phenomena involving frequencies $\omega \geq \sigma/\epsilon$, which are far in excess of optical frequencies, we may disregard this rapid process of charge relaxation in any conductor. Indeed if we wish to avoid getting bogged down with what happens during the first 10^{-18} seconds we must amend the law of charge conservation and replace it by $\operatorname{div} \mathbf{j} = 0$.

Noting that $D_t/j = \epsilon E_t/j$, too, is comparable to $\omega\epsilon/\sigma$, so that the displacement term may be dropped from Maxwell's equations, we conclude that Ohmic conductors (once charge relaxation has taken place) are governed by the equations

$$
\begin{aligned}
\mathbf{curl}\ \mathbf{H} &= \mathbf{j}, \\
\mathbf{j} &= \sigma \mathbf{E}, \\
\operatorname{div} \mathbf{B} &= 0, \\
\mathbf{curl}\ \mathbf{E} &= -\mathbf{B}_t.
\end{aligned} \tag{35.6}
$$

The first of these equations already ensures that div $\mathbf{j} = 0$. The system (35.6) must be supplemented by information regarding the magnetization $\mathbf{M} = \mathbf{B}/\mu_0 - \mathbf{H}$.

It is important to note that the system (35.6) holds in the interior, but not on the surface, of a conductor. For example, if the plates of a charged capacitor are connected by a conducting wire, the ensuing discharge may be calculated by applying (35.6) to the wire, but the surface charges $\pm Q(t)$ on the plates cannot be ignored along with the charge density q in the wire. On the contrary, the law of charge conservation, in the form $i(t) = dQ/dt$, is essential for calculating the discharge.

The passage of current through a conductor is accompanied by a thermal effect, which we shall consider in Chapter 15. Here we only quote the result: the scalar product

$$\mathbf{j} \cdot \mathbf{E}, \tag{35.7}$$

with dimensions of power per unit volume, acts (in the rest frame) as an additional source of heating density. That is, whether or not there are proper heating—or cooling—sources present, the integral $\int dt \int \mathbf{j} \cdot \mathbf{E} dV$ is an additional contribution to a conductor's energy. It is called *Joule heating*. But energy is subject to a law of conservation, or balance (the first law of thermodynamics). In matters of balance, as with money, one person's gain is another person's loss. The Joule heating term $\mathbf{j} \cdot \mathbf{E}$, particularly in the form j^2/σ, is therefore also referred to as *Ohmic loss*. Joule heating is responsible for the operation of all electrical heating elements and fuses.

Problem 35.1 A large sphere of radius b is made of material of conductivity σ and permittivity ϵ. At $t = 0$ a charge Q_0 is uniformly distributed over the surface of a small concentric sphere $r = a$. Determine how the charge Q on the inner sphere varies with time. If the temperature is initially and ultimately uniform and the sphere has a heat capacity C, find the ultimate change in temperature.

Solution For $a < r < b$ we have $D = Q_a/(4\pi r^2)$, $E = \epsilon D$ and $j = \sigma E$. Charge conservation at $r = a$ gives

$$\dot{Q}_a = -4\pi a^2 j = -4\pi a^2 \sigma E = \frac{\sigma}{\epsilon} Q_a.$$

The solution is

$$Q_a(t) = Q_0 e^{-\sigma t/\epsilon}.$$

The contribution of the Joule heating to the energy is

$$\int_0^\infty dt \int_a^b (\mathbf{j} \cdot \mathbf{E}) 4\pi r^2 \, dr = \int_0^\infty dt \int_a^b \sigma E^2 4\pi r^2 \, dr$$

$$= \frac{\sigma Q_0^2}{4\pi \epsilon^2} \int_0^\infty e^{-2\sigma t/\epsilon} \, dt \int_a^b \frac{dr}{r^2}$$

$$= \frac{Q_0^2}{8\pi \epsilon} \left(\frac{1}{a} - \frac{1}{b} \right).$$

The ultimate temperature change is the last expression, divided by the heat capacity C of the sphere.

---

36. Resistance

Consider now the steady flow of electric current through a stationary, conducting medium between two conductors, which we shall call the *electrodes*. The electrodes may be the two ends—or cross-sections—of a conducting wire, the inner and outer surfaces of a hollow, conducting cylinder, etc. According to (35.6), the following equations hold in the medium:

$$\operatorname{div} \mathbf{j} = 0,$$
$$\mathbf{j} = \sigma \mathbf{E}, \qquad (36.1)$$
$$\operatorname{\mathbf{curl}} \mathbf{E} = 0.$$

The last of these can be satisfied, as usual, by introducing an electric potential. As for the electrodes, we assume that the current leaves or enters them normally. According to $(36.1)_2$ we shall then have, on each electrode,

$$\mathbf{n} \times \mathbf{E} = 0. \qquad (36.2)$$

The requirement that the same current i leave one electrode and enter another is expressed by

$$\int j_n \, dS = \pm i, \qquad (36.3)$$

with the integral taken over the surface of the electrode.

Having found the distribution of $\mathbf{j}$ and $\mathbf{E}$ (or the electric potential) throughout the medium, we define its *resistance R* as

$$R = \frac{V}{i}, \qquad (36.4)$$

with the signs of the voltage $V = \int E_s \, ds$ between the electrodes, and of the current $i = \int j_n \, dS$ leaving or entering an electrode, chosen in such a way as to render R positive.

It is easy to see that the calculation of resistance requires the solution of a potential problem. But it is even more profitable to compare it with the calculation of the capacity of a dielectric capacitor. The latter problem, it will be recalled, was to solve the equations

$$\operatorname{div} \mathbf{D} = 0,$$
$$\mathbf{D} = \epsilon \mathbf{E}, \qquad (36.5)$$
$$\operatorname{\mathbf{curl}} \mathbf{E} = 0.$$

These were to be solved with the conditions that

$$\mathbf{n} \times \mathbf{E} = 0 \tag{36.6}$$

on each of the plates, and that the charges on the plates were

$$\int D_n \, dS = \pm Q. \tag{36.7}$$

The capacity was then defined by

$$C = \frac{Q}{V}. \tag{36.8}$$

If the conductivity σ of the resistor and the permittivity of the capacitor are distributed in the same way, the two problems are not merely similar; they are, mathematically speaking, the same problem. In particular, if σ and ϵ are constants, we have the simple relation

$$RC = \frac{\epsilon}{\sigma}. \tag{36.9}$$

Thus a parallel-plate resistor of cross-section S and length d has a resistance of

$$R = \frac{d}{\sigma S} \tag{36.10}$$

(cf. (28.3)). A coaxial resistor—a hollow cylinder with inner radius a, outer radius b and length L—has a resistance of (cf. (32.11))

$$R = \frac{\ln(b/a)}{2\pi \sigma L}. \tag{36.11}$$

Exercise

36.1. A spherical resistor consists of a hollow sphere with inner radius a and outer radius b, made of material with conductivity σ. Calculate its resistance. Use the answer to determine the capacity of a spherical capacitor which is filled with a dielectric of permittivity ϵ.

At a junction formed by several conductors, $\operatorname{div} \mathbf{j} = 0$ requires that the sum of the outgoing currents equals the sum of the ingoing ones. This is *Kirchhoff's first rule*:

$$\sum i = 0. \tag{36.12}$$

The currents i in this sum are to be taken with their 'correct algebraic signs'; this just means that all outgoing currents must be given one sign, and all ingoing currents, the opposite sign.

The entire Joule heating in the medium between the electrodes is

$$\int \mathbf{j} \cdot \mathbf{E} \, d^3x = -\int \mathbf{j} \cdot \mathbf{grad}\, V \, d^3x = -\int \mathrm{div}(V\mathbf{j}) \, d^3x. \qquad (36.13)$$

The last term can be transformed into an integral of Vj_n over the surface of the medium. Noting that the potential is constant on each electrode, that j_n vanishes at the boundary between the conducting medium and the insulating surroundings, and that the total current through the medium is given by (36.3), we find that the Joule heating can be simply expressed, in terms of the total current and the medium's resistance, as i^2R.

The mathematical equivalence between the problems of determining capacity and resistance extends, of course, to Thomson's theorem of Section 33. For any solenoidal distribution of the current density $\mathbf{j}$—a distribution satisfying $(36.1)_1$, but not necessarily $(36.1)_{2,3}$—with given total currents $\oint j_n \, dS$ issuing from, or entering into, a finite system of electrodes, the Joule heating,

$$W[\mathbf{j}] = \int \frac{j^2}{\sigma} \, d^3x, \qquad (36.14)$$

may be regarded as a functional of $\mathbf{j}$. By Thomson's theorem (Section 33) W will be minimal when $\mathbf{j} = \sigma \, \mathbf{grad}\, V$, with V constant on each electrode. In particular, j will be large in regions of high conductivity.

The foregoing variational formulation of the potential (or resistance) problem for Ohmic conductors does not correspond to the usual method of passing a current through a conducting medium. Rather than prescribe the total currents through the electrodes, we usually maintain each one at a given *potential*. Consider then the Joule heating in the form

$$W[V] = \int \sigma (\mathbf{grad}\, V)^2 \, d^3x. \qquad (36.15)$$

For a given distribution of $\sigma(\mathbf{x}) \geq 0$, we regard $W[V]$ as a functional of $V(\mathbf{x})$, the latter being subject only to the requirement of having prescribed constant values on the electrodes. We do *not* require V to satisfy $\mathrm{div}(\sigma \, \mathbf{grad}\, V) = 0$. Clearly, then, $W[V]$ will have no maximum, for we can let V oscillate wildly between the electrodes. It will, however, have a minimum—a positive one, unless the electrodes are all maintained at the same potential. In order to find it, we calculate the variation of $W[V]$. Taking account of the fact that $\delta V = 0$ on the electrodes, where V is prescribed, integration by parts gives

$$\delta W = -2 \int \delta V \, \mathrm{div}(\sigma \, \mathbf{grad}\, V) \, d^3x. \qquad (36.16)$$

Hence the minimum is obtained when V satisfies $\text{div}(\sigma \, \mathbf{grad} \, V) = 0$. This V is precisely the solution of the potential problem (36.1). The result (sometimes referred to as the principle of minimum dissipation) tells us that the actual distribution of electric potential throughout a conducting medium, in which a finite system of electrodes with prescribed potentials is embedded, is always the one for which the entire Joule heating is least[†]. Every electric toaster, for example, works in the worst possible way (a feature which is not usually advertised, and for which the manufacturer can hardly be blamed). Another consequence of this variational formulation is that the lines of electric field $\mathbf{E} = -\mathbf{grad} \, V$ are more concentrated in regions of *low* conductivity.

Exercises

36.2. A spherical hole of radius a is cut out of an infinite block of uniform conductivity. At large distances the current is uniform and in the z direction. Show that $V = A[r + a^3/(2r^2)] \cos \theta$ and that the lines of current flow are given by $(r^3 - a^3) \sin^2 \theta = \text{const} \times r$.

36.3. Prove that in a system of electrodes with zero total net current in an Ohmic medium, at least one electrode has current leaving everywhere, and one has current entering everywhere.

36.4. Prove that Green's reciprocal theorem $\sum i V' = \sum i' V$ holds for steady currents between electrodes in an Ohmic medium.

36.5. A closed curve c in the xy plane, lying entirely on the positive side of the y axis, is rotated through $180°$ about the y axis. The volume so formed is filled with a material of uniform conductivity σ, and electrodes are placed at the two plane ends. Prove that the resistance R is given by

$$\frac{1}{R} = \frac{\sigma}{\pi} \int \frac{dS}{x},$$

where the integration is over the area enclosed by c.

36.6. A thin spherical shell of radius a and thickness t is made of material with conductivity σ. Current enters and leaves by two small spherical electrodes of radius c whose centres are at the ends A, B of a diameter AOB. If i is the total current and P is a point on the shell such that the angle POA is θ, show that the current density at P is $i/(2\pi a t \sin \theta)$, and that the resistance of the shell is

$$\frac{1}{\pi \sigma t} \ln \cot \frac{c}{2a}.$$

36.7. Show that if a steady current with normal component j_n is flowing across the boundary between two conducting media, there must be a surface charge density

$$[\![\epsilon/\sigma]\!] j_n.$$

36.8. The space between two parallel conducting planes is filled with two slabs, one of thickness d_1, conductivity σ_1 and permittivity ϵ_1, and the other of thickness d_2, conductivity σ_2 and permittivity ϵ_2. Find the current density and the surface charge density on the boundary between the slabs when a voltage V is applied to the conducting planes.

[†]Kirchhoff (1848).

36.9. Twelve equal wires of resistance R are joined at their ends to form the edges of a cube. If current enters and leaves at opposite ends of the cube, show that the resistance is $5R/6$, and if it enters and leaves at two ends of one wire the resistance is $7R/12$.

36.10. A set of resistances $R_1, \ldots R_n$ are joined in series. Show that the total resistance equals $R = \sum R_a$, but if they are joined in parallel

$$\frac{1}{R} = \sum \frac{1}{R_a}.$$

36.11. In a uniform submarine cable there is a leak conductance (reciprocal resistance) G per unit length and the resistance of the cable is R per unit length. At a distance x along the cable the potential is V and the current is i. Show that

$$\frac{dV}{dx} = -Ri \quad \text{and} \quad \frac{di}{dx} = -GV.$$

Deduce a differential equation for V, and show that if the cable is of length l, the two ends being at potentials V_0 and 0, then

$$V = V_0 \frac{\sinh\left[\sqrt{GR}(l-x)\right]}{\sinh\sqrt{GR}l}.$$

Problem 36.1 In a plane-parallel *vacuum diode* electrons are emitted at zero velocity from a hot *cathode* at zero potential and accelerated over a distance d towards the parallel *anode*, which is maintained at positive potential V_0. A negative space charge of electrons builds up between the electrodes, until the electric field at the cathode vanishes. From then on a steady current i flows between the electrodes. If the electrode area A is large compared with d^2, so that edge effects can be neglected, show that $i = KV_0^{3/2}$, and determine the constant K.

Solution Let $x = 0$ be the cathode and $x = d$ the anode. From the energy integral and the initial conditions $v(0) = 0$ and $V(0) = 0$ we obtain

$$\tfrac{1}{2}mv^2(x) - eV(x) = 0.$$

The (convective) current density is

$$j = qv = (-\epsilon_0 \Delta V)\left(\frac{2eV}{m}\right)^{\frac{1}{2}}.$$

Note that j, like the electron charge density q, is negative: the current flows from the anode to the cathode. When j is constant

$$\Delta V = V''(x) = \frac{C}{V^{\frac{1}{2}}(x)}, \qquad C = -\frac{j}{\epsilon_0}\left(\frac{m}{2e}\right)^{\frac{1}{2}}.$$

Multiplication by $V'(x)$ leads to the first integral

$$\tfrac{1}{2}(V')^2 = 2CV^{\frac{1}{2}},$$

where the constant of integration has been determined by the boundary conditions at the cathode, which is at zero electric field $-V'$ and zero potential V. The differential equation $V' = 2C^{1/2}V^{1/4}$ has the integral

$$V^{\frac{3}{4}}(x) = \tfrac{3}{2}C^{\frac{1}{2}}x,$$

which satisfies the boundary condition $V(0) = 0$. Thus

$$V(x) = V_0\left(\frac{x}{d}\right)^{\frac{4}{3}}, \qquad \tfrac{3}{2}C^{\frac{1}{2}}d = V_0^{\frac{3}{4}}.$$

The current is given by Childs's formula

$$i = -jA = KV_0^{\frac{3}{2}}, \qquad K = \frac{4\epsilon_0 A}{9d^2}\left(\frac{2e}{m}\right)^{\frac{1}{2}}.$$

The space charge limited diode therefore constitutes a non-linear conductor.

---

37. Electromotive force

The foregoing considerations were based on Ohm's law $\mathbf{j} = \sigma\mathbf{E}$, which is a special case of the more general relation (35.1). We have already taken the line integral (cf. (35.2)) of that relation:

$$\int E_s\, ds = \int \frac{j_s}{\sigma}\, ds + \int \alpha\, d\vartheta. \tag{37.1}$$

The second, Thomson line integral on the right-hand side, taken with the *opposite sign*, is called the *Thomson electromotive force*, usually abbreviated to *Thomson emf*, and denoted $\mathcal{E}_T$. It is, of course, not a force; its units are those of electric field times distance, that is, volts. The notation is standard; the scalar $\mathcal{E}$ should not be confused with the electromotive intensity *vector* $\mathbf{\mathcal{E}}$. If we apply (37.1) to linear currents—which flow along wires, idealized as lines—the first term on the right-hand side becomes the product of current and resistance:

$$\int E_s\, ds = Ri - \mathcal{E}_T. \tag{37.2}$$

For a stationary conductor in a steady magnetic field, the integral on the left-hand side is independent of the path of integration, since **curl E** $= 0$; it is simply the voltage between the ends. In this case we therefore have

$$V_1 - V_2 = Ri - \mathcal{E}_T, \tag{37.3}$$

where R is the resistance between the points 1 and 2 along the current line.

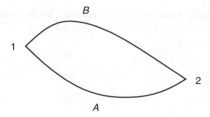

Fig. 37.1

A *thermocouple* is obtained by joining two different conductors as shown in Fig. 37.1. If the Thomson coefficients are constants, each Thomson emf is just $-\alpha \int d\vartheta$, and if we take the integrals in (37.2) around the whole loop (or apply (37.3) to each conductor), we obtain

$$(R_A + R_B)i + (\alpha_A - \alpha_B)(\vartheta_2 - \vartheta_1) = 0, \tag{37.4}$$

where i is counted as positive if it flows counter-clockwise. Thus a current will flow if the conductors are different ($\alpha_A \neq \alpha_B$) and the junctions are at different temperatures. This arrangement can be used for measuring any one of the quantities that appear in (37.4)—resistance, current, Thomson coefficient, temperature difference—if all the other quantities are known.

The Thomson emf is just minus the line integral of the Thomson term $\alpha \, \mathbf{grad} \, \vartheta$, which measures the effect of a temperature gradient in the constitutive relation (35.1) of a linear conductor of uniform composition. In a battery, or voltaic cell, the important inhomogeneity is that of chemical concentration. A constitutive theory of such cells leads to a concentration, or *battery*, emf $\mathcal{E}$, analogous to the Thomson emf. For a closed current loop, consisting of several elements, each with a resistance R, current i and emf $\mathcal{E}$, summation of relations like (37.3) gives

$$\sum (Ri - \mathcal{E}) = 0. \tag{37.5}$$

This is *Kirchhoff's second rule*.

Exercises

37.1. A capacitor is charged by means of a constant emf $\mathcal{E}$, the connecting wires having a resistance R. At $t = 0$ the emf is switched on. Show that the charge on the capacitor is $Q(t) = (1 - e^{-t/\tau})C\mathcal{E}$, where $\tau = RC$.

37.2. Show that, in a closed network containing batteries, the currents distribute themselves over the branches of the network in such a way that

$$\sum i_a (R_a i_a - 2\mathcal{E}_a)$$

is minimal.

37.3. AB is a uniform telephone wire. At some unknown point C of the wire there is a fault, that is, a resistance of unknown magnitude connecting C to earth. B is put to earth potential and an emf is applied at A. Next A is put to earth potential and the emf is applied at B. The total resistance is measured in each case. Show how to determine the position of C from these measurements.

38. The Faraday disc

The linear relation (35.1), as well as the simpler Ohm's law (35.3), are constitutive relations, which hold in the conductor's rest frame. If the conductor is moving relative to a given aether frame, we must state Ohm's law in the correct form

$$\mathbf{j}' = \sigma \mathbf{E}', \tag{38.1}$$

where the primed vectors refer to the rest frame, whose velocity $\dot{\mathbf{x}}$ is the instantaneous velocity of the conductor. The motion of macroscopic conductors in terrestrial experiments is always non-relativistic, so that we can apply Galilean transformations $\mathbf{j}' = \mathbf{j} - q\dot{\mathbf{x}} = \mathcal{J}$ and $\mathbf{E}' = \mathbf{E} + \dot{\mathbf{x}} \times \mathbf{B} = \mathcal{E}$, where $\mathcal{J}$ and $\mathcal{E}$ are the conduction current density and the electromotive intensity. Thus, in this non-relativistic approximation, Ohm's law becomes (cf. Exercise 23.2)

$$\mathcal{J} = \sigma \mathcal{E}. \tag{38.2}$$

If the charge density q vanishes, $\mathcal{J} = \mathbf{j}$, and we have

$$\mathbf{j} = \sigma (\mathbf{E} + \dot{\mathbf{x}} \times \mathbf{B}). \tag{38.3}$$

In order to investigate the effect of the $\dot{\mathbf{x}} \times \mathbf{B}$ term, we shall consider one of the well-known experiments of Faraday: the *Faraday disc*, also called the *unipolar inductor*, or the *homopolar generator*. It consists of a non-magnetic annular disc (a hollow cylinder of very short length) which rotates with constant angular velocity Ω in a uniform magnetic field $\mathbf{B}$, parallel to the axis.

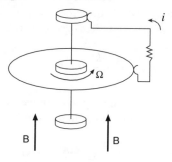

Fig. 38.1

Sliding contacts are attached to the inner and outer surfaces of the disc, and connected through an external resistance (Fig. 38.1). The disc, of conductivity σ, has inner and outer radii a and b, and thickness L.

We assume the inner and outer surfaces, and the sliding contacts, to be highly conducting. Then $r = a$ and $r = b$ are equipotentials, and the electric field in the disc is radial. Since $\dot{\mathbf{x}} \times \mathbf{B}$ is also radial, so are $\mathcal{E} = \mathbf{E} + \dot{\mathbf{x}} \times \mathbf{B}$ and $\mathbf{j} = \sigma \mathcal{E}$. According to (38.3), we have, in this steady case,

$$\operatorname{div} \mathbf{j} = 0,$$
$$\mathbf{j} = \sigma (\mathbf{E} + \dot{\mathbf{x}} \times \mathbf{B}), \tag{38.4}$$
$$\operatorname{\mathbf{curl}} \mathbf{E} = 0.$$

These equations are the same as (36.1), with $\mathbf{j} = \sigma \mathbf{E}$ replaced by $\mathbf{j} = \sigma \mathcal{E}$. From them we obtain, respectively,

$$j = \frac{i}{2\pi r L},$$
$$j = \sigma (E + \Omega r B), \tag{38.5}$$
$$\int_a^b E \, dr - V_e = 0,$$

where V_e is the voltage across the external resistance. Hence

$$
\begin{aligned}
V_e &= \int_a^b \left(\frac{i}{2\pi \sigma L r} - \Omega r B \right) dr \\
&= \frac{i}{2\pi \sigma L} \ln \frac{b}{a} - \tfrac{1}{2} \Omega B (b^2 - a^2) \\
&= Ri - \frac{\Omega}{2\pi} B\pi (b^2 - a^2),
\end{aligned} \tag{38.6}
$$

where R is the resistance (36.11) of the disc.

We consider two extreme cases. If the circuit is open—or the external resistance infinite—the current will vanish, and (38.6) gives the open-circuit voltage as

$$V_{OC} = -\frac{\Omega}{2\pi} B\pi (b^2 - a^2). \tag{38.7}$$

If, on the other hand, the terminals are shorted—or the external resistance is zero—V_e will vanish, and (38.6) gives the short-circuit current as

$$i_{SC} = \frac{V_{OC}}{R}. \tag{38.8}$$

For typical values of $B = 1$ tesla, $\sigma = 6 \times 10^7$ siemen m^{-1}, $\Omega = 3600$ rpm, $a = 1$ cm, $b = 10$ cm and $L = 1$ mm, V_{OC} is about 2 volt and i_{SC} about 3×10^5 amp. Homopolar generators are therefore typically high-current, low-voltage devices.

In the foregoing discussion we have assumed—in going from (38.2) to (38.3)—that there was no charge density q in the conductor. But this assumption, which was based on the rapid charge relaxation in metallic conductors at rest, does not strictly apply to *moving* conductors. Indeed the ratio of the convection current density $q\dot{\mathbf{x}} = \dot{\mathbf{x}} \operatorname{div} \mathbf{D}$ to $\sigma\mathbf{E}$ is of the order $\Omega\epsilon/\sigma$, which is extremely small. Thus (38.3) is quite accurate. We have tacitly made another assumption, that the magnetic field inside the disc was the same as outside. This, too, can be easily justified, so long as $(\Omega b)^2 \ll c^2$.

The reader must already have noticed the similarity between Faraday's disc and Wilson's experiment (Section 34). We have mentioned, at the end of Section 34, the controversy regarding the correct form of the aether relations in a moving linear dielectric; which—in the language of those days—was about the question of 'whether moving bodies do, or do not, carry the aether along with them'. A similar, older controversy began as soon as Faraday published his disc experiment. It raged over the question of 'whether the magnetic field lines do, or do not, rotate with the material'. Faraday himself maintained that they did not, and in this he anticipated Lorentz.

To us, accustomed as we are to regarding Lorentz's aether relations as an unambiguous, clearly stated and independent principle of electromagnetism, these questions seem, not only vague and confusing, but unnecessary. Nevertheless, the controversies that began with Faraday's disc experiment persisted for many decades, and Lorentz's (1895) views did not win universal acceptance until 1913, when H.A. Wilson, together with Marjorie Wilson, repeated his 1904 experiment (Section 34), this time with a *non-conducting magnetizable medium*, which they created by embedding a large number of steel balls in a dielectric wax.

Problem 38.1   An electrically neutral conducting sphere of radius a is placed in a uniform magnetic field $\mathbf{B}$ and made to rotate with angular velocity $\mathbf{\Omega}$ around an axis that is parallel to the field. Find the electric potential inside and outside the sphere.

Solution   The mathematics of this problem is similar to that of Problem 34.1, although the physics is not. After charge relaxation $\mathcal{E} = \mathbf{E} + \mathbf{v} \times \mathbf{B}$ vanishes inside the conducting sphere. With $\mathbf{v} = \mathbf{\Omega} \times \mathbf{r}$ and θ denoting the polar angle measured from $\mathbf{B}$ we have

$$\begin{aligned}
rE_r &= \mathbf{r} \cdot \mathbf{E} \\
&= -\mathbf{r} \cdot \mathbf{v} \times \mathbf{B} \\
&= -(\mathbf{\Omega} \cdot \mathbf{B})r^2 \sin^2 \theta \\
&= -\tfrac{2}{3}(\mathbf{\Omega} \cdot \mathbf{B})r^2[1 - P_2(\cos\theta)],
\end{aligned}$$

$$\operatorname{div}\mathbf{E} = -\operatorname{div}\mathbf{v} \times \mathbf{B} = \mathbf{v} \cdot \mathbf{curl}\,\mathbf{B} - \mathbf{B} \cdot \mathbf{curl}\,\mathbf{v} = -2(\mathbf{\Omega} \cdot \mathbf{B}),$$
$$q = \operatorname{div}\mathbf{D} = \operatorname{div}\epsilon_0\mathbf{E} = -2\epsilon_0(\mathbf{\Omega} \cdot \mathbf{B}).$$

The potential inside the sphere satisfies

$$\Delta V = \operatorname{div}\mathbf{grad}\,V = -\operatorname{div}\mathbf{E} = 2(\mathbf{\Omega} \cdot \mathbf{B}).$$

This has the particular solution

$$V_0(r) = \tfrac{1}{3}(\mathbf{\Omega} \cdot \mathbf{B})(r^2 - a^2).$$

Thus V must be of the form

$$V = V_0(r) + \sum A_n r^n P_n(\cos\theta),$$

which includes an undetermined constant term A_0. Except for this constant term we can determine V because we know $\mathbf{E}$ inside the sphere: $E_r = -\partial V/\partial r$ gives

$$-\tfrac{2}{3}(\mathbf{\Omega} \cdot \mathbf{B})r(P_0 - P_2) = -\tfrac{2}{3}(\mathbf{\Omega} \cdot \mathbf{B})r - \sum n A_n r^{n-1} P_n,$$

from which we obtain

$$A_2 = -\tfrac{1}{3}(\mathbf{\Omega} \cdot \mathbf{B}).$$

A_0 is still undetermined, and the remaining A's all vanish.

Outside, the potential has the form

$$V_e = \sum \frac{B_n}{r^{n+1}} P_n(\cos\theta),$$

which corresponds to zero at infinity. In order to satisfy the jump condition $\mathbf{n} \times [\![\mathbf{E}]\!] = 0$ we require that V be continuous at $r = a$. Thus

$$B_0 = A_0 a, \qquad B_2 = A_2 a^5 = -\tfrac{1}{3}(\mathbf{\Omega} \cdot \mathbf{B})a^5,$$

and the remaining B's all vanish.

In order to determine A_0 (and $B_0 = A_0 a$) we apply the second jump condition $\sigma = \mathbf{n} \cdot [\![\mathbf{D}]\!]$ (the use of σ to denote the surface charge density is harmless because we shall not need the conductivity in this problem) at $r = a$:

$$\sigma = \frac{\mathbf{r} \cdot \mathbf{D}_e - \mathbf{r} \cdot \mathbf{D}}{r}$$

$$= -\epsilon_0 \frac{\partial V_e}{\partial r} - \epsilon_0 E_r$$

$$= \epsilon_0 \sum \frac{(n+1)B_n}{a^{n+2}} P_n + \tfrac{2}{3}\epsilon_0(\mathbf{\Omega} \cdot \mathbf{B})a(P_0 - P_2)$$

$$= \epsilon_0 \left(\frac{B_0}{a^2} + \frac{3B_2}{a^4} P_2\right) + \tfrac{2}{3}\epsilon_0(\mathbf{\Omega} \cdot \mathbf{B})a(1 - P_2).$$

The total surface charge is

$$\oint \sigma \, dS = 4\pi\epsilon_0 \left(B_0 + \tfrac{2}{3}(\mathbf{\Omega} \cdot \mathbf{B})a^3\right),$$

since $\oint P_2 \, d\Omega = 0$. The total volume charge is

$$\int q \, d^3x = \frac{4\pi}{3} a^3 q = -4\pi\epsilon_0 \tfrac{2}{3}(\mathbf{\Omega} \cdot \mathbf{B})a^3.$$

The sum $4\pi\epsilon_0 B_0$ of these must vanish because the sphere is neutral. Hence $B_0 = 0$ and $A_0 = B_0/a = 0$. Thus, finally, the potential inside the sphere is

$$V = V_0(r) + A_2 r^2 P_2 = \tfrac{1}{3}(\boldsymbol{\Omega}\cdot\mathbf{B})(r^2 - a^2) - \tfrac{1}{3}(\boldsymbol{\Omega}\cdot\mathbf{B})r^2 P_2(\cos\theta).$$

Outside there is a quadrupole potential

$$V_e = \frac{B_2}{r^3}P_2 = -\tfrac{1}{3}(\boldsymbol{\Omega}\cdot\mathbf{B})a^2\left(\frac{a}{r}\right)^3 P_2(\cos\theta).$$

Exercises

38.1. In a Faraday disc with contacts open $\mathcal{J} = 0$. Hence show that there is a constant volume charge density $-2\epsilon_0\boldsymbol{\Omega}\cdot\mathbf{B}$. If the disc is electrically neutral, prove that the inner and outer surfaces $r = a$ and $r = b$ have surface charge densities $\sigma_a = -\epsilon_0 a\boldsymbol{\Omega}\cdot\mathbf{B}$ and $\sigma_b = \epsilon_0 b\boldsymbol{\Omega}\cdot\mathbf{B}$.

38.2. During the time interval $t_1 \le t \le t_2$ the magnetic flux through a closed linear circuit of resistance R changes by the amount $\Delta\Phi$. Find the total charge flowing through the circuit during this time interval.

38.3. A locomotive is moving with velocity v. Show that if the two rails are insulated from the ground, there is a voltage between them equal to vlB, where l is the distance between the rails and B is the vertical component of the earth's magnetic field.

---

39. The skin effect

A stationary Ohmic conductor is governed by eqns (35.6). With $\mathbf{j}$ eliminated, these take the form

$$\mathbf{curl\ H} = \sigma\mathbf{E},$$
$$\mathrm{div}\,\mathbf{B} = 0, \qquad\qquad (39.1)$$
$$\mathbf{curl\ E} = -\mathbf{B}_t.$$

We have already discussed solutions of this system for steady fields, when $\mathbf{B}_t = 0$. Now we wish to consider variable fields that are still sufficiently weak for the conduction current to be linear in the electromotive intensity. This introduces a complication, because $\mathcal{J}$ may fail to keep up with $\mathcal{E}$ when the latter changes rapidly. We shall therefore replace $\mathcal{J} = \sigma\mathcal{E}$ by a more general, history-dependent, linear relation of the form

$$\mathcal{J}(t) = \int_0^\infty f(\tau)\mathcal{E}(t - \tau)\,d\tau, \qquad\qquad (39.2)$$

where $f(\tau)$ must, of course, be such that the integral converges. This is the same as saying that electromotive intensities in the distant past must have a vanishing effect. Without an assumption of this kind, experimental physics would become impossible, for the physicist is usually ignorant of the *complete* history of any material he uses. He therefore works on the assumption that, if a material has any memory of the past, it is a *fading memory*.

The assumption of fading memory relates to the *distant* past. It does not mean that matter is forgetful. A piece of magnetized steel may retain vivid recollection of the processes to which it was subjected when it was forged. Its fading memory in this case is with regard to the aeons it has spent underground before being mined, or with regard to its creation through nuclear reactions in a supernova progenitor, long before the solar system was formed.

If we substitute the right-hand side of (39.2) (with $\mathcal{E} = \mathbf{E}$ for a stationary conductor) for $\sigma\mathbf{E}$, the system (39.1) remains linear, but it ceases to be a system of partial differential equations. We therefore resolve the fields by a Fourier expansion into *monochromatic* components, each of which depends on the time through the factor $e^{-i\omega t}$. For such monochromatic fields, the history-dependent Ohm's law (39.2) reduces to $\mathcal{J} = \sigma(\omega)\mathcal{E}$, where the frequency-dependent, monochromatic conductivity $\sigma(\omega)$ is given by

$$\sigma(\omega) = \int_0^\infty f(\tau)e^{i\omega\tau}\,d\tau. \tag{39.3}$$

Thus $\sigma(\omega)$ is the Fourier transform of the recollection function $f(t)$. In terms of the monochromatic components of $\mathbf{E}$ and $\mathbf{H} = \mathbf{B}/\mu_0$, the equations $(39.1)_{1,3}$ now become

$$\mathbf{curl\ H} = \sigma(\omega)\mathbf{E}, \qquad \mathbf{curl\ E} = i\omega\mu_0\mathbf{H}. \tag{39.4}$$

If $\sigma(\omega)$ is uniform throughout the conductor, $\operatorname{div}\sigma\mathbf{E} = \sigma\operatorname{div}\mathbf{E} = 0$ is a consequence of $(39.4)_1$. Similarly, $\operatorname{div}\mathbf{H} = 0$ follows from $(39.4)_2$. If we eliminate $\mathbf{H}$ between the two equations (39.4), we obtain

$$\mathbf{curl}^2\,\mathbf{E} = k^2\mathbf{E}, \qquad k^2 = i\omega\mu_0\sigma. \tag{39.5}$$

It is immediately verified that $\mathbf{H}$, too, satisfies the same equation.

We shall apply (39.5) to a conducting wire of circular cross-section. Using cylindrical coordinates (r, ϕ, z), we seek a solution with $\mathbf{E} = (0, 0, E(r))$. Outside the wire, the electric field will be uniform and equal to the field $E(a)$ at the surface $r = a$ of the wire. Inside, $E(r)$ will be the solution of

$$\frac{1}{r}\frac{d}{dr}\left(r\frac{dE}{dr}\right) + k^2E = 0, \qquad k = \frac{1+i}{\delta}, \qquad \delta = \sqrt{\frac{2}{\omega\mu_0\sigma}}, \tag{39.6}$$

which is finite at $r = 0$. The form of the solution (known as a Bessel function) can be obtained in the following way.

Near the axis, we expand $E(r)$ as a power series $\sum a_n r^n$, which we substitute in (39.6). The result is

$$\sum_{n=1} n^2 a_n r^{n-2} + k^2 \sum_{n=0} a_n r^n = 0. \tag{39.7}$$

It follows that $a_1 = 0$ and $a_m = -k^2 a_{m-2}/m^2$ for $m = 1, 2, \ldots$. The expansion therefore contains only even powers, and the electric field near the axis has the form

$$E = \text{const.} \times \left[1 - \tfrac{1}{2} i (r/\delta)^2 - \tfrac{1}{16} (r/\delta)^4 \right] e^{-i\omega t}. \tag{39.8}$$

The amplitude of E, and with it that of the current density, increases away from the axis of the wire as $1 + \tfrac{1}{8}(r/\delta)^4$. The constant in (39.8) is determined by the amplitude of the electric field at $r = a$ (say), or by the total current corresponding to the current density $\mathbf{j} = \sigma \mathbf{E}$. The approximate formula (39.8) is valid for $r \ll \delta$. For low frequencies, such that $a \ll \delta$, it holds up to the boundary.

Near the boundary $r = a$ of the wire, we neglect the curvature of the boundary and regard it as plane. Then (39.6)$_1$ becomes $E''(r) + k^2 E = 0$, and the solution that remains finite for $r < a$ is

$$E = \text{const.} \times e^{-(a-r)/\delta} e^{i(a-r)/\delta} e^{-i\omega t}. \tag{39.9}$$

This approximation at the boundary is valid if the frequency is large enough so that $\delta \ll a$. The electric field, and with it the current density, is then confined to a thin layer of thickness δ near the surface of the wire. This is the *skin effect*.

Since the effective cross-section of the conducting wire is reduced, its resistance increases at high frequencies. The same is true of the Joule heating at a given total current. A cylindrical conductor of radius a and length L has an ordinary resistance, at zero frequency, of $R_0 = L/(\pi a^2 \sigma)$. At high frequency, when $\delta \ll a$, we may regard the current as flowing through the annular region of thickness δ near the surface. The resistance is then

$$R(\omega) = \frac{L}{2\pi a \delta \sigma} = \frac{a}{2\delta} R_0 \gg R_0. \tag{39.10}$$

In the limit of very high $\omega\sigma$ we have $\delta \to 0$: the current becomes a surface current and the electromagnetic field ($\mathbf{E}$ and $\mathbf{B}$) vanishes inside the conductor.

11

Linear magnets

40. Magnetic effects of currents

In the absence of magnetization, $\mathbf{H} = \mathbf{B}/\mu_0$, and the equations governing a steady system of currents are

$$\operatorname{div} \mathbf{B} = 0,$$
$$\operatorname{curl} \mathbf{B} = \mu_0 \mathbf{j}. \tag{40.1}$$

If we regard $\mathbf{j}$ as given, these equations determine $\mathbf{B}$. The first is satisfied by introducing a vector potential

$$\mathbf{B} = \operatorname{curl} \mathbf{A}. \tag{40.2}$$

The potential $\mathbf{A}$ is subject to a gauge transformation, and this can be chosen in such a way that the condition

$$\operatorname{div} \mathbf{A} = 0 \tag{40.3}$$

be satisfied. Such a vector potential is said to satisfy the *Coulomb gauge*.

If we now combine (40.2) and (40.3) with the remaining equation (40.1)$_2$ we obtain the equation

$$\Delta \mathbf{A} = -\mu_0 \mathbf{j}, \tag{40.4}$$

where $\Delta \mathbf{A} = \operatorname{\mathbf{grad}} \operatorname{div} \mathbf{A} - \operatorname{\mathbf{curl}}^2 \mathbf{A}$. The Cartesian components of $\Delta \mathbf{A}$ are $(\Delta A_x, \Delta A_y, \Delta A_z)$. The Cartesian components of (40.4) are therefore three Poisson equations. Hence the regular solution of (40.4) is (cf. (27.3))

$$\mathbf{A} = \frac{\mu_0}{4\pi} \int \frac{\mathbf{j} \, d^3x}{r}, \tag{40.5}$$

where r is the length of the vector $\mathbf{r}$ from the volume element d^3x to the point at which $\mathbf{A}$ is evaluated. In order to get $\mathbf{B}$, we must take the curl; note that the curl only operates on r. The result is

$$\mathbf{B} = \frac{\mu_0}{4\pi} \int \mathbf{grad}\, \frac{1}{r} \times \mathbf{j}\, d^3x = \frac{\mu_0}{4\pi} \int \frac{\mathbf{j} \times \mathbf{r}\, d^3x}{r^3}. \tag{40.6}$$

For a linear current (the adjective refers to a current flowing along a line, which represents a thin wire, not to a linear functional relation) we replace the vector $\mathbf{j}\, d^3x$ by the vector $i\,\mathbf{ds}$, where $\mathbf{ds}$ is an element of the line, in the direction of $\mathbf{j}$:

$$\mathbf{A} = \frac{\mu_0 i}{4\pi} \oint \frac{\mathbf{ds}}{r}, \qquad \mathbf{B} = \frac{\mu_0 i}{4\pi} \oint \frac{\mathbf{ds} \times \mathbf{r}}{r^3}. \tag{40.7}$$

Equation $(40.7)_2$ is the formula of Biot and Savart. Since $\mathbf{H} = \mathbf{B}/\mu_0$, it expresses the current potential of linear currents as

$$\mathbf{H} = \frac{i}{4\pi} \oint \frac{\mathbf{ds} \times \mathbf{r}}{r^3}. \tag{40.8}$$

The right-hand sides of (40.7)–(40.8) extend over all current lines. If there are several current loops, every product $i \oint \mathbf{ds} \cdots$ becomes a sum.

Problem 40.1   A current i flows in a circular wire of radius a, as in Fig. 40.1. Find the current potential $\mathbf{H}$ at a point P on the axis.

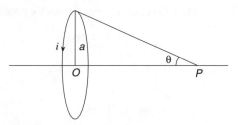

Fig. 40.1

Solution   All elements ds of the wire are at the same distance $r = \sqrt{(a^2 + z^2)}$ from P. Since $\mathbf{ds}$ is normal to $\mathbf{r}$ each element of the wire contributes a $d\mathbf{H}$ of magnitude $i\,ds/(4\pi r^2)$. In order to get the component along the axis we must multiply by a/r. The component normal to the axis vanishes by symmetry. Thus

$$H_P = \frac{i}{4\pi} \frac{2\pi a}{r^2} \frac{a}{r} = \frac{i}{2} \frac{a^2}{r^3} = \frac{1}{2} i \frac{a^2}{(a^2 + z^2)^{\frac{3}{2}}}. \tag{40.9}$$

The vector potential $\mathbf{A}$ of a current loop is given by $(40.7)_1$ in terms of a line integral. It can, alternatively, be written in terms of a surface integral. In order to find this form, we note that Stokes's theorem,

$$\oint \mathbf{ds} \cdot \mathbf{a} = \int \mathbf{dS} \cdot (\nabla \times \mathbf{a}) = \int (\mathbf{dS} \times \nabla) \cdot \mathbf{a},$$

which holds for any $\mathbf{a}$, can be written as the transformation formula

$$\oint \mathbf{ds} = \int \mathbf{dS} \times \nabla. \qquad (40.10)$$

We use this formula to transform the vector potential $(40.7)_1$:

$$\mathbf{A} = \frac{\mu_0 i}{4\pi} \oint \mathbf{ds} \frac{1}{r} = \frac{\mu_0 i}{4\pi} \int \mathbf{dS}' \times \nabla' \frac{1}{r}. \qquad (40.11)$$

The last integral is over any surface spanned by the current loop. We shall of course choose a surface which does not include the point P at which $\mathbf{A}$ is to be evaluated. The primes indicate that ∇' differentiates with respect to the integration coordinates. The distance $r = |\mathbf{x} - \mathbf{x}'|$ from dS' to the point $P(x, y, z)$, at which $\mathbf{A}$ is evaluated, depends on the coordinates of P as well. Indeed, when applied to any function of r, $\nabla = -\nabla'$. Since $\nabla(1/r) = -\mathbf{r}/r^3$, where $\mathbf{r}$ is the vector from dS to P, we have

$$\mathbf{A} = \frac{\mu_0 i}{4\pi} \int \frac{\mathbf{dS} \times \mathbf{r}}{r^3}. \qquad (40.12)$$

If P is at a great distance from the current loop, the vector potential becomes

$$\mathbf{A} = \frac{\mu_0}{4\pi} \frac{\mathbf{m} \times \mathbf{r}}{r^3}, \qquad (40.13)$$

where

$$\mathbf{m} = i \int \mathbf{dS} \qquad (40.14)$$

is *the magnetic moment* of the current loop[†]. Its direction is fixed by the sense of the current i and the conventions of Stokes's theorem. We note that the integral $\int \mathbf{dS}$ in (40.14) is over *any* surface spanned by the current loop; all such integrals have the same value, because $\oint \mathbf{dS}$, taken over any *closed* surface, vanishes.

The formula (40.13) is the magnetic analogue of the formula

$$V = \frac{1}{4\pi\epsilon_0} \frac{\mathbf{d} \cdot \mathbf{r}}{r^3}. \qquad (40.15)$$

(cf. (14.8)) for the potential of an electric dipole moment $\mathbf{d}$.

[†] In Section 20 the magnetic moment with respect to O was defined as $\frac{1}{2} \int \mathbf{r}' \times \mathbf{j}\, d^3x$, where $\mathbf{r}'$ is the vector from O to d^3x. For a linear current loop this is $\frac{1}{2} i \oint \mathbf{r}' \times \mathbf{ds}$. By the transformation (40.10) this becomes $-\frac{1}{2} i \int (\mathbf{dS}' \times \nabla') \times \mathbf{r}'$, which can easily be reduced to (40.14).

In Chapter 15 we shall prove that the magnetic field exerts on a volume element d^3x in which a current is flowing a force $\mathbf{j} \times \mathbf{B} \, d^3x$. The resultant force is therefore

$$\mathbf{f} = \int \mathbf{j} \times \mathbf{B} \, d^3x. \tag{40.16}$$

Generally, $\mathbf{B}$ in this equation is the total magnetic field, which includes the contribution (40.6) of the current density $\mathbf{j}$ on which it is acting. In the limit of a linear current the self-field becomes infinite, but its contribution to the force (40.16) vanishes by symmetry, because the self-field lines near a linear current element $i\mathbf{ds}$ are circles enclosing the line element. Thus, for a linear current loop, the resultant force becomes

$$\mathbf{f} = i \oint \mathbf{ds} \times \mathbf{B}, \tag{40.17}$$

and in this formula the self-field is to be left out. The resultant torque, with respect to a point P, is

$$\mathbf{t} = i \oint \mathbf{r}' \times (\mathbf{ds} \times \mathbf{B}), \tag{40.18}$$

where $\mathbf{r}'$ is the vector from P to $\mathbf{ds}$. The resultant force $\mathbf{f}$ will vanish for a homogeneous $\mathbf{B}$, since $\oint \mathbf{ds} = 0$. In that case, the torque must be independent of the reference point P: it becomes a couple. This is easily verified:

$$\mathbf{t} = i \oint [\mathbf{ds}(\mathbf{r}' \cdot \mathbf{B}) - \mathbf{B}(\mathbf{r}' \cdot \mathbf{ds})].$$

For a homogeneous $\mathbf{B}$ the last term vanishes because $\oint \mathbf{r}' \cdot \mathbf{ds}$ is the line integral of $\mathbf{r}' = \frac{1}{2} \mathbf{grad}'(r')^2$. Using the transformation formula (40.10), the torque becomes

$$\mathbf{t} = i \int \mathbf{dS}' \times \nabla'(\mathbf{r}' \cdot \mathbf{B}) = i \int \mathbf{dS} \times \mathbf{B},$$

which is manifestly independent of the reference point P. In terms of the magnetic moment (40.14) of the loop,

$$\mathbf{t} = \mathbf{m} \times \mathbf{B}. \tag{40.19}$$

If $\mathbf{B}$ is *inhomogeneous* the resultant force will no longer vanish (and the formula (40.19) will be only an *approximation*, in terms of an average $\mathbf{B}$, for the torque (40.18)). Using the transformation formula (40.10) in (40.17),

$$\mathbf{f} = i \int (\mathbf{dS} \times \nabla) \times \mathbf{B}.$$

Now

$$(\mathbf{dS} \times \nabla) \times \mathbf{B} - \mathbf{dS} \times (\nabla \times \mathbf{B}) = (\mathbf{dS} \cdot \nabla)\mathbf{B} - \mathbf{dS}(\nabla \cdot \mathbf{B})$$

and $\nabla \cdot \mathbf{B} = \text{div } \mathbf{B} = 0$. Thus, finally,

$$\mathbf{f} = i \int (\mathbf{dS} \cdot \mathbf{grad})\mathbf{B} + i \int \mathbf{dS} \times \text{curl } \mathbf{B}. \qquad (40.20)$$

For a small loop

$$\mathbf{f} = (\mathbf{m} \cdot \mathbf{grad})\mathbf{B} + \mathbf{m} \times \text{curl } \mathbf{B}. \qquad (40.21)$$

The second term will vanish for any steady vacuum field, for which $\mathbf{B} = \mu_0 \mathbf{H}$ and curl $\mathbf{H} = 0$; we have already agreed to leave out the self-field of the linear current loop.

Problem 40.2   Find the electromagnetic field arising from the motion of a small magnet of moment $\mathbf{m}$ at the origin, which is rotating about its centre with an angular velocity Ω.

Solution   From the formula (40.13) for $\mathbf{A}$, we obtain

$$\mathbf{B} = \text{curl } \mathbf{A} = \text{curl } \frac{\mu_0}{4\pi} \frac{\mathbf{m}(t) \times \mathbf{r}}{r^3} = \frac{\mu_0}{4\pi} \frac{3(\mathbf{m} \cdot \mathbf{n})\mathbf{n} - \mathbf{m}}{r^3},$$

where $\mathbf{n} = \mathbf{r}/r$ is the unit vector along $\mathbf{r}$. In order to calculate the electric field we use $d\mathbf{m}/dt = \Omega \times \mathbf{m}$. Thus

$$\mathbf{E} = -\mathbf{A}_t = -\frac{\mu_0}{4\pi} \frac{(\Omega \times \mathbf{m}) \times \mathbf{r}}{r^3}$$

$$= \frac{1}{4\pi \epsilon_0 c^2} \frac{\mathbf{n} \times (\Omega \times \mathbf{m})}{r^2}$$

$$= \frac{1}{4\pi \epsilon_0 c^2} \frac{(\mathbf{m} \cdot \mathbf{n})\Omega - (\Omega \cdot \mathbf{n})\mathbf{m}}{r^2}.$$

The formula (40.13) for $\mathbf{A}$ was a consequence of the static equation curl $\mathbf{H} = \mathbf{j}$. In the present case the neglected displacement term $\mathbf{D}_t = \epsilon_0 \mathbf{E}_t = -\epsilon_0 \mathbf{A}_{tt}$ is of order $(\Omega r/c)^2$ times curl $\mathbf{H} = \mu_0^{-1} \text{curl}^2 \mathbf{A}$. Hence the foregoing results for $\mathbf{B}$ and $\mathbf{E}$ are only valid in the region near the magnet, where $(\Omega r)^2 \ll c^2$. Actually, the obliquely rotating magnet emits electromagnetic waves (Chapter 12, Problem 45.1).

Exercises

40.1.   A current i flows in a circular wire $r = a$, $z = 0$ in cylindrical coordinates. Show that the vector potential at any point (r, ϕ, z) is given by

$$A_r = A_z = 0, \qquad A_\phi = \frac{\mu_0}{4\pi} i \int_0^{2\pi} \frac{a \cos \psi \, d\psi}{(a^2 + r^2 + z^2 - 2ar \cos \psi)^{\frac{1}{2}}}.$$

If the loop is small show that A_ϕ is approximately $\frac{1}{4}\mu_0 i a^2 r/(r^2 + z^2)^{\frac{3}{2}}$, in agreement with (40.13).

40.2.   A coil of n turns of wire of radius a rests with its plane vertical and in the magnetic meridian. The horizontal component of the earth's magnetic field is B. Show that when a current i flows around the coil a small compass needle placed at the centre of the coil is deflected through an angle θ where $\tan \theta = \frac{1}{2} \mu_0 ni / (aB)$.

40.3. [D. J. Griffiths, *Introduction to Electrodynamics*, 2nd ed., Prentice Hall, 1989] A current i flows in a plane wire loop. Part of the loop is in a region (shaded in Fig. 40.2) in which a uniform magnetic field **B**, perpendicular to the plane of the loop, prevails. Show that the loop experiences a resultant force ilB, where l is the distance from P to Q. What is the direction of the force if **B** in Fig. 40.2 points into the page?

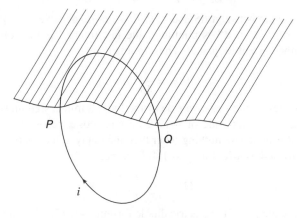

Fig. 40.2

40.4. Calculate the torque required for driving the homopolar generator (Faraday's disc of Section 38).

---

41. The scalar magnetic potential

With linear currents the system (40.1) is homogeneous, except for singularities of **curl H** on the lines along which currents are flowing. It is then possible to introduce a scalar magnetic potential Ω, such that

$$\mathbf{H} = - \mathbf{grad}\, \Omega, \qquad \Delta \Omega = 0. \qquad (41.1)$$

The scalar potential Ω is often more convenient than the vector potential **A**, because scalars can be summed more easily—when there are several current loops—than vectors. But the magnetic scalar potential is not as convenient as the electric potential V, because it is many-valued: in traversing a complete circuit, $\oint d\Omega = - \oint \mathbf{H} \cdot \mathbf{ds}$, and this equals $-i$, not zero, whenever the circuit encloses a current i.

In order to find the scalar magnetic potential of a single current loop, we calculate the magnetic field **B** from the magnetic potential (40.11):

$$\mathbf{B} = \nabla \times \mathbf{A} = -\frac{\mu_0 i}{4\pi} \int \nabla \times (\mathbf{dS} \times \nabla) \frac{1}{r}. \qquad (41.2)$$

The latter integrand is $[\mathbf{dS}(\nabla \cdot \nabla) - \nabla(\mathbf{dS} \cdot \nabla)](1/r)$. But $\nabla \cdot \nabla(1/r) = 0$. Hence

$$\mathbf{H} = \frac{i}{4\pi}\nabla \int (\mathbf{dS} \cdot \nabla)\frac{1}{r} = -\nabla\frac{i}{4\pi}\int \frac{\mathbf{dS} \cdot \mathbf{r}}{r^3}. \tag{41.3}$$

Thus, according to (41.1), the scalar magnetic potential of a current loop is given by

$$\Omega = \frac{i}{4\pi}\int \frac{\mathbf{dS} \cdot \mathbf{r}}{r^3} = \frac{i}{4\pi}\int \frac{dS\cos\theta}{r^2}, \tag{41.4}$$

where θ is the angle between $\mathbf{dS}$ and $\mathbf{r}$. If P is on the positive side of the current loop—the side to which $\mathbf{dS}$ is pointing—the last integral is the solid angle ω_P subtended by the current loop at P. Thus, in this case,

$$\Omega_P = \frac{i}{4\pi}\omega_P. \tag{41.5}$$

If, however, P lies on the negative side of the loop, a minus sign must be inserted in front of the right-hand side of (41.5), because $\cos\theta$ becomes negative. This embarrassment is of course nothing other than the many-valued nature of Ω.

Far away from the current loop, (41.4) becomes

$$\Omega = \frac{1}{4\pi}\frac{\mathbf{m} \cdot \mathbf{r}}{r^3}, \tag{41.6}$$

an even closer analogue of the electric dipole formula (40.15). If we apply (41.5) to the current loop of Fig. 40.1, we obtain

$$\Omega_P = \tfrac{1}{2}i(1 - \cos\theta). \tag{41.7}$$

From this and $H = -\partial\Omega/\partial z$ we again arrive at (40.9).

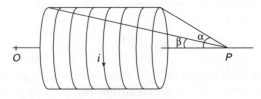

Fig. 41.1

If we now lay a large number of circular coils side by side, as in Fig. 41.1, they form a *solenoid*. In practice, we achieve this by winding a single wire in a helical shape. If each turn carries the same current i and there are n turns per unit length, Ω_P can be found by integration:

$$\begin{aligned}
\Omega_P &= \int \tfrac{1}{2}i(1 - \cos\theta)n\,dz \\
&= -\tfrac{1}{2}ni\int_\alpha^\beta (1 - \cos\theta)a\sin^{-2}\theta\,d\theta \\
&= \tfrac{1}{2}nia\left(\tan\tfrac{1}{2}\alpha - \tan\tfrac{1}{2}\beta\right).
\end{aligned} \tag{41.8}$$

From this we obtain

$$H = -\frac{\partial \Omega}{\partial z} = \tfrac{1}{2}ni(\cos\beta - \cos\alpha). \tag{41.9}$$

If the solenoid is infinite, we set $\alpha = \pi$ and $\beta = 0$. Then

$$H = ni. \tag{41.10}$$

This result can be obtained much more quickly by regarding the infinite solenoid as a cylinder with a surface current $K = ni$ flowing on its surface. A zero field outside, and a uniform field $\mathbf{H}$ in the direction of the axis inside the solenoid, clearly provide a solution of (40.1). It remains to satisfy the jump condition $\mathbf{n} \times [\![\mathbf{H}]\!] = \mathbf{K}$, and this is (41.10). This method is the analogue of the direct method by which we have, in Section 26, found the electric field between two infinite planes carrying surface charges $\pm\sigma$.

The foregoing analogy between magnetic fields of currents and electrostatics goes much further. The magnetic potential Ω satisfies (41.1), and one can apply to $\Delta\Omega = 0$ all the methods of Chapter 8, this time with the boundary conditions

$$\mathbf{n} \cdot [\![\mathbf{B}]\!] = 0, \qquad \mathbf{n} \times [\![\mathbf{H}]\!] = \mathbf{K}. \tag{41.11}$$

The uniqueness theorems of Section 27 can also be used in magnetic problems.

Consider, for example, the following magnetostatic problem, which arises in discussions of the magnetic confinement of a plasma: within a simply connected region (like the interior of a sphere) it is desired to set up a given, static magnetic field $\mathbf{B}$. Can this be achieved by suitable surface currents on the boundary S of the region, like the seam on a tennis ball?

We first note that the given field inside S has a given normal component B_n on S, and this must be continuous according to $(41.11)_1$. If there are to be no currents outside S, the external field $\mathbf{B}_e = \mu_0 \mathbf{H}_e$ must satisfy div $\mathbf{B}_e = 0$ and curl $\mathbf{H}_e = 0$. It follows that $\mathbf{H}_e = -\mathbf{grad}\,\Omega$ and $\Delta\Omega = 0$, with $\partial\Omega/\partial n = -B_n/\mu_0$ given on S. Since this Neumann problem has a unique regular solution, $\mathbf{B}_e = -\mu_0\,\mathbf{grad}\,\Omega$ is determined, and so is the required surface current $\mathbf{K} = \mathbf{n} \times [\![\mathbf{H}]\!] = \mathbf{n} \times (\mathbf{B}_e - \mathbf{B})/\mu_0$ on S.

Problem 41.1 Two equal circular coils of radius a are placed opposite one another, a distance $2b$ apart. They carry the same current in the same direction. Show that at the point midway between the two centres the first three derivatives of the field vanish if $2b = a$. Such an arrangement which gives an approximately uniform field is known as a Helmholtz coil.

Solution The field on the axis of a single coil is

$$B(z) = \tfrac{1}{2}\mu_0 i a^2 f(z), \qquad f(z) = \frac{1}{r^3}, \qquad r = (a^2 + z^2)^{\frac{1}{2}}.$$

At a point on the axis between the two coils, at a distance z from one of them,

$$B(z) = \tfrac{1}{2}\mu_0 i a^2 [f(z) + f(2b - z)], \qquad B(b) = \tfrac{1}{2}\mu_0 i a^2 2 f(b),$$
$$B'(z) = \tfrac{1}{2}\mu_0 i a^2 [f'(z) - f'(2b - z)], \qquad B'(b) = 0,$$
$$B''(z) = \tfrac{1}{2}\mu_0 i a^2 [f''(z) + f''(2b - z)], \qquad B''(b) = \tfrac{1}{2}\mu_0 i a^2 2 f''(b),$$
$$B'''(z) = \tfrac{1}{2}\mu_0 i a^2 [f'''(z) - f'''(2b - z)], \qquad B'''(b) = 0.$$

We need only show that $f''(b) = 0$ when $2b = a$.

$$f'(z) = -\frac{3z}{r^5}, \qquad f''(z) = -\frac{3}{r^5} + \frac{15z^2}{r^7} = \frac{12z^2 - 3a^2}{r^7}.$$

For $2b = a$, $f''(b)$ indeed vanishes. The field at the point midway between the centres is then $B = 8\mu_0 i / (5^{3/2} a)$.

Exercises

41.1. An infinite straight wire whose cross-section is a circle of radius a carries a uniform current i. Using cylindrical coordinates, show that the magnetic field is given by

$$r > a: \quad H_r = 0, \quad H_\theta = i/(2\pi r), \quad H_z = 0,$$
$$r < a: \quad H_r = 0, \quad H_\theta = ir/(2\pi a^2), \quad H_z = 0.$$

41.2. A current i is flowing along a circular wire of radius a. Find the scalar magnetic potential Ω (at points on or off the axis) by using (41.7) and the device of Exercise 29.5.

41.3. Show that the current potential at the centre of a square coil of side $2a$ carrying a current i is $i\sqrt{2}/(\pi a)$.

41.4. Verify that a uniform field $\mathbf{B}$ is derivable from the vector potential $\mathbf{A} = \tfrac{1}{2}\mathbf{B} \times \mathbf{x}$.

41.5. Show that very close to a wire carrying a current i the lines of $\mathbf{B}$ are circular and deduce that at small distances r from the wire $B = \mu_0 i / (2\pi r)$. Use this result to show that two similar loops of wire carrying currents in the same direction, placed nearly in coincidence, attract one another. This is the reverse of the electrical case where like charges repel.

41.6. A small current loop is placed at a distance r from an infinite straight current i. Prove that, in addition to a couple, it also experiences a force $\mu_0 im/(2\pi r^2)$, where m is the projection of the moment upon the shortest distance between the loop and the current.

41.7. Show that the force exerted on a wire carrying a current i_1 due to a current i_2 in a second wire may be expressed in the form

$$\frac{\mu_0 i_1 i_2}{4\pi} \oint \oint \frac{\mathbf{ds}_1 \times (\mathbf{ds}_2 \times \mathbf{r})}{r^3},$$

where $\mathbf{ds}_1$ and $\mathbf{ds}_2$ are elements of the two wires and $\mathbf{r}$ is the vector from $\mathbf{ds}_2$ to $\mathbf{ds}_1$. Calculate the force between two parallel wires.

41.8. Show that the formula of the last exercise may be written

$$\frac{\mu_0 i_1 i_2}{4\pi} \oint \oint (\mathbf{ds}_1 \cdot \mathbf{ds}_2) \, \mathbf{grad} \, \frac{1}{r} - \frac{\mu_0 i_1 i_2}{4\pi} \oint \oint (\mathbf{ds}_1 \cdot \mathbf{grad} \, \frac{1}{r}) \, \mathbf{ds}_2.$$

Verify by integrating with respect to s_1 that the second term is zero, so that the force is

$$\frac{\mu_0 i_1 i_2}{4\pi} \oint \oint (\mathbf{ds}_1 \cdot \mathbf{ds}_2) \, \mathbf{grad} \, \frac{1}{r}.$$

This means that each pair of elements ds_1 and ds_2 at an angle θ may be regarded as contributing a force $\mu_0 i_1 i_2 \cos\theta \, ds_1 ds_2 / (4\pi r^2)$.

41.9. Use $\Omega = \mathbf{m} \cdot \mathbf{r} / (4\pi r^3)$ and $\mathbf{H} = -\mathbf{grad}\,\Omega$ to calculate the magnetic field $\mathbf{B}$ of a small current loop.

41.10. Use the result of the last exercise and $\mathbf{t} = \mathbf{m} \times \mathbf{B}$ to show that the couple exerted by a small loop with magnetic moment $\mathbf{m}'$ on another small loop with magnetic moment $\mathbf{m}$ is

$$\frac{\mu_0}{4\pi} \left[\frac{3(\mathbf{m} \times \mathbf{r})(\mathbf{m}' \cdot \mathbf{r})}{r^5} - \frac{\mathbf{m} \times \mathbf{m}'}{r^3} \right],$$

where $\mathbf{r}$ is the vector from $\mathbf{m}'$ to $\mathbf{m}$.

41.11. In the previous exercise why is the couple exerted on m by m' not equal and opposite to the couple exerted on m' by m?

41.12. Two small loops are free to rotate about their centres which are fixed. Show that the positions of stable equilibrium are those in which the magnetic moments are pointing in the same direction along the line joining the loop centres.

42. Inductance

Consider the integral

$$U = \int \frac{B^2}{2\mu_0} d^3x = \frac{1}{2} \int \mathbf{H} \cdot \mathbf{B} \, d^3x. \tag{42.1}$$

In the identity

$$\text{div } \mathbf{A} \times \mathbf{H} = \mathbf{H} \cdot \text{curl } \mathbf{A} - \mathbf{A} \cdot \text{curl } \mathbf{H} \tag{42.2}$$

we substitute $\mathbf{B} = \text{curl } \mathbf{A}$ and $\text{curl } \mathbf{H} = \mathbf{j}$, and obtain

$$\text{div } \mathbf{A} \times \mathbf{H} = \mathbf{H} \cdot \mathbf{B} - \mathbf{A} \cdot \mathbf{j}. \tag{42.3}$$

If we integrate this and note that the surface integral at infinity vanishes in accordance with the behaviour of $\mathbf{A}$ and $\mathbf{H}$ (cf. $(40.7)_1$ and (40.8)), we get

$$U = \tfrac{1}{2} \int \mathbf{A} \cdot \mathbf{j} \, d^3x = \tfrac{1}{2} \sum i_a \oint \mathbf{A} \cdot \mathbf{ds}_a, \tag{42.4}$$

where the sum extends over all the linear current loops, because $\int \mathbf{j} \, d^3x = \sum i_a \oint \mathbf{ds}_a$. We denote by

$$\mathbf{A}_b = \frac{\mu_0}{4\pi} i_b \oint \frac{\mathbf{ds}_b}{r} \tag{42.5}$$

the vector potential due to the bth current. Of course $\mathbf{A} = \sum \mathbf{A}_b$. Then

$$\begin{aligned} U &= \tfrac{1}{2} \sum i_a \oint \mathbf{A} \cdot \mathbf{ds}_a \\ &= \frac{\mu_0}{8\pi} \sum i_a i_b \oint \oint \frac{\mathbf{ds}_a \cdot \mathbf{ds}_b}{r} \\ &= \tfrac{1}{2} \sum L_{ab} i_a i_b, \end{aligned} \tag{42.6}$$

where the last two sums are of course double sums. The coefficients

$$L_{ab} = \frac{\mu_0}{4\pi} \oint \oint \frac{\mathbf{ds}_a \cdot \mathbf{ds}_b}{r} \tag{42.7}$$

are the *coefficients of inductance*. They are manifestly symmetric: $L_{ba} = L_{ab}$. Since $U > 0$, the matrix L is positive definite. In particular, the coefficients of *self-inductance* L_{aa} are all positive, and the coefficients of *mutual inductance* L_{ab} satisfy $L_{ab}^2 \leq L_{aa} L_{bb}$. Another way of writing U is

$$\begin{aligned} U &= \sum \tfrac{1}{2} i_a \oint \mathbf{A} \cdot \mathbf{ds}_a \\ &= \sum \tfrac{1}{2} i_a \int (\mathbf{curl\ A})_n \, dS_a \\ &= \tfrac{1}{2} \sum i_a \Phi_a, \end{aligned} \tag{42.8}$$

where

$$\Phi_a = \int B_n \, dS_a \tag{42.9}$$

is the magnetic flux through the ath circuit. Clearly

$$\Phi_a = \sum L_{ab} i_b. \tag{42.10}$$

Mutual inductance coefficients are usually calculated from this formula, rather than from the Neumann formula (42.7).

The units of L_{ab} are those of magnetic flux, divided by current. The SI unit weber amp^{-1} is called the *henry*.

Problem 42.1   Two coils, one of m turns and one of n turns, are wound on a circular ring of arbitrary cross-section. Show that the coefficients of inductance are

$$L_{11} = \frac{m^2 \mu_0}{2\pi} \int \frac{dS}{s}, \quad L_{22} = \frac{n^2 \mu_0}{2\pi} \int \frac{dS}{s}, \quad L_{12} = \frac{mn \mu_0}{2\pi} \int \frac{dS}{s},$$

where s is the distance of the element dS of the cross-section from the axis of the ring. Note that $L_{11} L_{22} = L_{12}^2$. If the cross-section is a circle of radius a with its centre at a distance b from the axis, show that

$$L_{12} = \mu_0 mn[b - \sqrt{(b^2 - a^2)}].$$

Solution   A single coil of n turns, closely wound on a ring of any cross-section, in which a current i is flowing, may be regarded as a surface current of density $K = ni/(2\pi s)$, where s is the distance from the axis of the ring. Using cylindrical coordinates, the current potential $\mathbf{H} = (0, ni/(2\pi s), 0)$ satisfies the equations $\oint \mathbf{H} \cdot d\mathbf{s} = \int j_n \, dS$ and div $\mathbf{B} = 0$, as well as the jump condition $\mathbf{n} \times [\![\mathbf{H}]\!] = \mathbf{K}$. The flux of $\mathbf{B}$ through the coil is

$$n \int B_n \, dS = \frac{n^2 \mu_0 i}{2\pi} \int \frac{dS}{s},$$

and this equals i times the self-inductance of the coil. Thus when two coils, one of m turns and one of n turns, are closely wound on the same ring, we obtain

$$L_{11} = \frac{m^2 \mu_0}{2\pi} \int \frac{dS}{s}, \quad L_{22} = \frac{n^2 \mu_0}{2\pi} \int \frac{dS}{s}, \quad L_{12} = \frac{mn \mu_0}{2\pi} \int \frac{dS}{s}.$$

If the cross-section is a circle of radius a with its centre at a distance b from the axis

$$\int \frac{dS}{s} = \int_{b-a}^{b+a} \frac{2\sqrt{a^2 - (s - b)^2} \, ds}{s} = 2 \int_{-a}^{a} \frac{\sqrt{a^2 - x^2} \, dx}{b + x}.$$

The integral can be evaluated by substituting $t = \sqrt{[(a - x)/(a + x)]}$. It has the value

$$\int_{-a}^{a} \frac{\sqrt{a^2 - x^2} \, dx}{b + x} = \pi[b - \sqrt{(b^2 - a^2)}]. \tag{1}$$

Exercises

42.1.   Show that the coefficient of mutual inductance between a circle of radius a and an infinite straight line in the same plane is

$$L_{12} = \mu_0 \left(d - \sqrt{d^2 - a^2}\right).$$

where d is the shortest distance from the centre of the circle to the straight line.

42.2. Show that the coefficient of mutual inductance between a long straight wire and a coplanar equilateral triangular loop of wire is

$$L_{12} = \frac{\mu_0}{3^{\frac{1}{2}} \pi} [(a + b) \ln(1 + a/b) - a],$$

where a is the height of the triangle and b is the distance from the straight wire to the side of the triangle parallel to and nearest it.

42.3. Two equal circular loops of radius a lie opposite each other, a distance c apart. Show that the coefficient of mutual inductance is

$$L_{12} = \tfrac{1}{2} \mu_0 a^2 \int_0^{2\pi} \frac{\cos \psi \, d\psi}{(c^2 + 2a^2 - 2a^2 \cos \psi)^{\frac{1}{2}}}.$$

If c is very large, show that $L_{12} = \tfrac{1}{2} \mu_0 \pi a^4 / c^3$.

---

43. Paramagnetic and diamagnetic materials

Magnetic materials, or magnets, are analogous to dielectrics and present similar phenomena. There is nothing, in principle, to prevent a material from being both dielectric and magnetic. Common magnets, however, show no polarization[†]. The simplest among them are linear: their response is expressed by the constitutive relations

$$\mathbf{M}' = \frac{\chi_B}{\mu_0} \mathbf{B}', \qquad \mathbf{P}' = 0, \tag{43.1}$$

where, as usual, the primes refer to the material rest frame. We call such materials *linear magnets*. The coefficient χ_B is called the *magnetic susceptibility*.

In a linear magnet at rest (in an aether frame) we can leave out the primes and write

$$\mu_0 \mathbf{M} = \chi_B \mathbf{B}. \tag{43.2}$$

The aether relation $\mathbf{H} = \mathbf{B}/\mu_0 - \mathbf{M}$ can be written in the form

$$\mathbf{B} = \mu \mathbf{H}, \tag{43.3}$$

where

$$\mu = \frac{\mu_0}{1 - \chi_B} \tag{43.4}$$

is the *magnetic permeability*. Unlike the dielectric susceptibility, which is always positive, χ_B may be of either sign. Materials with positive χ_B are called *paramagnetic*; those with negative χ_B are called *diamagnetic*. Whether positive or negative,

---

[†]Since common magnets, like iron, are conductors, charge relaxation (at suboptical frequencies) results in $\mathcal{E} = 0$. Even if $\epsilon \neq \epsilon_0$, $\mathbf{P} = (\epsilon - \epsilon_0)\mathcal{E}$ will then still vanish.

values of $|\chi_B|$ range from 10^{-9} to 10^{-4}. There is certainly no danger of the permeability according to (43.4) blowing up as $\chi_B \to 1$.

In the older literature the magnetic susceptibility is defined differently, viz.:

$$\mathbf{M} = \chi_H \mathbf{H}. \tag{43.5}$$

In terms of χ_H, the permeability is

$$\mu = (1 + \chi_H)\mu_0. \tag{43.6}$$

This is analogous to $\epsilon = (1 + \chi)\epsilon_0$ (cf. (32.2)). The newer χ_B and the older χ_H are related by

$$(1 - \chi_B)(1 + \chi_H) = 1. \tag{43.7}$$

The two obviously have the same sign. Indeed they are equal to first order in either of them, and to distinguish between them would be a refinement that is not justified by the accuracy with which magnetic susceptibilities are usually measured.

The equations governing a magnet are

$$\begin{aligned}
\mathbf{curl}\,\mathbf{H} &= \mathbf{j}, & \mathbf{n} \times [\![\mathbf{H}]\!] &= \mathbf{K}, \\
\mathrm{div}\,\mathbf{B} &= 0, & \mathbf{n} \cdot [\![\mathbf{B}]\!] &= 0, \\
\mathbf{H} &= \mathbf{B}/\mu_0 - \mathcal{M}.
\end{aligned} \tag{43.8}$$

For permanent magnets—materials in which $\mathbf{M}$ is a given, non-zero vector field—these equations have non-trivial regular solutions, even in the absence of any currents. For a linear—diamagnetic or paramagnetic—material, eqns (43.8) have only trivial regular solutions when $\mathbf{j} = \mathbf{K} = 0$ everywhere. A linear magnet can therefore only act as an *electromagnet*. If it fills all of the space around the circuits, the magnetic field is given by the formulae of Sections 40 and 41, provided we replace μ_0 by μ everywhere.

If the currents flow along lines, or on surfaces, we have $\mathbf{curl}\,\mathbf{H} = 0$ almost everywhere. We can then introduce a magnetic potential, as in (41.1), and seek solutions of Laplace's equation $\Delta\Omega = 0$. The methods for solving electrostatic problems with linear dielectrics can all be applied, but now, of course, with the aether relation $\mathbf{B} = \mu\mathbf{H}$ (instead of $\mathbf{D} = \epsilon\mathbf{E}$) and the boundary conditions (41.11).

In applying the complex potential method we recognize that the notation $i = \sqrt{-1}$ is firmly established in mathematics. We therefore denote a current by I. For a linear current I flowing along the z axis, the current potential—using polar coordinates (r, θ) in the xy plane—is

$$\mathbf{H} = (0, I/(2\pi r)). \tag{43.9}$$

The corresponding scalar potential is

$$\Omega = -2I\theta, \tag{43.10}$$

which is the real part of

$$w = 2iI \ln z = -2I\theta + 2iI \ln r. \qquad (43.11)$$

The complex potential $w = 2iI \ln z$ therefore describes a current I flowing through the origin perpendicular to the xy plane. If the current flows through the point $z_0 = (x_0, y_0)$, the appropriate complex potential is

$$w = 2iI \ln(z - z_0). \qquad (43.12)$$

We shall dispense with a formulation of Thomson's theorem for magnetic fields. It can easily be done, and one of its corollaries is that, in linear magnetic media, the field lines of the solenoidal field $\mathbf{B}$ will be more concentrated in regions of high permeability. In magnetic machinery, where high magnetic flux is desirable, iron cores are therefore placed inside the current coils.

In dealing with electricity the experimenter always uses the electric field $\mathbf{E}$, rather than the (partial) charge potential, or the displacement, $\mathbf{D}$. That is because voltage, the line integral of $\mathbf{E}$, can be measured, but (free) electric charge cannot: there is no 'displacement meter'. In magnetic problems, however, it is the (free) electric current, the line integral of $\mathbf{H}$, that is measurable. In situations of simple symmetry such formulae as $H = i/(2\pi r)$ (straight wire) or $H = ni$ (solenoid) are much simpler than the corresponding formulae for $\mathbf{B}$, which often depend on the ill-known, or uncontrollable, magnetization. Hence the experimenter frequently prefers to use the (partial) current potential $\mathbf{H}$, rather than the magnetic field $\mathbf{B}$.

Problem 43.1 A sphere of radius a and permeability μ is placed in a uniform magnetic field $\mathbf{B}_0$. Determine the field inside the sphere.

Solution   Outside the sphere we assume a superposition of $\mathbf{B}_0$ and a dipole field (the first term of $\Omega = \sum A_n r^{-n-1} P_n(\cos\theta)$):

$$\mathbf{B}_e = \mathbf{B}_0 + \alpha \left(\frac{a}{r}\right)^3 [3(\mathbf{n} \cdot \mathbf{B}_0)\mathbf{n} - \mathbf{B}_0],$$

where $\mathbf{n} = \mathbf{r}/r$. Inside, we assume a uniform field (the first non-constant term of $\Omega = \sum B_n r^n P_n$)

$$\mathbf{B} = \beta \mathbf{B}_0.$$

At $r = a$ the jump condition $\mathbf{n} \cdot [\![\mathbf{B}]\!] = 0$ gives

$$1 + 2\alpha = \beta.$$

The jump condition $\mathbf{n} \times [\![\mathbf{H}]\!] = 0$ gives

$$\frac{1 - \alpha}{\mu_0} = \frac{\beta}{\mu}.$$

Hence $\alpha = (\mu - \mu_0)/(2\mu_0 + \mu)$ and $\beta = 3\mu/(2\mu_0 + \mu)$. Thus the field inside the sphere is uniform and independent of the radius:

$$\mathbf{B} = \frac{3\mu}{2\mu_0 + \mu}\mathbf{B}_0.$$

Problem 43.2 A tiny loop of magnetic moment **m** lies at the centre of a sphere of radius a and permeability μ. Find the magnetic field inside and outside the sphere.

Solution Assume a superposition of a dipole field and a constant field (in the direction of **m**) inside the sphere, and a dipole field outside:

$$\mathbf{H}_e = \frac{\alpha}{4\pi}\frac{3(\mathbf{m}\cdot\mathbf{n})\mathbf{n} - \mathbf{m}}{r^3},$$

$$\mathbf{H} = \frac{1}{4\pi}\left[\frac{3(\mathbf{m}\cdot\mathbf{n})\mathbf{n} - \mathbf{m}}{r^3} + \beta\frac{\mathbf{m}}{a^3}\right].$$

Near the centre **H** reduces to the correct expression (Exercise 41.9). The jump condition $\mathbf{n} \times [\![\mathbf{H}]\!] = 0$ at $r = a$ requires the relation $\alpha = 1 - \beta$ between α and β. The jump condition $\mathbf{n} \cdot [\![\mathbf{B}]\!] = 0$, together with $\mathbf{B}_e = \mu_0\mathbf{H}_e$ and $\mathbf{B} = \mu\mathbf{H}$, yields the second relation, $2\alpha\mu_0 = (2 + \beta)\mu$. Thus

$$\alpha = \frac{3\mu}{\mu + 2\mu_0}, \qquad \beta = -\frac{2(\mu - \mu_0)}{\mu + 2\mu_0}.$$

The magnetic field is

$$\mathbf{B}_e = \frac{3\mu\mu_0}{4\pi(\mu + 2\mu_0)}\frac{3(\mathbf{m}\cdot\mathbf{n})\mathbf{n} - \mathbf{m}}{r^3},$$

$$\mathbf{B} = \frac{\mu}{4\pi}\left[\frac{3(\mathbf{m}\cdot\mathbf{n})\mathbf{n} - \mathbf{m}}{r^3} - \frac{2(\mu - \mu_0)}{\mu + 2\mu_0}\frac{\mathbf{m}}{a^3}\right].$$

Problem 43.3 A torus of soft iron with permeability μ has n turns of wire closely and evenly wound around it (Fig. 43.1). A current i flows in the wire. Show that at points inside the torus $B = \mu ni/(2\pi r)$, where r is the distance from the axis of the torus. A small air-gap is formed in the iron by cutting away a thin sector bounded by two planes through the axis which make a small angle α with each other. Show that B in the gap is reduced from the foregoing value by the factor $1 + (\mu/\mu_0 - 1)\alpha/(2\pi)$.

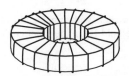

Fig. 43.1

Solution   Before the gap is formed the field is determined as in Problem 42.1, except that the torus is now made of material with permeability μ. Thus $H = ni/(2\pi r)$ and $B = \mu ni/(2\pi r)$, where r is the distance from the axis. After the gap is made the new field B' is again continuous. Since H is B'/μ in the iron and B'/μ_0 in the gap, $\oint \mathbf{H} \cdot \mathbf{ds} = ni$ takes the form

$$(2\pi - \alpha)r\frac{B'}{\mu} + \alpha r\frac{B'}{\mu_0} = ni.$$

Thus

$$\left[1 + \left(\frac{\mu}{\mu_0} - 1\right)\frac{\alpha}{2\pi}\right]B' = B.$$

Exercises

43.1. Show that the lines of **B** are refracted at a change of medium, and that if the angles made with the normal are θ_1 and θ_2, as in Fig. 43.2, then

$$\mu_1 \cot \theta_1 = \mu_2 \cot \theta_2.$$

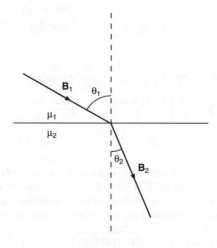

Fig. 43.2

43.2. A paramagnetic spherical shell with inner and outer radii a and b is placed in a uniform field $\mathbf{H}_0$. Show that the field $\mathbf{H}_i$ in the inner cavity $r \le a$ is uniform and parallel to $\mathbf{H}_0$, and that

$$\frac{H_i}{H_0} = \frac{9\mu\mu_0}{9\mu\mu_0 + 2(\mu - \mu_0)^2(1 - a^3/b^3)}.$$

43.3. A current i flows in a straight wire parallel to a semi-infinite block of material of permeability μ. Show that the magnetic field is given by an image system similar to that of Exercise 33.5.

43.4. A current i flows in a straight wire parallel to a circular cylinder of permeability μ. Show that inside the cylinder $\mathbf{B}$ is $2\mu/(\mu + \mu_0)$ times as large as it would be if the cylinder were removed, and is everywhere in the same direction.

12

Radiation

44. The wave equations

The second pair of Maxwell's equations can always be satisfied by introducing the potentials $\mathbf{A}$ and V:

$$\mathbf{B} = \text{curl } \mathbf{A},$$
$$\mathbf{E} = -\mathbf{A}_t - \text{grad } V. \tag{44.1}$$

In an aether frame, the first pair of Maxwell's equations is (cf. (13.1))

$$\text{div } \epsilon_0 \mathbf{E} = q,$$
$$\text{curl } \mathbf{B}/\mu_0 - \epsilon_0 \mathbf{E}_t = \mathbf{j}, \tag{44.2}$$

where q and $\mathbf{j}$ are the *total* charge and current densities, including those of polarization and magnetization. For linear dielectrics and magnets, we may replace ϵ_0 and μ_0 by ϵ and μ, respectively, and then q and $\mathbf{j}$ are the free charge and current densities; we shall then also have to replace, in the following equations, the squared speed of light in vacuum $c^2 = (\epsilon_0\mu_0)^{-1}$ by $(\epsilon\mu)^{-1}$. If we substitute (44.1) in (44.2) we obtain

$$\Delta V - \frac{1}{c^2}V_{tt} + \frac{\partial}{\partial t}\left(\text{div } \mathbf{A} + \frac{1}{c^2}V_t\right) = -\frac{q}{\epsilon_0},$$
$$\Delta \mathbf{A} - \frac{1}{c^2}\mathbf{A}_{tt} - \text{grad}\left(\text{div } \mathbf{A} + \frac{1}{c^2}V_t\right) = -\mu_0\mathbf{j}. \tag{44.3}$$

The potentials are still subject to a gauge transformation, and this can be chosen so that they satisfy the *Lorenz gauge condition*

$$\text{div } \mathbf{A} + \frac{1}{c^2}V_t = 0. \tag{44.4}$$

Equations (44.3) then become

$$\Delta V - \frac{1}{c^2} V_{tt} = -\frac{q}{\epsilon_0},$$
$$\Delta \mathbf{A} - \frac{1}{c^2} \mathbf{A}_{tt} = -\mu_0 \mathbf{j}. \tag{44.5}$$

These are inhomogeneous wave equations for V and for the Cartesian components of $\mathbf{A}$. If the time derivatives are discarded, the Lorenz gauge (44.4) reduces to the Coulomb gauge (40.3); (44.5)$_1$ reduces to Poisson's equation (26.3); and (44.5)$_2$ reduces to (40.4).

The strangest claims have been made regarding the solutions of the inhomogeneous wave equations (44.5), usually because of careless mathematics. We shall therefore treat these equations in a mathematically responsible manner. This requires a little patience.

Each of the four components of (44.5) is of the form

$$\Delta u - \frac{1}{c^2} u_{tt} = -g, \tag{44.6}$$

which is a linear partial differential equation for u. Solutions of this *inhomogeneous* equation can be obtained from any *particular* one by adding solutions of the *homogeneous* equation with $g = 0$. This is, of course, the analogue of the statement that solutions of Poisson's equation $\Delta V = -q/\epsilon_0$ can be obtained from any particular one by adding harmonic functions. But in the present case there are initial conditions to consider.

We begin with a discussion of the homogeneous wave equation,

$$\Delta u - \frac{1}{c^2} u_{tt} = 0. \tag{44.7}$$

Solutions of this equation that have the form $u(\mathbf{x}, t) = f(\mathbf{x})g[h(\mathbf{x}) \pm ct]$ are called *progressive waves*, or *travelling waves*. The argument $h(\mathbf{x}) \pm ct$ of g is called the *phase* of the wave; the surfaces $h(\mathbf{x}) = $ const. are called the *wave fronts*. Solutions of the form $u(\mathbf{x}, t) = f(\mathbf{x})g(t)$ are called *standing waves*. The most important example of progressive waves is given by the spatially decreasing spherical waves: if we look for solutions of (44.7) that depend only on $r = |\mathbf{x}|$ and t, that is $u = u(r, t)$, we find $\Delta u = u_{rr} + (2/r)u_r = (ru)_{rr}/r = (1/c^2)u_{tt}$, or $(ru)_{rr} - (1/c^2)(ru)_{tt} = 0$. It is easy to show that the general solution of the latter equation is

$$u(r, t) = \frac{f_1(r - ct)}{r} + \frac{f_2(r + ct)}{r}, \tag{44.8}$$

where f_1 and f_2 are arbitrary. The first term is an outgoing, the second an incoming, spherical wave. Of course we may choose to define r as the distance from any point P; then (44.8) is still the general solution, and it describes waves which emanate from, or converge to, P. These spherical waves play a role which is similar to that of the fundamental harmonic function $1/r$ in electrostatics.

Let $r = |\mathbf{y} - \mathbf{x}|$, the distance from $\mathbf{x}$ to $\mathbf{y}$. Then, for any ψ and f,

$$u(\mathbf{x}, t) = \int \psi(\mathbf{y}) \frac{f(r - ct)}{r} \, d^3 y \qquad (44.9)$$

is a superposition of spherical waves, and is therefore a solution of the wave equation (44.7). We choose $f(\lambda)$ to be a non-negative function that vanishes outside the interval $-\epsilon < \lambda < \epsilon$, and for which

$$\int_{-\infty}^{\infty} f(\lambda) \, d\lambda = 1.$$

The contributions to the integral in (44.9) are then only from a spherical shell of thickness 2ϵ around $r = ct$. Letting ϵ tend to zero and passing to the limit, so that f becomes a delta function, we obtain

$$u = ct \int_{|\mathbf{y} - \mathbf{x}| = ct} \psi(\mathbf{y}) \, d\omega, \qquad (44.10)$$

where $d\omega$ is an element of solid angle. Any point on the sphere $|\mathbf{y} - \mathbf{x}| = ct$ can be written as $\mathbf{y} = \mathbf{x} + \mathbf{n}ct$, where $\mathbf{n}$ is a unit radial vector. We denote

$$M(t)\psi = \frac{1}{4\pi} \int \psi(\mathbf{x} + \mathbf{n}ct) \, d\omega. \qquad (44.11)$$

This is a mean of ψ over the sphere of radius ct around $\mathbf{x}$. It is therefore a function of $\mathbf{x}$ and t, and its value at $t = 0$ is $\psi(\mathbf{x})$. We now compute its time derivative:

$$\begin{aligned}
(M(t)\psi)_t &= \frac{c}{4\pi} \int \mathbf{n} \cdot \mathbf{grad}\, \psi \, d\omega \\
&= \frac{1}{4\pi c t^2} \int_{|\mathbf{y} - \mathbf{x}| = ct} \mathbf{n} \cdot \mathbf{grad}\, \psi(\mathbf{y}) \, dS \\
&= \frac{1}{4\pi c t^2} \int_{|\mathbf{y} - \mathbf{x}| \leq ct} \Delta \psi(\mathbf{y}) \, d^3 y, \qquad (44.12)
\end{aligned}$$

where we have used $dS = r^2 \, d\omega = c^2 t^2 \, d\omega$, as well as Gauss's theorem. Since the volume in the last integral goes as $c^3 t^3$, $(M(t)\psi)_t$ vanishes at $t = 0$.

According to (44.10) and (44.11), the function

$$u(\mathbf{x}, t) = t M(t)\psi, \qquad (44.13)$$

for any ψ, is a solution of the homogeneous wave equation (44.7). Since $u_t = t(M(t)\psi)_t + M(t)\psi$, u satisfies the initial conditions

$$u(\mathbf{x}, 0) = 0, \qquad u_t(\mathbf{x}, 0) = \psi(\mathbf{x}). \qquad (44.14)$$

With this u, let $v = u_t$. Then v, too, satisfies the wave equation and has the initial values $v(\mathbf{x}, 0) = u_t(\mathbf{x}, 0) = \psi(\mathbf{x})$ and $v_t(\mathbf{x}, 0) = u_{tt}(\mathbf{x}, 0) = c^2 \Delta u(\mathbf{x}, 0) = 0$. Since ψ is arbitrary, this proves that, for any φ, the function

$$u(\mathbf{x}, t) = (tM(t)\varphi)_t \tag{44.15}$$

is a solution of (44.7) with the initial values

$$u(\mathbf{x}, 0) = \varphi(\mathbf{x}), \qquad u_t(\mathbf{x}, 0) = 0. \tag{44.16}$$

By superposing these solutions we conclude that

$$u(\mathbf{x}, t) = (tM(t)\varphi)_t + tM(t)\psi \tag{44.17}$$

is a solution with initial values

$$u(\mathbf{x}, 0) = \varphi(\mathbf{x}), \qquad u_t(\mathbf{x}, 0) = \psi(\mathbf{x}). \tag{44.18}$$

We now return to the inhomogeneous wave equation and, using an idea known as *Duhamel's principle*, we seek a particular solution with zero initial conditions. Having found it, we can always make it conform to the initial conditions (44.18) by adding (44.17).

Consider the following initial value problem at $t = \tau$:

$$\Delta v - \frac{1}{c^2} v_{tt} = 0, \qquad v(\mathbf{x}, \tau) = 0, \qquad v_t(\mathbf{x}, \tau) = g(\mathbf{x}, \tau). \tag{44.19}$$

According to (44.13) and (44.14), the solution of (44.19), which we denote by $v(\mathbf{x}, t; \tau)$, is

$$v(\mathbf{x}, t; \tau) = (t - \tau)M(t - \tau)g. \tag{44.20}$$

We now prove that

$$u(\mathbf{x}, t) = c^2 \int_0^t v(\mathbf{x}, t; \tau)\, d\tau \tag{44.21}$$

is a solution of the inhomogeneous equation (44.6) with zero initial conditions. For

$$u_t = c^2 v(\mathbf{x}, t; t) + c^2 \int_0^t v_t(\mathbf{x}, t; \tau)\, d\tau$$

$$= c^2 \int_0^t v_t(\mathbf{x}, t; \tau)\, d\tau, \tag{44.22}$$

since $v(\mathbf{x}, t; t) = 0$. Clearly, $u(\mathbf{x}, 0) = u_t(\mathbf{x}, 0) = 0$. Furthermore,

$$u_{tt} = c^2 v_t(\mathbf{x}, t; t) + c^2 \int_0^t v_{tt}(\mathbf{x}, t; \tau)\, d\tau$$

$$= c^2 g(\mathbf{x}, t) + c^2 \int_0^t v_{tt}\, d\tau, \tag{44.23}$$

and

$$\Delta u = c^2 \int_0^t \Delta v \, d\tau. \tag{44.24}$$

Hence $\Delta u - (1/c^2)u_{tt} = -g$.

In order to obtain a more illuminating form of this solution, we substitute (44.20) in (44.21) and recall the definition (44.11) of M:

$$u(\mathbf{x}, t) = c^2 \int_0^t \frac{d\tau(t - \tau)}{4\pi} \int g(\mathbf{x} + \mathbf{n}c(t - \tau), \tau) \, d\omega.$$

With a change of variable from τ to $r = c(t - \tau)$, we find

$$u(\mathbf{x}, t) = \frac{1}{4\pi} \int_0^{ct} r \, dr \int g(\mathbf{x} + \mathbf{n}r, t - r/c) \, d\omega,$$

or

$$u(\mathbf{x}, t) = \frac{1}{4\pi} \int_{r \le ct} \frac{g(\mathbf{x}', t - r/c) \, d^3x'}{r}, \tag{44.25}$$

where r is the distance from $\mathbf{x}$ to the integration point $\mathbf{x}'$, and the integral extends over the sphere of radius ct around $\mathbf{x}$.

In the form given by (44.25), the solution is easy to visualize. Unlike the solution of Poisson's equation, each volume element d^3x' makes its contribution $g(\mathbf{x}', t - r/c) \, d^3x'/r$ at the *retarded* time $t - r/c$, rather than at t; and the amount of retardation is precisely that which a signal, travelling at the velocity of light c, would need to traverse the distance r from $\mathbf{x}'$ to $\mathbf{x}$. In addition, only sources g that are within a distance ct from $\mathbf{x}$ contribute to the integral. Those that lie beyond the sphere $r = ct$, which may be called the *horizon* at time t, make no contribution—except at a later time, when they rise, so to speak, above the horizon.

The solutions of equations (44.5) for potentials that satisfy zero initial conditions are then

$$\mathbf{A}(\mathbf{x}, t) = \frac{\mu_0}{4\pi} \int_{r \le ct} \frac{\mathbf{j}(\mathbf{x}', t - r/c) \, d^3x'}{r},$$

$$V(\mathbf{x}, t) = \frac{1}{4\pi\epsilon_0} \int_{r \le ct} \frac{q(\mathbf{x}', t - r/c) \, d^3x'}{r}. \tag{44.26}$$

The fields (44.1) that are derived from these potentials involve **curl A**, $\mathbf{A}_t$ and **grad** V. Thus they, too, will vanish initially.

Anyone who wishes to regard the current density $\mathbf{j}$ or the charge density q as 'causes', and the potentials $\mathbf{A}$ or V that satisfy zero initial conditions, as well as the Lorenz gauge (44.4), as 'effects', must accept (44.26) as a demonstration, indeed a mathematical proof, of *the principle of causality*, according to which 'the cause must precede the effect'. As a mere restatement of a proved result this platitude is certainly correct, though hardly a principle.

For a finite system (such that $\mathbf{j}$ and q vanish outside a finite region of space), the potentials (44.26), which are zero initially, will continue to vanish until such time as is required for the horizon to reach the system. If d is the distance from $\mathbf{x}$ to the nearest part of the system, this time is $t_1 = d/c$. The *whole* system will affect the potentials after the time $t_2 = D/c$, where D is the distance to the furthest part of the system. After t_2 the whole system lies within the horizon, and the limitation $r \le ct$ on the volume integrals of (44.26) becomes irrelevant.

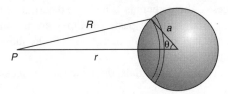

Fig. 44.1

As an example, we consider a conducting sphere of radius a, on which the uniform charge density $\sigma(t)$ is made to vary so that $\sigma = 0$ for $t < 0$ and $\sigma(t) = \sigma_0 \sin \omega t$ for $t > 0$. At a distance r $(r > a)$ from the centre of the sphere, the electric potential (assumed to vanish up to $t = 0$) will vanish for $ct < r - a$. Later, it will be given by

$$V = \frac{1}{4\pi\epsilon_0} \int_{R \le ct} \frac{\sigma(t - R/c)\, dS}{R}. \tag{44.27}$$

The contribution to the integral from a ring (Fig. 44.1) between R and $R + dR$ will be $\sigma_0 \sin \omega(t - R/c) 2\pi a \sin \theta a\, d\theta / R$, where $\theta(R)$ is given by $R^2 = a^2 + r^2 - 2ar \cos \theta$. Noting that $R\, dR = ar \sin \theta\, d\theta$, we obtain

$$V = \frac{a\sigma_0}{2\epsilon_0 r} \int \sin \omega(t - R/c)\, dR. \tag{44.28}$$

In the last integral, the lower limit is $r - a$, and the upper limit is the smaller of ct and $r + a$. An easy calculation gives

$$
\begin{aligned}
V &= \frac{ac\sigma_0}{2\epsilon_0 \omega r}\left[1 - \cos \omega\left(t - \frac{r - a}{c}\right)\right], \quad r - a < ct < r + a, \\
V &= \frac{ac\sigma_0}{\epsilon_0 \omega r} \sin \frac{\omega a}{c} \sin \omega\left(t - \frac{r}{c}\right), \quad r + a < ct.
\end{aligned}
\tag{44.29}
$$

We note that both expressions are of the form $f(t - r/c)/r$. Each one is therefore an outgoing spherical wave.

Often the formulae (44.26) are offered without the constraint $r \leq ct$ on the integration domain, that is, in the form

$$
\mathbf{A}(\mathbf{x}, t) = \frac{\mu_0}{4\pi} \int \frac{\mathbf{j}(\mathbf{x}', t - r/c)\, d^3x'}{r},
$$

$$
V(\mathbf{x}, t) = \frac{1}{4\pi\epsilon_0} \int \frac{q(\mathbf{x}', t - r/c)\, d^3x'}{r},
$$

(44.30)

and the claim is made that these *retarded potentials*[†] are the 'appropriate' solutions of (44.5), because they satisfy the principle of causality. Now, it is true that they are solutions, because they are of the form (44.9). But they correspond to definite non-zero initial values of the potentials, which are determined by the values of the sources at $(\mathbf{x}', -r/c)$, that is, in the past. Thus they are not the same 'effects' as previously defined, and it is only after this change in the meaning of 'effect' that they satisfy the causality 'principle'.

Along with these retarded potentials, there are the *advanced potentials*,

$$
\mathbf{A}(\mathbf{x}, t) = \frac{\mu_0}{4\pi} \int \frac{\mathbf{j}(\mathbf{x}', t + r/c)\, d^3x'}{r},
$$

$$
V(\mathbf{x}, t) = \frac{1}{4\pi\epsilon_0} \int \frac{q(\mathbf{x}', t + r/c)\, d^3x'}{r},
$$

(44.31)

which are also solutions, of the form (44.9) with $f(r + ct)$ instead of $f(r - ct)$, but they depend on the sources at $(\mathbf{x}', t + r/c)$, that is, in the *future*. They are often rejected as 'unphysical', because they seem to violate the causality 'principle'. But since even the initial conditions of the advanced potentials involve the sources at $(\mathbf{x}', r/c)$, that is, in the future, they are not at all the kind of 'effect' we have been discussing until now. Thus they *do not*, indeed *cannot*, violate the causality 'principle', because it does not apply to them.

The advanced potentials are indeed solutions of Maxwell's equations, but in order to evaluate them we must, according to (44.31), know the distribution of the sources in the future. We may have this knowledge, at least for some of the sources; for example, when we possess next week's television programme. If we do not, the advanced potentials are merely useless, not wrong, and certainly not 'unphysical'.

Problem 44.1   A uniform surface current $\mathbf{K}(t)$ is made to flow along an infinite plane. There is no electromagnetic field prior to $t = 0$. Find the electromagnetic field on either side of the plane for $t \geq 0$.

Solution   At a point P, which is at a distance x from the plane, the contribution to $\mathbf{A}$ from the ring of radius s and width ds shown in Fig. 44.2 is

$$
\frac{\mu_0}{4\pi} \frac{2\pi s\, ds\, \mathbf{K}(t - r/c)}{r}.
$$

[†]Lorenz (1867).

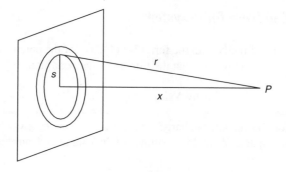

Fig. 44.2

Since $s^2 = r^2 - x^2$ we have $s\,ds = r\,dr$, so that the contribution is

$$\tfrac{1}{2}\mu_0 \mathbf{K}(t - r/c)\,dr.$$

Thus

$$\mathbf{A}(x, t) = \tfrac{1}{2}\mu_0 \int_x^{ct} \mathbf{K}(t - r/c)\,dr = \tfrac{1}{2}\mu_0 c \int_0^{t-x/c} \mathbf{K}(y)\,dy,$$

provided that $t > x/c$. Prior to that $\mathbf{A}(x, t) = 0$. The fields are

$$\mathbf{E} = -\mathbf{A}_t = -\tfrac{1}{2}\mu_0 c \mathbf{K}(t - x/c),$$
$$\mathbf{B} = \operatorname{curl} \mathbf{A} = \left(0, -\frac{\partial A_z}{\partial x}, \frac{\partial A_y}{\partial x}\right)$$
$$= \left[0, \tfrac{1}{2}\mu_0 K_z(t - x/c), -\tfrac{1}{2}\mu_0 K_y(t - x/c)\right]$$
$$= \frac{1}{c}\mathbf{n} \times \mathbf{E},$$

where $\mathbf{n}$ is the unit normal that points away from the plane. Both fields, which depend on x and t through $t - x/c$, are transverse plane waves propagating away from the plane. Like $\mathbf{K}$, $\mathbf{E}$ is parallel to the plane. It is continuous across the plane (at $x = 0$). The magnetic field $\mathbf{B}$, too, is parallel to the plane, and the jump condition $\mathbf{n} \times [\![\mathbf{H}]\!] = \mathbf{K}$ is satisfied because $\mathbf{n}$ changes sign on crossing the plane.

Exercises

44.1. Calculate the electric potential at a distance r ($r > a$) from the centre of a sphere of radius a, which is suddenly given a uniformly distributed surface charge Q.

44.2. At $t = 0$ a current i suddenly begins to flow in an infinite, straight wire. Determine the electromagnetic field.

45. Radiation from finite systems

Instead of working directly with the formulae (44.26) for the potentials $\mathbf{A}$ and V, it often proves convenient to note that the Lorenz gauge condition,

$$\operatorname{div}\mathbf{A}/\mu_0 + \epsilon_0 V_t = 0, \tag{45.1}$$

has the same form as the law of charge conservation. It can therefore be satisfied by introducing two vectors $\mathbf{Z}^e$ and $\mathbf{Z}^m$, similar to the charge and current potentials $\mathbf{D}$ and $\mathbf{H}$:

$$\mathbf{A} = \mu_0(\mathbf{Z}_t^e - \operatorname{\mathbf{curl}} \mathbf{Z}^m), \qquad V = -\epsilon_0^{-1} \operatorname{div} \mathbf{Z}^e. \tag{45.2}$$

The vectors $\mathbf{Z}^e$ and $\mathbf{Z}^m$ are called *Hertz vectors*. Since they are potentials for the potentials $\mathbf{A}$ and V, they are also called *superpotentials*. In terms of these, the fields (44.1) become

$$\begin{aligned}
\mathbf{B} &= \mu_0 \operatorname{\mathbf{curl}} \mathbf{Z}_t^e - \mu_0 \operatorname{\mathbf{curl}}^2 \mathbf{Z}^m, \\
\mathbf{E} &= -\mu_0 \mathbf{Z}_{tt}^e + \epsilon_0^{-1} \operatorname{\mathbf{grad}} \operatorname{div} \mathbf{Z}^e + \mu_0 \operatorname{\mathbf{curl}} \mathbf{Z}_t^m,
\end{aligned} \tag{45.3}$$

and substitution in the first pair (44.2) of Maxwell's equations gives

$$\operatorname{div}\!\left(\Delta\mathbf{Z}^e - \frac{1}{c^2}\mathbf{Z}_{tt}^e\right) = q,$$

$$-\!\left(\Delta\mathbf{Z}^e - \frac{1}{c^2}\mathbf{Z}_{tt}^e\right)_t + \operatorname{\mathbf{curl}}\left(\Delta\mathbf{Z}^m - \frac{1}{c^2}\mathbf{Z}_{tt}^m\right) = \mathbf{j}. \tag{45.4}$$

Consider now the wave equations

$$\Delta\mathbf{Z}^e - \frac{1}{c^2}\mathbf{Z}_{tt}^e = -\mathbf{p},$$

$$\Delta\mathbf{Z}^m - \frac{1}{c^2}\mathbf{Z}_{tt}^m = \mathbf{m}, \tag{45.5}$$

where $\mathbf{p}$ and $\mathbf{m}$, the *Hertz source vectors*, are given vector fields. The solutions of (45.5), at time t at a point P, which correspond to zero initial conditions, are

$$\mathbf{Z}^e = \frac{1}{4\pi}\int_{r\leq ct}\frac{\mathbf{p}(\mathbf{x}, t-r/c)\, d^3x}{r},$$

$$\mathbf{Z}^m = -\frac{1}{4\pi}\int_{r\leq ct}\frac{\mathbf{m}(\mathbf{x}, t-r/c)\, d^3x}{r}, \tag{45.6}$$

where r is the magnitude of the vector $\mathbf{r}$ from the integration point $\mathbf{x}$ to P. From (45.4) and (45.5) it is clear that, for any pair of Hertz source vectors $\mathbf{p}$ and $\mathbf{m}$, the potentials (45.2) and the fields (45.3) that result from (45.6) are the solutions corresponding to the sources

$$q = -\operatorname{div} \mathbf{p}, \qquad \mathbf{j} = \mathbf{p}_t + \operatorname{curl} \mathbf{m}. \tag{45.7}$$

Of course we know that, conversely, any charge-current distribution can be represented in this form (cf. (20.2)). Indeed the Hertz source vector $\mathbf{p}$ is often referred to as a *Hertz dipole*, and the source vector $\mathbf{m}$, as a *Hertz magnetic moment*; more precisely, $\mathbf{p}$ and $\mathbf{m}$ are *densities* (cf. (20.7)–(20.8)).

A simple example is provided by the Hertz source vectors $\mathbf{m} = 0$ and

$$\mathbf{p}(\mathbf{x}, t) = \mathbf{d}(t)\delta(\mathbf{x}), \tag{45.8}$$

where $\delta(\mathbf{x})$ is the three-dimensional delta function. Since $\mathbf{p}$ is the dipole moment density and $\delta(\mathbf{x})$ is a unit density concentrated at the origin, (45.8) represents a time-dependent dipole $\mathbf{d}(t)$ at the origin. At any point with position vector $\mathbf{r}$ that, at time t, is within the horizon $r \le ct$, (45.6) give the solutions

$$\mathbf{Z}^e(\mathbf{r}, t) = \frac{1}{4\pi r}\mathbf{d}(t - r/c), \qquad \mathbf{Z}^m = 0. \tag{45.9}$$

Exercise

45.1. Let $\mathbf{n} = \mathbf{r}/r$. Prove the following identities for $\mathbf{d}(t - r/c)$:

$$\operatorname{div} \mathbf{d} = -\frac{1}{c}\mathbf{n} \cdot \dot{\mathbf{d}},$$

$$\operatorname{div} \frac{1}{r}\mathbf{d} = -\frac{1}{r^2}\mathbf{n} \cdot \mathbf{d} - \frac{1}{cr}\mathbf{n} \cdot \dot{\mathbf{d}},$$

$$\operatorname{curl} \mathbf{d} = -\frac{1}{c}\mathbf{n} \times \dot{\mathbf{d}}, \tag{45.10}$$

$$\operatorname{curl} \frac{1}{r}\mathbf{d} = -\frac{1}{r^2}\mathbf{n} \times \mathbf{d} - \frac{1}{cr}\mathbf{n} \times \dot{\mathbf{d}},$$

$$\operatorname{grad}(\mathbf{r} \cdot \mathbf{d}) = \mathbf{d} - \frac{r}{c}(\mathbf{n} \cdot \dot{\mathbf{d}})\mathbf{n},$$

where $\mathbf{d}$ and its derivatives are, of course, evaluated at the retarded time $t - r/c$. Hence prove that the electromagnetic potentials (45.2) and fields (45.3) that follow from (45.9) are:

$$\mathbf{A} = \frac{\mu_0}{4\pi r}\dot{\mathbf{d}},$$

$$V = \frac{1}{4\pi \epsilon_0}\left(\frac{1}{r^2}\mathbf{n} \cdot \mathbf{d} + \frac{1}{cr}\mathbf{n} \cdot \dot{\mathbf{d}}\right),$$

$$\mathbf{B} = -\frac{\mu_0}{4\pi}\left(\frac{1}{r^2}\mathbf{n} \times \dot{\mathbf{d}} + \frac{1}{cr}\mathbf{n} \times \ddot{\mathbf{d}}\right), \tag{45.11}$$

$$\mathbf{E} = \frac{1}{4\pi \epsilon_0}\left\{\frac{1}{r^3}[3(\mathbf{n} \cdot \mathbf{d})\mathbf{n} - \mathbf{d}] + \frac{1}{cr^2}[3(\mathbf{n} \cdot \dot{\mathbf{d}})\mathbf{n} - \dot{\mathbf{d}}]\right\}$$

$$+ \frac{\mu_0}{4\pi r}[(\mathbf{n} \cdot \ddot{\mathbf{d}})\mathbf{n} - \ddot{\mathbf{d}}].$$

Near the dipole the electric field $(45.11)_4$ is dominated by its first term, which is an electrostatic field corresponding to the instantaneous value of the dipole moment $\mathbf{d}(t)$ (since the retardation vanishes at $r = 0$). Far away from the dipole, the fields are

$$\mathbf{E} = \frac{\mu_0}{4\pi} \frac{\mathbf{n} \times [\mathbf{n} \times \ddot{\mathbf{d}}(t - r/c)]}{r},$$

$$\mathbf{B} = \frac{1}{c}\mathbf{n} \times \mathbf{E}. \tag{45.12}$$

In any given direction $\mathbf{n}$, the electric field in the far zone is an outgoing spherical wave. It is *transverse*, its direction being normal to $\mathbf{n}$, and its amplitude is angle-dependent. The magnetic field, too, is a transverse outgoing spherical wave; it is normal to both $\mathbf{n}$ and $\mathbf{E}$. The far zone is therefore also called the *wave zone*, or the *radiation zone*.

In setting up the link between electromagnetism and thermodynamics we shall introduce (in Section 54) a basic hypothesis: a body in an electromagnetic field is subject to an additional energy flux $\mathcal{E} \times \mathcal{H}$. This means that, if $\mathbf{n}$ denotes the inward pointing unit normal, the body's energy increases at the additional rate $\oint \mathcal{E} \times \mathcal{H} \cdot \mathbf{n} \, dS$, where the integral is over the surface of the body. The adjective 'additional' means that this increase is to be added to the other sources of energy increase, which are work and heat. For a body at rest $\mathcal{E} = \mathbf{E} + \mathbf{v} \times \mathbf{B} = \mathbf{E}$ and $\mathcal{H} = \mathbf{H} - \mathbf{v} \times \mathbf{D} = \mathbf{H}$, and the extra energy flux is $\mathbf{E} \times \mathbf{H}$, which is called the *Poynting vector*.

For the fields (45.12)

$$\mathbf{E} \times \mathbf{H} = \frac{\mathbf{E} \times \mathbf{B}}{\mu_0} = \frac{E^2}{\mu_0 c}\mathbf{n} = \frac{\mu_0}{4\pi} \frac{\sin^2\theta}{4\pi c r^2}\ddot{d}^2\mathbf{n}, \tag{45.13}$$

where θ is the angle between $\ddot{\mathbf{d}}$ and $\mathbf{n}$ (the polar angle measured from $\ddot{\mathbf{d}}$). If we therefore place in the far zone, where the fields are given by (45.12), a target $dS = r^2 \, d\Omega$ that is normal to $\mathbf{n}$, the amount of energy that enters it per unit time is

$$\mathbf{E} \times \mathbf{H} \cdot \mathbf{n} \, dS = \frac{1}{4\pi\epsilon_0} \frac{\ddot{d}^2 \sin^2\theta}{c^3} \frac{d\Omega}{4\pi}. \tag{45.14}$$

This may be regarded as the power radiated by the dipole into the solid angle $d\Omega$. The angular distribution of the radiation is given by the factor $\sin^2\theta$: nothing in the fore and aft directions, and maximal power in the directions perpendicular to $\ddot{\mathbf{d}}$ (Fig. 45.1).

By integrating (45.14) over all angles we obtain *Larmor's formula* for the total power:

$$P = \oint \mathbf{E} \times \mathbf{H} \cdot \mathbf{n} \, dS = \frac{1}{4\pi\epsilon_0} \frac{2}{3c^3}\ddot{d}^2. \tag{45.15}$$

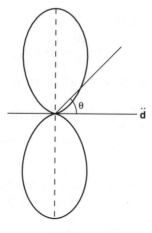

Fig. 45.1

Exercise

45.2. Carry out the analogous calculations for a magnetic dipole, represented by the Hertz source vector

$$\mathbf{m}(\mathbf{x}, t) = \mathbf{d}(t)\delta(\mathbf{x}).$$ (45.16)

Prove the identity

$$\mathbf{curl}\,(\mathbf{r} \times \mathbf{d}) = -2\mathbf{d} - \frac{r}{c}\mathbf{n} \times (\mathbf{n} \times \dot{\mathbf{d}}).$$ (45.17)

Show that the total power radiated is

$$P = \frac{\mu_0}{4\pi}\frac{2}{3c^3}\ddot{d}^2.$$ (45.18)

Problem 45.1 A small spinning body has a magnetic moment $\mathbf{d}$, inclined at an angle χ to the angular velocity $\boldsymbol{\Omega}$. If the spin-down rate $\dot{\Omega}$ is slow ($\dot{\Omega} \ll \Omega^2$), show that

$$\frac{\mu_0}{4\pi}\frac{2\Omega^3}{3c^3}d^2 \sin^2 \chi = -I\dot{\Omega},$$

where I is the moment of inertia about the axis of rotation. Such an oblique rotator provides a crude model for a pulsar[‡], and the foregoing relation is used to estimate its magnetic moment. The latter determines the magnetic field near the pulsar (Problem 40.2).

Solution We substitute

$$\dot{\mathbf{d}} = \boldsymbol{\Omega} \times \mathbf{d}, \qquad \ddot{\mathbf{d}} = \dot{\boldsymbol{\Omega}} \times \mathbf{d} + \boldsymbol{\Omega} \times (\boldsymbol{\Omega} \times \mathbf{d}) \approx \boldsymbol{\Omega} \times (\boldsymbol{\Omega} \times \mathbf{d})$$

[‡]A compact star, not usually visible, that sends out rapid radio signals at regular intervals.

(since $\dot{\Omega} \ll \Omega^2$) in the formula (45.18) for the radiated power:

$$P = \frac{\mu_0}{4\pi} \frac{2\Omega^4}{3c^3} (d \sin \chi)^2.$$

This energy is radiated at the expense of the kinetic energy $\frac{1}{2} I \Omega^2$. Hence

$$\frac{\mu_0}{4\pi} \frac{2\Omega^4}{3c^3} (d \sin \chi)^2 = -I \Omega \dot{\Omega}.$$

We now apply our results for the Hertz electric dipole to the radiation from a linear antenna, represented by the segment $-L/2 \le z \le L/2$ on the z axis. Let

$$i = i_0 \cos \frac{\pi z}{L} \sin \omega t \qquad (45.19)$$

be an alternating current flowing in the antenna. If S is the cross-section of the antenna, the law of charge conservation gives

$$\frac{1}{S} \frac{\partial i}{\partial z} = -q_t, \qquad (45.20)$$

from which we obtain the charge density:

$$q(z, t) = \frac{i_0 \pi}{S \omega L} \sin \frac{\pi z}{L} \cos \omega t. \qquad (45.21)$$

At $t = 0$, q has a maximum at $z = L/2$ and a minimum at $z = -L/2$; at $t = \pi/(2\omega)$, q vanishes everywhere; and at $t = \pi/\omega$, the distribution of q is the same, except for sign, as at $t = 0$.

The antenna has the dipole moment

$$d(t) = \int zq S \, dz = \frac{2i_0 L}{\pi \omega} \cos \omega t \qquad (45.22)$$

in the z direction. If we regard it as a point dipole, the power it radiates is, according to (45.15),

$$P = \frac{1}{4\pi \epsilon_0} \frac{2}{3c^3} \left[\frac{2i_0 \omega L}{\pi} \cos \omega (t - r/c) \right]^2. \qquad (45.23)$$

Averaged over a period $2\pi/\omega$, this becomes

$$\bar{P} = \frac{1}{4\pi \epsilon_0} \frac{4\omega^2 L^2 i_0^2}{3\pi^2 c^3}. \qquad (45.24)$$

The approximation involved in regarding the antenna as a point dipole is justified when $\omega L/(2\pi c)$, which is the length of the antenna divided by the wavelength $\lambda = 2\pi c/\omega$, is small.

46. Radiation from a moving point charge

In order to calculate the potential V for a moving point charge in accordance with
$(44.26)_2$, we first take account of the retardation by integrating with respect to t'
with the delta function $\delta(t' - t + |\mathbf{x} - \mathbf{x}'|/c)$:

$$V(\mathbf{x}, t) = \frac{1}{4\pi\epsilon_0} \int \int \frac{q(\mathbf{x}', t')\delta(t' - t + |\mathbf{x} - \mathbf{x}'|/c)\, dt'\, d^3 x'}{|\mathbf{x} - \mathbf{x}'|}. \tag{46.1}$$

In this integral we now substitute the charge density

$$q(\mathbf{x}, t) = e\delta[\mathbf{x} - \mathbf{r}_0(t)] \tag{46.2}$$

that corresponds to a point charge with charge e which is moving along the trajectory
$\mathbf{r}_0(t)$. The space integration gives

$$V(\mathbf{x}, t) = \frac{1}{4\pi\epsilon_0} \int \frac{e\delta[t' - t + |\mathbf{x} - \mathbf{r}_0(t')|/c]\, dt'}{|\mathbf{x} - \mathbf{r}_0(t')|}. \tag{46.3}$$

As before, we denote by

$$\mathbf{r}(\mathbf{x}, t) = \mathbf{x} - \mathbf{r}_0(t) \tag{46.4}$$

the vector from the position $\mathbf{r}_0(t)$ of the point charge to $\mathbf{x}$. Also, let

$$t' + \frac{r(\mathbf{x}, t')}{c} = \tau. \tag{46.5}$$

We regard this as (implicitly) determining t' as a function $t'(\mathbf{x}, \tau)$ of $\mathbf{x}$ and τ. By
differentiation at fixed $\mathbf{x}$ we obtain

$$(1 - \mathbf{n} \cdot \mathbf{v}/c)\frac{\partial t'}{\partial \tau} = 1, \tag{46.6}$$

where

$$\mathbf{n} = \frac{\mathbf{r}(\mathbf{x}, t')}{r(\mathbf{x}, t')}, \qquad \mathbf{v} = \dot{\mathbf{r}}_0(t'). \tag{46.7}$$

We can now use (46.6) to change the variable of integration in (46.3) from t' to τ:

$$V(\mathbf{x}, t) = \frac{1}{4\pi\epsilon_0} \int \frac{e\delta(\tau - t)\, d\tau}{r(\mathbf{x}, t')(1 - \mathbf{n} \cdot \mathbf{v}/c)} = \frac{1}{4\pi\epsilon_0} \frac{e}{r - \mathbf{r} \cdot \mathbf{v}/c}, \tag{46.8}$$

where the denominator in the final result is to be taken at the value $t'(\mathbf{x}, t)$ of t' that
satisfies (46.5) with $\tau = t$:

$$t' + \frac{r(\mathbf{x}, t')}{c} = t \quad \text{or} \quad r(\mathbf{x}, t') = c(t - t'). \tag{46.9}$$

The requirement that the point charge lies within the horizon $r \le ct$ of $\mathbf{x}$ at time t is satisfied when $t'(\mathbf{x}, t)$ is positive.

A similar calculation for $\mathbf{A}$, starting with $(44.26)_1$ and

$$\mathbf{j}(\mathbf{x}, t) = q(\mathbf{x}, t)\mathbf{v}(t) = e\delta[\mathbf{x} - \mathbf{r}_0(t)]\dot{\mathbf{r}}_0(t), \tag{46.10}$$

leads to the result

$$\mathbf{A}(\mathbf{x}, t) = \frac{\mu_0}{4\pi} \frac{e\mathbf{v}}{r - \mathbf{r} \cdot \mathbf{v}/c}, \tag{46.11}$$

where, again, all quantities on the right-hand side are evaluated at the 'retarded time' t' of (46.9).

The expressions (46.8) and (46.11) are called the *Liénard–Wiechert potentials*. From them we obtain the electromagnetic fields by differentiation with respect to x, y, z and t. But the Liénard–Wiechert potentials are expressed in terms of $\mathbf{x} = (x, y, z)$ and $t'(\mathbf{x}, t)$; hence we must first find the partial derivatives of t' and then use the chain rule.

Exercises

46.1. Derive the following formulae for the derivatives of $t'(\mathbf{x}, t)$:

$$\frac{\partial t'}{\partial t} = \frac{1}{1 - \mathbf{n} \cdot \mathbf{v}/c}, \qquad \mathbf{grad}\, t' = -\frac{\mathbf{n}/c}{1 - \mathbf{n} \cdot \mathbf{v}/c}. \tag{46.12}$$

46.2. Prove that the fields (44.1) for a moving point charge are:

$$\mathbf{E} = \frac{e}{4\pi\epsilon_0} \frac{1 - v^2/c^2}{r^2(1 - \mathbf{n} \cdot \mathbf{v}/c)^3}(\mathbf{n} - \mathbf{v}/c)$$

$$+ \frac{\mu_0}{4\pi} \frac{e}{r(1 - \mathbf{n} \cdot \mathbf{v}/c)^3}\mathbf{n} \times [(\mathbf{n} - \mathbf{v}/c) \times \dot{\mathbf{v}}], \tag{46.13}$$

$$\mathbf{B} = \frac{1}{c}\mathbf{n} \times \mathbf{E}.$$

We emphasize, again, that all the quantities appearing on the right-hand sides of (46.12) and (46.13) are to be taken at the retarded time t' of (46.9).

For an unaccelerated point charge, the second term of $\mathbf{E}$ vanishes, and the fields become

$$\mathbf{E} = \frac{e}{4\pi\epsilon_0} \frac{1 - v^2/c^2}{r^2(1 - \mathbf{n} \cdot \mathbf{v}/c)^3}(\mathbf{n} - \mathbf{v}/c), \qquad \mathbf{B} = \frac{1}{c}\mathbf{n} \times \mathbf{E}. \tag{46.14}$$

In the rest-frame, they reduce to

$$\mathbf{E} = \frac{e}{4\pi\epsilon_0} \frac{\mathbf{n}}{r^2}, \qquad \mathbf{B} = 0. \tag{46.15}$$

Indeed (46.14) and (46.15) are related by the Lorentz transformation formulae (12.11)–(12.12). That they do not follow from each other by applying a Galilean transformation is obvious from the appearance of $1 - v^2/c^2$ in $(46.14)_1$. The Lorentz transformation is superior to the Galilean on the basis of experimental evidence, but this knowledge is not contained in our present calculations. How then does this apparent preference for the Lorentz transformation come about? The reason is that both (46.14) and the electrostatic solution (46.15) assume the aether relations, and we have shown in Section 12 that every transformation from one aether frame to another must be a Lorentz transformation.

The contribution of the *velocity term* (46.14) to the power $\mathbf{E} \times \mathbf{H} \cdot \mathbf{n} \, dS = r^2 E^2 \, d\Omega/(\mu_0 c)$ into the solid angle $d\Omega$ vanishes as r^{-2}. We therefore consider the *acceleration term*

$$\mathbf{E} = \frac{\mu_0}{4\pi} \frac{e}{r(1 - \mathbf{n} \cdot \mathbf{v}/c)^3} \mathbf{n} \times [(\mathbf{n} - \mathbf{v}/c) \times \dot{\mathbf{v}}],$$

$$\mathbf{B} = \frac{1}{c} \mathbf{n} \times \mathbf{E}. \tag{46.16}$$

Even by itself, the acceleration term is quite complicated. We shall therefore discuss a few special cases:

(a) *For low velocities* (but arbitrary accelerations),

$$\mathbf{E} \times \mathbf{H} \cdot \mathbf{n} \, dS = \frac{e^2}{4\pi\epsilon_0} \frac{\dot{v}^2 - \dot{v}_n^2}{c^3} \frac{d\Omega}{4\pi} = \frac{e^2 \dot{v}^2}{4\pi\epsilon_0 c^3} \sin^2\theta \frac{d\Omega}{4\pi}, \tag{46.17}$$

where θ is the angle between $\mathbf{n}$ and $\dot{\mathbf{v}}$. The angular distribution of (46.17) is the same as that of electric dipole radiation (cf. (45.14)) or the radiation from a linear antenna, and is shown in Fig. 45.1.

The total power is obtained by integrating over all angles (cf. (45.15)):

$$P = \frac{1}{4\pi\epsilon_0} \frac{2e^2 \dot{v}^2}{3c^3}. \tag{46.18}$$

This formula, like (45.15)—which it resembles—is also called *Larmor's formula*.

(b) *For the case of an acceleration that is parallel to the velocity*,

$$\mathbf{E} \times \mathbf{H} \cdot \mathbf{n} \, dS = \frac{e^2 \dot{v}^2}{4\pi\epsilon_0 c^3} \frac{\sin^2\theta}{(1 - \beta\cos\theta)^6} \frac{d\Omega}{4\pi}, \tag{46.19}$$

where θ is now the angle between $\mathbf{n}$ and the velocity $\mathbf{v}$, and $\beta = v/c$. This is symmetric about $\mathbf{v}$ and vanishes in the fore and aft directions, but now the denominator causes the lobes of Fig. 45.1 to be displaced towards the forward direction (Fig. 46.1).

For relativistic speeds, when $\beta \approx 1$, the displacement is quite pronounced, and most of the radiation takes place in a narrow cone about the forward

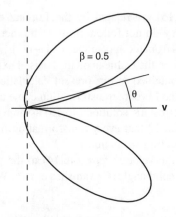

Fig. 46.1

direction. A rapidly moving electron, which is decelerated by interaction with matter, emits radiation of this kind, called *bremsstrahlung* (braking radiation).

(c) *When the acceleration is normal to the velocity*, as in circular motion,

$$\mathbf{E} \times \mathbf{H} \cdot \mathbf{n} \, dS$$

$$= \frac{e^2 \dot{v}^2}{4\pi \epsilon_0 c^3} \left[\frac{1}{(1 - \beta \cos \theta)^4} - \frac{(1 - \beta^2) \sin^2 \theta \cos^2 \phi}{(1 - \beta \cos \theta)^6} \right] \frac{d\Omega}{4\pi}, \qquad (46.20)$$

where θ is, as before, the angle between $\mathbf{n}$ and $\mathbf{v}$, and ϕ is the azimuthal angle of $\mathbf{n}$ relative to the plane formed by $\mathbf{v}$ and $\dot{\mathbf{v}}$; specifically, $v_n = v \cos \theta$ and $\dot{v}_n = \dot{v} \sin \theta \cos \phi$. This is symmetric with respect to the latter plane; and in that plane, it vanishes along the two directions for which $\cos \theta = \beta = v/c$. Since the second term in the square brackets is positive, the expression has an upper limit given by the first term. The latter has its maximum in the forward direction (for which the second term vanishes). The radiation therefore has a maximum in the forward direction. For relativistic speeds, this becomes a sharp ray. Fig. 46.2 shows the angular distribution in the plane formed by $\mathbf{v}$ and $\dot{\mathbf{v}}$.

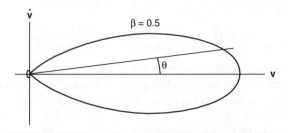

Fig. 46.2

Radiation emitted by an electron in circular motion is called *synchrotron radiation*. At relativistic speeds the sharply peaked synchrotron radiation sweeps around like a turning motorcycle's headlight.

The formulae (46.17), (46.19) and (46.20) for $\mathbf{E} \times \mathbf{H} \cdot \mathbf{n} dS$ determine the energy crossing dS per unit time. During a time interval the energy crossing dS is $\mathbf{n} dS \cdot \int dt \mathbf{E} \times \mathbf{H}$. By the remark following (46.13) the integrand $\mathbf{E} \times \mathbf{H}$ is a function of the retarded time t'. In order to effect the integration, we make the substitution $dt = dt'(\partial t'/\partial t)^{-1}$. According to (46.12)$_1$, this is equivalent to multiplying the expressions on the right-hand sides of (46.19)–(46.20) by $1 - \beta \cos \theta$, in which case they become energy rates per unit t', rather than t.

Thus, when the acceleration is parallel to the velocity, the rate per unit t' is

$$\frac{e^2 \dot{v}^2}{4\pi \epsilon_0 c^3} \frac{\sin^2 \theta}{(1 - \beta \cos \theta)^5} \frac{d\Omega}{4\pi}. \tag{46.21}$$

The total power P' radiated by the particle per unit t' is obtained by integrating over the angles $(d\Omega = \sin \theta d\theta d\phi)$:

$$P' = \frac{1}{4\pi \epsilon_0} \frac{2e^2 \dot{v}^2}{3c^3} \gamma^6, \tag{46.22}$$

where $\gamma = (1 - \beta^2)^{-1/2}$.

When the acceleration is normal to the velocity, the rate per unit t' is

$$\frac{e^2 \dot{v}^2}{4\pi \epsilon_0 c^3} \frac{(1 - \beta \cos \theta)^2 - (1 - \beta^2) \sin^2 \theta \cos^2 \phi}{(1 - \beta \cos \theta)^5} \frac{d\Omega}{4\pi}. \tag{46.23}$$

Integration over the angles yields

$$P' = \frac{1}{4\pi \epsilon_0} \frac{2e^2 \dot{v}^2}{3c^3} \gamma^4. \tag{46.24}$$

In the low-velocity approximation the multiplying factor $1 - \beta \cos \theta$ is unity, and the rate (46.17), per unit t, is indistinguishable from the rate per unit t'. Larmor's formula (46.18) is of course the common limit of (46.22) and (46.24) when $\beta = 0$ (and $\gamma = 1$).

Problem 46.1 An electron with initial velocity v_0 which would, if undisturbed, pass at a distance b from a nucleus of atomic number Z, describes a hyperbolic orbit under the attraction of the nucleus. Assuming that the motion is non-relativistic, find the total energy radiated in the encounter.

Solution In non-relativistic motion the electron radiates at the rate given by Larmor's formula (46.18). The total energy radiated will be

$$\int P \, dt = \frac{1}{4\pi \epsilon_0} \frac{2e^2}{3c^3} \int \dot{v}^2 \, dt.$$

The acceleration, determined from $m\dot{\mathbf{v}} = -Ze^2\mathbf{r}/(4\pi\epsilon_0 r^3)$, is

$$\dot{\mathbf{v}} = -\alpha\frac{\mathbf{r}}{r^3}, \qquad \alpha = \frac{Ze^2}{4\pi\epsilon_0 m}.$$

Motion under any central force is governed by the law of areas

$$r^2\dot{\phi} = h = bv_0.$$

For motion along the hyperbola

$$\frac{l}{r} = 1 + \epsilon\cos\phi$$

we obtain

$$P\,dt = \frac{1}{4\pi\epsilon_0}\frac{2e^2\alpha^2}{3c^3}\frac{dt}{r^4} = \frac{1}{4\pi\epsilon_0}\frac{2e^2\alpha^2}{3c^3l^2}(1+\epsilon\cos\phi)^2\frac{d\phi}{h}.$$

The latus rectum l, the eccentricity ϵ and the angle 2χ between the asymptotes are given by the well-known relations

$$\alpha l = h^2 = b^2v_0^2, \qquad \tan\chi = \sqrt{(\epsilon^2-1)} = \frac{l}{b} = \frac{bv_0^2}{\alpha}.$$

Hence

$$\int P\,dt = \frac{1}{4\pi\epsilon_0}\frac{2e^2\alpha^2}{3c^3l^2h}\int_{-(\pi-\chi)}^{(\pi-\chi)}(1+\epsilon\cos\phi)^2\,d\phi$$

$$= \frac{e^2}{4\pi\epsilon_0}\frac{2\alpha^4}{3c^3h^5}[(\pi-\chi)(2+\sec^2\chi)+3\tan^2\chi],$$

where ϵ has been replaced by its value $\sec\chi$. In terms of the initial velocity v_0 and the impact parameter b

$$\int P\,dt = \frac{2}{3c^3}\left(\frac{e^2}{4\pi\epsilon_0}\right)^5\frac{Z^4}{m^4b^5v_0^5}[(\pi-\chi)(2+\sec^2\chi)+3\tan^2\chi], \tag{1}$$

where χ is given by

$$\tan\chi = mv_0^2\Big/\left(\frac{Ze^2}{4\pi\epsilon_0 b}\right).$$

The last equation shows that $\tan\chi$ is (twice) the ratio of the electron's initial kinetic energy to the potential energy it would have at a distance b from the nucleus. If this is small the orbit is almost parabolic, and the square bracket in eqn (1) is approximately 3π.

Although we have not formulated a law of energy balance for radiating charged particles, we expect the foregoing calculation to become inconsistent if Z, v_0 and b are such that the radiated energy (1) exceeds $\frac{1}{2}mv_0^2$. The electron must then end up in a bound orbit of negative energy around the nucleus: it must be captured.

13

Electromagnetic wave propagation

47. Monochromatic plane waves

In Chapter 12 we treated the *generation* of electromagnetic waves, which was governed by inhomogeneous wave equations. The *propagation* of electromagnetic waves is characterized by homogeneous wave equations. This does not mean that we are going to set all charges and currents equal to zero, because that would limit us to materials without response. The homogeneity follows rather from the assumptions we shall make about the material, and from the structure of the equations.

We shall confine the discussion to electromagnetic waves that are sufficiently weak for the material response to be linear[†]. In addition to the linear relations $\mathbf{P} = \epsilon_0 \chi \mathcal{E}$ and $\mu_0 \mathcal{M} = \chi_B \mathbf{B}$ we shall also assume Ohm's law $\mathcal{J} = \sigma \mathcal{E}$. In matter which is stationary (with respect to an aether frame), we are thus led to the following system of linear equations:

$$\begin{aligned} \text{div}\, \epsilon \mathbf{E} &= q, & \mathbf{curl}\, \mathbf{H} &= \sigma \mathbf{E} + (\epsilon \mathbf{E})_t, \\ \text{div}\, \mu \mathbf{H} &= 0, & \mathbf{curl}\, \mathbf{E} &= -(\mu \mathbf{H})_t. \end{aligned} \tag{47.1}$$

We recall, from the discussion of the initial value problem in Section 24, that with the present set of equations, in which the displacement term—the time derivative $\mathbf{D}_t$—is not neglected, we must read $(47.1)_1$ from right to left. The jump conditions involving the surface charge and current densities are to be read in a similar way. The *remaining* equations and jump conditions are clearly homogeneous.

We further recall, from the discussion of the skin effect in Section 39, that the linear relation $\mathcal{J} = \sigma \mathcal{E}$ ceases to be instantaneous at high frequencies, because it cannot keep up with the rapid changes in the electromagnetic field. The same thing happens with the response relations $\mathbf{P} = \epsilon_0 \chi \mathcal{E}$ and $\mu_0 \mathcal{M} = \chi_B \mathbf{B}$. As a result, Maxwell's equations cease to be partial differential relations. We therefore resolve the fields by a Fourier expansion into monochromatic components, each of which depends on the time through the factor $e^{-i\omega t}$, and replace the instantaneous response

---

[†]The theory of waves in which non-linear response is important is called *non-linear optics*.

relations and Ohm's law by the appropriate relations for the monochromatic components (cf. Section 39). From $(47.1)_{2,4}$ we are thus led to the following equations for the monochromatic components of **E** and **H**:

$$\mathbf{curl\ H} = (\sigma - i\omega\epsilon)\mathbf{E}, \qquad \mathbf{curl\ E} = i\omega\mu\mathbf{H}. \qquad (47.2)$$

In an insulator $\sigma = 0$. Real dielectrics, however, do have finite, even if small, conductivities. It is customary to replace $\sigma - i\omega\epsilon$ by $-i\omega\epsilon$, where the new ϵ is complex. Denoting its real and imaginary parts by ϵ' and ϵ'', we have complex permittivity

$$\epsilon = \epsilon' + i\epsilon'', \qquad \epsilon'' = \sigma(\omega)/\omega, \qquad (47.3)$$

and $(47.2)_1$ becomes $\mathbf{curl\ H} = -i\omega\epsilon\mathbf{E}$. Indeed handbooks on material properties never mention the conductivity of dielectrics. Instead, they list ϵ'', or the angle $\delta = \tan^{-1}(\epsilon''/\epsilon')$, at specified frequencies.

Metals, too, can of course be described by (47.3); for them, the imaginary part of the complex 'permittivity' ϵ is large compared with the real part. Often, however, $\sigma - i\omega\epsilon$ is replaced in discussions of metals by a complex $\sigma = \sigma' + i\sigma''$, where the (normally small) imaginary part σ'' is related to the ordinary dielectric susceptibility. With this notation, $(47.2)_1$ becomes $\mathbf{curl\ H} = \sigma\mathbf{E}$.

We have already introduced, in Section 45, the assumption that a body, or any part of a body, receives in an electromagnetic field an extra energy flux $\mathcal{E} \times \mathcal{H}$. The total additional energy rate is $\oint \mathcal{E} \times \mathcal{H} \cdot \mathbf{n}\, dS$, where $\mathbf{n}$ is the unit *inward* normal. According to Gauss's theorem this is equal to $-\int \mathrm{div}\, \mathcal{E} \times \mathcal{H}\, dV$ (the minus sign is due to $\mathbf{n}$ pointing inward). In stationary matter (where $\mathcal{E} = \mathbf{E}$ and $\mathcal{H} = \mathbf{H}$) the additional rate of energy increase is therefore $-\mathrm{div}\, \mathbf{E} \times \mathbf{H}$ per unit volume. It is called the rate of *absorption* (of radiation per unit volume). Positive absorption does not necessarily mean that the material (at the point considered) is warming up: the absorption may be counteracted by cooling (heat loss) or by negative mechanical power.

In order to calculate the rate of absorption in a monochromatic field of frequency ω we use the vector identity

$$-\mathrm{div}\, \mathbf{E} \times \mathbf{H} = \mathbf{E} \cdot \mathbf{curl\ H} - \mathbf{H} \cdot \mathbf{curl\ E}$$

and substitute for the curls from (47.2). Using the notation (47.3), we have $\mathbf{curl\ H} = -i\omega\epsilon\mathbf{E}$ and $\mathbf{curl\ E} = i\omega\mu\mathbf{H}$. We shall (formally) allow for a complex magnetic permeability, $\mu = \mu' + i\mu''$. It is important to notice that, in calculating the absorption, we are forming products of quantities (e.g., **E** and **curl H**), which are in turn given by the real parts of complex numbers (e.g., **E** and $-i\omega\epsilon\mathbf{E}$), and that each one of these fluctuates with a frequency ω. If $a(t)$ and $b(t)$ are complex, and each one is proportional to $e^{-i\omega t}$, the product of their real parts is $(a + a^*)(b + b^*)/4$ (where a^* denotes the complex conjugate of a), which is a sum of four terms. Two of these fluctuate (with a frequency 2ω), and the remaining two are constant. The time average of the product is therefore $(ab^* + ba^*)/4$, which is $\Re(\frac{1}{2}ab^*)$. This result (in which the product was only assumed to satisfy the distributive law of multiplication) applies, in particular, to scalar and vector products of complex vectors. Using it, we find that, in a monochromatic field, the average rate of absorption (per unit volume) is

$$\frac{1}{2}\omega\big(\epsilon''|E|^2 + \mu''|H|^2\big). \tag{47.4}$$

We have seen that a non-zero ϵ'' is to be expected, because it is merely an alternative notation for a non-zero conductivity. It should also be noted that the real part ϵ' of the permittivity behaves quite differently from the static permittivity (which corresponds to $\omega = 0$): it is quite common for ϵ' to vanish, and even to become negative in some frequency ranges. The complex permeability μ, on the other hand, has been introduced purely formally.

In order to assess the relative roles of magnetization and polarization, we estimate the two terms of the bound current density $\mathbf{curl}\ \mathbf{M} + \mathbf{P}_t$. If ℓ is the scale over which the fields vary, then $\mathbf{curl}\ \mathbf{M}$ is of the order of $\chi_B B/(\mu_0\ell)$, and $\mathbf{P}_t$ is of the order of $\omega\epsilon_0\chi E$, or $\omega^2\epsilon_0\chi\ell B$ (since $\mathbf{curl}\ \mathbf{E} = -\mathbf{B}_t$). Noting that $c/\omega = \lambda_0/2\pi$, where λ_0 is the wavelength corresponding to the frequency ω in vacuum, we find that the ratio of $\mathbf{curl}\ \mathbf{M}$ to $\mathbf{P}_t$ is $(\lambda_0/2\pi\ell)^2(\chi_B/\chi)$. In macroscopic samples, at optical frequencies, the first factor is quite small. As for the second factor, we have no reason to suppose that χ_B at high frequencies is significantly different from the tiny, static values that it has in para- and diamagnets (less than 10^{-4})[†]. Compared with polarization, magnetization is therefore quite insignificant in optical phenomena. In what follows, we shall take μ to be real and close to μ_0. The rate of absorption will then reduce to the first term of (47.4), which in view of $(47.3)_2$ is just the average Ohmic loss (cf. Section 35). A medium in which ϵ'' vanishes is called *transparent*.

We shall now consider solutions of (47.2) that have the dependence $e^{i\mathbf{k}\cdot\mathbf{x}}$ on the coordinates $\mathbf{x} = (x, y, z)$. For any vector field $\mathbf{f}$ with this spatial dependence, $\mathrm{div}\,\mathbf{f} = i\mathbf{k}\cdot\mathbf{f}$ and $\mathbf{curl}\ \mathbf{f} = i\mathbf{k}\times\mathbf{f}$. Equations (47.2), with the notation (47.3), then become

$$\mathbf{k}\times\mathbf{H} = -\omega\epsilon\mathbf{E}, \qquad \mathbf{k}\times\mathbf{E} = \omega\mu\mathbf{H}. \tag{47.5}$$

Solutions $\mathbf{E}$ and $\mathbf{H}$ of this linear, homogeneous system will vanish unless

$$k^2 = \omega^2\epsilon\mu. \tag{47.6}$$

In vacuum, where $\sigma = 0$, $\epsilon = \epsilon_0$ and $\mu = \mu_0$, $\mathbf{k}$ is a real vector, but in a medium this need not be so, and $\mathbf{k}$ may be complex. As usual, we shall write $\mathbf{k} = \mathbf{k}' + i\mathbf{k}''$. Then (47.6) becomes

$$k'^2 - k''^2 + 2i\mathbf{k}'\cdot\mathbf{k}'' = \omega^2(\epsilon' + i\epsilon'')\mu. \tag{47.7}$$

[†] We are concerned with weak electromagnetic fields. A weak, variable field is one of the common methods of demagnetizing ferromagnets, thus reducing their permeability to $\mu \approx \mu_0$.

This may be solved for the real and imaginary parts of $\mathbf{k}$. If $\epsilon'' = 0$ (so that the medium is transparent) and $\epsilon'\mu > 0$, $\mathbf{k}$ is real and has magnitude

$$k = \omega\sqrt{\epsilon\mu} = \frac{n\omega}{c}, \qquad (47.8)$$

where $n = c\sqrt{\epsilon\mu}$ is the *refractive index* of the medium. According to (47.5), the electric and magnetic fields are in this case in a plane normal to $\mathbf{k}$ and are also perpendicular to each other. They form a plane wave with the dependence $e^{i(\mathbf{k}\cdot\mathbf{x}-\omega t)}$; and this wave has the phase velocity $\omega/k = c/n$.

If $\mathbf{k}$ is complex, $i\mathbf{k}\cdot\mathbf{x} = i\mathbf{k}'\cdot\mathbf{x} - \mathbf{k}''\cdot\mathbf{x}$, and we see from the spatial dependence $e^{i\mathbf{k}'\cdot\mathbf{x}}e^{-\mathbf{k}''\cdot\mathbf{x}}$ that the planes of constant phase are normal to $\mathbf{k}'$, whereas those of constant amplitude are normal to $\mathbf{k}''$. The fields themselves are, in general, constant on neither. Such solutions are not really plane waves; they are called *inhomogeneous plane waves*.

In the special case in which $\mathbf{k}'$ and $\mathbf{k}''$ are parallel, the solution *is* a plane wave, although a damped one. If $\mathbf{l}$ is a unit vector in the common direction of $\mathbf{k}'$ and $\mathbf{k}''$, we have $\mathbf{k} = k\mathbf{l}$, where k satisfies (47.6). We denote

$$k = (n + i\kappa)\frac{\omega}{c}, \qquad (47.9)$$

with real n and κ. They are, respectively, called the *refractive index* and the *absorption coefficient* of the medium. With this notation, (47.7) becomes

$$n^2 - \kappa^2 + 2in\kappa = c^2(\epsilon' + i\epsilon'')\mu. \qquad (47.10)$$

By separating real and imaginary parts, and solving the two resulting equations, we obtain

$$n^2 = \tfrac{1}{2}\mu c^2(\sqrt{\epsilon'^2 + \epsilon''^2} + \epsilon'),$$
$$\kappa^2 = \tfrac{1}{2}\mu c^2(\sqrt{\epsilon'^2 + \epsilon''^2} - \epsilon'). \qquad (47.11)$$

The wave, which has the dependence $e^{i(nx-ct)\omega/c}e^{-\kappa x\omega/c}$, is damped if $\kappa > 0$. According to (47.10), this will always be the case in an absorbing medium, because $\kappa \neq 0$ if $\epsilon'' \neq 0$. In metals, where ϵ' is small compared with $\epsilon'' = \sigma/\omega$,

$$n = \kappa = c\sqrt{\frac{\mu\sigma}{2\omega}} = \frac{c}{\omega\delta}, \qquad (47.12)$$

where δ is the skin depth (cf. $(39.6)_3$). Damping can also occur, *without* any absorption of radiation, in a transparent medium: if $\epsilon'' = 0$ and ϵ' is negative, (47.10) gives $n = 0$ and $\kappa = c\sqrt{-\epsilon'\mu}$.

Problem 47.1   An electromagnetic wave exerts Lorentz forces on the electrons and nuclei of matter through which it propagates. Assuming that the displacements $\mathbf{x}$ of the nuclei can be neglected on account of their greater masses, and that the electrons, which have non-relativistic velocities, are acted upon by restoring forces $-m\omega_0^2\mathbf{x}$ and damping forces $-m\gamma\dot{\mathbf{x}}$, show that the (complex) permittivity is

$$\epsilon = \epsilon_0 + \frac{n_e e^2}{m(\omega_0^2 - i\gamma\omega - \omega^2)},$$

where n_e is the number of electrons per unit volume.

Solution Since in an electromagnetic wave $E = cB$ and the motion is non-relativistic, we may neglect the magnetic part $e\dot{\mathbf{x}} \times \mathbf{B}$ of the Lorentz force. The equation of motion of an electron is

$$m\ddot{\mathbf{x}} = e\mathbf{E} - m\omega_0^2\mathbf{x} - m\gamma\dot{\mathbf{x}}.$$

The monochromatic component (varying as $e^{-i\omega t}$) of this equation is

$$m(-\omega^2 + \omega_0^2 - i\gamma\omega)\mathbf{x} = e\mathbf{E},$$

with the solution

$$\mathbf{x} = \frac{e\mathbf{E}}{m(\omega_0^2 - i\gamma\omega - \omega^2)}.$$

According to (20.7) the polarization $\mathbf{P}$ is the dipole moment per unit volume, that is, $n_e e\mathbf{x}$. The permittivity ϵ follows from $\mathbf{D} = \epsilon_0\mathbf{E} + \mathbf{P} = \epsilon\mathbf{E}$. Thus

$$\epsilon = \epsilon_0 + \frac{n_e e^2}{m(\omega_0^2 - i\gamma\omega - \omega^2)}.$$

If the electrons are *free*, we set the restoring force equal to zero. Then $\omega_0^2 = 0$, and ϵ becomes, at low frequency ω, purely imaginary:

$$\epsilon = \frac{i n_e e^2}{m\gamma\omega}.$$

According to (47.3)$_2$ this corresponds to a real conductivity

$$\sigma = \frac{n_e e^2}{m\gamma}.$$

Generally, at sufficiently high frequencies, we have

$$\epsilon = \epsilon_0(1 - \omega_p^2/\omega^2),$$

where

$$\omega_p^2 = \frac{n_e e^2}{m\epsilon_0}$$

is the square of the *plasma frequency*. Because of the factor n_e, ω_p is proportional to the square root of the density. In the ionosphere the plasma frequency is a few MHz, just above the AM radio band. Radio waves of lower frequency cannot penetrate the ionosphere, and are reflected because for them ϵ becomes negative.

Exercise

47.1. Prove that, in the case of damped plane waves in a transparent medium, the time average of the Poynting vector $\mathbf{E} \times \mathbf{H}$ vanishes.

---

48. Reflection and refraction

In a transparent medium with $\epsilon' > 0$, according to the results of the last section, a plane monochromatic wave which is propagating in the direction $\mathbf{l}$ is characterized by

$$\mathbf{k} = k\mathbf{l}, \qquad k = \omega\sqrt{\epsilon\mu} = \frac{n\omega}{c},$$

$$\mathbf{H} = \sqrt{\frac{\epsilon}{\mu}}\mathbf{l} \times \mathbf{E},$$

(48.1)

and the fields have the dependence $e^{i(\mathbf{k}\cdot\mathbf{x} - \omega t)}$ on the coordinates[†].

Consider now such a wave which impinges normally on the plane boundary between two different media, both of which are at rest (with respect to an aether frame). Since $\mathbf{l}$ is normal to the boundary, $\mathbf{E}$ and $\mathbf{H}$ in the incident wave are both parallel to the boundary. Both must therefore be continuous. Can the wave then simply proceed, without any change, through the interface into the second medium? Clearly not, because $(48.1)_3$ cannot be satisfied with the same fields if ϵ/μ is discontinuous.

We must therefore conclude that, if anything does go through into the second medium, there must be a corresponding disturbance in the first medium. In other words, *transmission* must result in *reflection*. Is it possible (conversely) for an incident wave to be reflected without anything passing through? Obviously yes, and this situation constitutes a perfect mirror.

We now turn to the more general case of oblique incidence from a transparent medium (designated 1) into a second medium (designated 2), which is not necessarily transparent. We denote the propagation vectors and frequencies of the incident, transmitted and reflected waves (respectively) by $\mathbf{k}^{(i)}, \mathbf{k}^{(t)}, \mathbf{k}^{(r)}, \omega^{(i)}, \omega^{(t)}, \omega^{(r)}$, and choose the coordinates so that the (stationary) surface of separation becomes the plane $z = 0$. On $z = 0$, the x and y components of $\mathbf{E}$ and $\mathbf{H}$ must be continuous. Since the fields have the dependence $e^{i(\mathbf{k}\cdot\mathbf{x} - \omega t)}$, these conditions can only be satisfied if the three arguments $\mathbf{k} \cdot \mathbf{x} - \omega t$ are all equal there:

$$xk_x^{(r)} + yk_y^{(r)} - \omega^{(r)}t = xk_x^{(i)} + yk_y^{(i)} - \omega^{(i)}t,$$

$$xk_x^{(t)} + yk_y^{(t)} - \omega^{(t)}t = xk_x^{(i)} + yk_y^{(i)} - \omega^{(i)}t.$$

(48.2)

[†]The expression $\sqrt{(\mu/\epsilon)}$ has the dimensions of resistance. In vacuum, it has the value $\sqrt{(\mu_0/\epsilon_0)} \approx$ 377 ohms, which is called *the impedance of free space*. More properly, it is the impedance of an aether frame (cf. (12.6)), or simply the resistance of the aether.

These equations must hold all over the boundary at all times, that is, for all x, y and t. It follows, first, that the three frequencies must be equal[†]. Similarly, we must have $k_x^{(t)} = k_x^{(r)} = k_x^{(i)}$ and $k_y^{(t)} = k_y^{(r)} = k_y^{(i)}$. We define the *plane of incidence* as the plane that contains the (real) propagation vector $\mathbf{k}^{(i)}$ of the incident wave and the normal to the surface of separation, and choose the coordinates so that this is the xz plane. By this choice $k_y^{(i)} = 0$. It follows that $k_y^{(t)}$ and $k_y^{(r)}$, too, must vanish. This means that all three waves have their propagation vectors in the same plane, and we have the situation shown in Fig. 48.1.

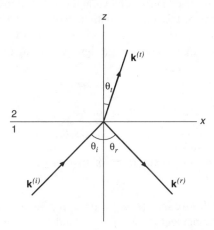

Fig. 48.1

We have not yet made any use of the equality of the x components of the $\mathbf{k}$'s. In the transparent medium 1, we have (cf. Fig. 48.1)

$$k_x^{(i)} = (\omega/c)n_1 \sin \theta_i, \qquad k_x^{(r)} = (\omega/c)n_1 \sin \theta_r, \qquad (48.3)$$

where $n_1 = c\sqrt{\epsilon_1 \mu_1}$. The equality $k_x^{(r)} = k_x^{(i)}$ gives the *law of reflection*

$$\theta_r = \theta_i. \qquad (48.4)$$

If the second medium, too, is transparent,

$$k_x^{(t)} = (\omega/c)n_2 \sin \theta_t, \qquad (48.5)$$

and the equality $k_x^{(t)} = k_x^{(i)}$ gives *Snell's law of refraction*

$$n_2 \sin \theta_t = n_1 \sin \theta_i. \qquad (48.6)$$

[†]That is because the surface of separation has been assumed stationary. If it were moving, we would have to apply the jump conditions on $z(t)$, rather than on $z = 0$; the resulting relations (called the *Doppler relations*) between the ω's would then involve the normal speed of the surface.

So far, our discussion has merely concerned itself with directions. In order to deal with the amplitudes, we must apply the jump conditions, which require the continuity of the x and y components of $\mathbf{E}$ and $\mathbf{H} = \mathbf{k} \times \mathbf{E}/(\omega\mu)$ across $z = 0$. Since the k_y's all vanish, the jump conditions are seen to fall into two groups: one, which involves E_y (the component perpendicular to the plane of incidence) and $H_x = -k_z E_y/(\omega\mu)$; and the other, involving E_x and E_z (the components in the plane of incidence) and $H_y = (k_z E_x - k_x E_z)/(\omega\mu)$. The plane formed by the propagation and electric vectors is called the *plane of polarization* of the wave. The foregoing two groups are therefore waves polarized perpendicular and parallel to the plane of incidence.

In order to implement the jump conditions we need the z components of the $\mathbf{k}$'s . Since the k_y's all vanish, $k_z^2 = k^2 - k_x^2 = \omega^2 \epsilon\mu - k_x^2$, and we obtain k_z by taking the square root with the appropriate sign. According to (48.3), together with the equality of the k_x's, we have

$$
\begin{aligned}
k_z^{(i)} &= \omega\sqrt{\epsilon_1\mu_1}\cos\theta_i = (\omega/c)n_1\cos\theta_i, \\
k_z^{(r)} &= -\omega\sqrt{\epsilon_1\mu_1}\cos\theta_i = -(\omega/c)n_1\cos\theta_i, \\
k_z^{(t)} &= \omega\sqrt{\epsilon_2\mu_2 - \epsilon_1\mu_1\sin^2\theta_i} = (\omega/c)\sqrt{c^2\epsilon_2\mu_2 - n_1^2\sin^2\theta_i}.
\end{aligned}
\tag{48.7}
$$

In the following formulae we shall set $\mu_1 = \mu_2 = \mu_0$. Taking, first, the case in which the electric field in the incident wave is perpendicular to the plane of incidence, we have $E_y^{(i)} + E_y^{(r)} = E_y^{(t)}$ and $k_z^{(i)} E_y^{(i)} + k_z^{(r)} E_y^{(r)} = k_z^{(t)} E_y^{(t)}$. The solution of these equations gives *Fresnel's formulae* (we omit the subscript y):

$$
\begin{aligned}
E^{(r)} &= \frac{n_1\cos\theta_i - \sqrt{\epsilon_2/\epsilon_0 - n_1^2\sin^2\theta_i}}{n_1\cos\theta_i + \sqrt{\epsilon_2/\epsilon_0 - n_1^2\sin^2\theta_i}} E^{(i)}, \\
E^{(t)} &= \frac{2n_1\cos\theta_i}{n_1\cos\theta_i + \sqrt{\epsilon_2/\epsilon_0 - n_1^2\sin^2\theta_i}} E^{(i)}.
\end{aligned}
\tag{48.8}
$$

If the second medium, too, is transparent, we can make use of Snell's law $n_1 \sin\theta_i = n_2 \sin\theta_t$. Fresnel's formulae then become

$$
E^{(r)} = \frac{\sin(\theta_t - \theta_i)}{\sin(\theta_t + \theta_i)} E^{(i)}, \qquad E^{(t)} = \frac{2\cos\theta_i \sin\theta_t}{\sin(\theta_t + \theta_i)} E^{(i)}.
\tag{48.9}
$$

A similar calculation can be carried out for the case in which the electric vector in the incident wave is *in* the plane of incidence. The magnetic vector is then perpendicular to this plane, and Fresnel's formulae (in terms of $H = H_y$) are

$$H^{(r)} = \frac{(\epsilon_2/\epsilon_0)\cos\theta_i - n_1\sqrt{\epsilon_2/\epsilon_0 - n_1^2\sin^2\theta_i}}{(\epsilon_2/\epsilon_0)\cos\theta_i + n_1\sqrt{\epsilon_2/\epsilon_0 - n_1^2\sin^2\theta_i}}H^{(i)},$$

$$H^{(t)} = \frac{2(\epsilon_2/\epsilon_0)\cos\theta_i}{(\epsilon_2/\epsilon_0)\cos\theta_i + n_1\sqrt{\epsilon_2/\epsilon_0 - n_1^2\sin^2\theta_i}}H^{(i)}.$$

(48.10)

If the second medium, too, is transparent, these formulae become

$$H^{(r)} = \frac{\tan(\theta_i - \theta_t)}{\tan(\theta_i + \theta_t)}H^{(i)},$$

$$H^{(t)} = \frac{\sin 2\theta_i}{\sin(\theta_t + \theta_i)\cos(\theta_i - \theta_t)}H^{(i)}.$$

(48.11)

Each of the waves is associated with an energy flux, the time average of which is $\Re(\frac{1}{2}\mathbf{E} \times \mathbf{H}^*)$. In order to obtain the z component, which gives the amount of energy crossing unit area of the surface of separation per unit time, we must multiply by $\cos\theta$. The ratio R of the normal components of the average energy flux in the reflected and incident waves is called the *reflectivity*. Similarly, the corresponding ratio T of the transmitted and incident waves is called the *transmissivity*.

We shall first calculate R and T (from (48.9) and (48.11)) for the case in which both media are transparent. For an incident wave with the electric vector perpendicular to the plane of incidence (we indicate this by the subscript $\perp$) we have

$$R_\perp = \frac{\sin^2(\theta_i - \theta_t)}{\sin^2(\theta_i + \theta_t)}, \qquad T_\perp = \frac{\sin 2\theta_i \sin 2\theta_t}{\sin^2(\theta_i + \theta_t)}.$$

(48.12)

We note that $R_\perp + T_\perp = 1$, as expected. For an incident wave with the electric vector in the plane of incidence (subscript $\parallel$), we have

$$R_\parallel = \frac{\tan^2(\theta_i - \theta_t)}{\tan^2(\theta_i + \theta_t)}, \qquad T_\parallel = \frac{\sin 2\theta_i \sin 2\theta_t}{\sin^2(\theta_i + \theta_t)\cos^2(\theta_i - \theta_t)}.$$

(48.13)

Again, $R_\parallel + T_\parallel = 1$.

For normal incidence the distinction between (48.12) and (48.13) disappears, and either of them reduces to

$$R = \frac{(n-1)^2}{(n+1)^2}, \qquad T = \frac{4n}{(n+1)^2},$$

(48.14)

where $n = n_2/n_1$. In all cases, when n_2 approaches n_1, the reflectivity goes to zero and the transmissivity goes to 1.

In the special case when $\theta_i + \theta_t = \pi/2$, that is, when the transmitted and reflected rays are perpendicular to each other, $\tan(\theta_i + \theta_t)$ becomes infinite. According to

$(48.13)_1$, $R_\parallel$ vanishes in this case. The angle of incidence for which this special case arises is given by

$$\tan \theta_i = n = \frac{n_2}{n_1}, \tag{48.15}$$

which follows from $\sin \theta_t = \sin(\pi/2 - \theta_i) = \cos \theta_i$, together with Snell's law (48.6). This θ_i is called the *Brewster angle*. Thus, when light is incident under this angle, the electric vector in the reflected wave will have no component in the plane of incidence. Such a reflected wave is said to be *totally polarized*; the Brewster angle is therefore also called the (incidence) *angle of total polarization.* .

It may be noted that, according to $(48.12)_1$ and $(48.13)_1$, $R_\parallel < R_\perp$ (except when $\theta_i = 0$ or $\pi/2$). Natural light (an equal mixture of waves with **E** in every direction perpendicular to the propagation vector) is therefore always partially polarized by reflection: the reflected wave will have its electric vector predominantly normal to the plane of incidence.

If $n_2 > n_1$, the second medium is said to be optically denser than the first medium. In this case, according to Snell's law (48.6), $\sin \theta_t < \sin \theta_i$; hence $\theta_t < \theta_i$. If, on the other hand, the second medium is optically less dense than the first, $\sin \theta_t$ becomes unity (so that the refracted wave is propagated along the surface of separation) when

$$\sin \theta_i = n = \frac{n_2}{n_1}. \tag{48.16}$$

This angle of incidence is called the angle of *total reflection*.

Exercise

48.1. Show that $R_\parallel = R_\perp = 1$ at the angle of total reflection.

If (for $n_2 < n_1$) θ_i exceeds the angle of total reflection, $k_z^{(t)}$ becomes imaginary (cf. (48.7)), and the fields in the second medium undergo damping. Since the second medium is transparent, this damping is not accompanied by absorption. It is easily verified that the Poynting vector, though not zero, has a vanishing time average. Energy therefore flows to and fro, but there is no lasting flow.

A modern application of total reflection is found in *optical fibres*. These are tubes made of highly transparent material (which is optically denser than the surroundings). Electromagnetic waves with propagation vectors making a small angle with the axis of the tube are then bounced off the walls by total reflection, that is, with no loss of energy.

The foregoing discussion of reflectivity and transmissivity has been confined to the case in which both media are transparent. If the second medium is not transparent, R and T must be calculated from (48.8) and (48.10). The general formulae are rather unwieldy. For normal incidence ($\theta_i = 0$), however, the real and imaginary parts of the complex $\mathbf{k}^{(t)}$ are both in the z direction. The wave in medium 2 then becomes homogeneous, and we can use (47.9).

Exercise

48.2. Show that, for normal incidence,

$$R = \frac{(n-1)^2 + \kappa^2}{(n+1)^2 + \kappa^2}, \tag{48.17}$$

where $n = n_2/n_1$ and $\kappa = \kappa_2/n_1$.

We have seen that, in metals, n and κ are of the order of the wavelength, divided by the skin depth (cf. (47.12)). So long as the frequency is not too high, n^2 and κ^2 are therefore large, and $R \approx 1 - 2/n$ is close to 1. At optical frequencies the simple formulae (47.12) break down, because $\omega\epsilon$ is no longer small compared with σ, but n^2 and κ^2 are usually still large. Metals are therefore good reflectors, even at optical frequencies.

Problem 48.1 Light falls normally on a slab of transparent material, which is bounded by two parallel planes. Find the total fraction of energy reflected, and the total transmitted. It is necessary to take account of the multiple reflections that take place at each boundary.

Solution First, we note that the reflection and transmission coefficients from medium 1 to medium 2 at normal incidence are the same as from medium 2 to medium 1. This is because, according to (48.14)

$$R(1/n) = \frac{(1/n - 1)^2}{(1/n + 1)^2} = \frac{(n-1)^2}{(n+1)^2} = R(n),$$

$$T(1/n) = \frac{4/n}{(1/n + 1)^2} = \frac{4n}{(n+1)^2} = T(n),$$

so that it does not matter whether n in (48.14) denotes n_2/n_1 or n_1/n_2. Thus, at normal incidence, transmission into the slab, or out of it, takes place with the same coefficient T. Similarly, at normal incidence, reflection from the slab, or from the air, takes place with the same coefficient R.

The total fraction of energy reflected from a slab includes, first of all, the fraction R that has not entered the slab. Next, the fraction TRT which has undergone transmission, reflection and transmission. Proceeding in this way, we arrive at the infinite sum

$$R + TRT + TRRRT + \cdots = R + T(R + R^3 + R^5 + \cdots)T.$$

The geometric series has the sum $R/(1 - R^2)$. Noting that $T = 1 - R$, we obtain

$$R + T\frac{R}{1 - R^2}T = R + T\frac{R}{(1 - R)(1 + R)}T$$

$$= R + \frac{R(1 - R)}{1 + R}$$

$$= \frac{2R}{1 + R}$$

$$= \frac{(n - 1)^2}{n^2 + 1}.$$

The total fraction transmitted must of course be one minus the total fraction reflected, but, as a check, we calculate it by a similar procedure. This leads to the sum

$$TT + TRRT + TRRRRT + \cdots = T(1 + R^2 + R^4 + \cdots)T$$

$$= \frac{T^2}{1 - R^2}$$

$$= \frac{1 - R}{1 + R}$$

$$= \frac{2n}{n^2 + 1}.$$

For $n = 1.5$, as in the case of a glass slab in air, the total fraction transmitted is 12/13, and the total reflected is 1/13. One can therefore check one's own appearance while looking through a shop window.

Exercises

48.3. Show that if a straight, transparent pipe is to act as an optical fibre for any light ray incident on one end, its index of refraction (relative to air) must satisfy $n^2 > 2$.

48.4. An optical fibre in the form of a cylindrical pipe of diameter d is bent into a semi-circle. Show that the least inner radius of curvature (before a possible leak develops) is $r = d/(n - 1)$.

---

49. Conditions at metallic surfaces

We shall now examine in more detail the case in which the second medium is a metal, with a conductivity σ that is large compared with $\omega\epsilon$. In this case we have (cf. (39.4))

$$\text{curl } \mathbf{H} = \sigma\mathbf{E}, \qquad \text{curl } \mathbf{E} = i\omega\mathbf{B}. \tag{49.1}$$

In a *perfect conductor* ($\sigma \to \infty$), $\mathbf{E} = \text{curl } \mathbf{H}/\sigma$ must vanish. Along with $\mathbf{E}$, any variable magnetic field ($\omega \neq 0$) must vanish as well (according to (49.1)$_2$). From the jump conditions $\mathbf{n} \times [\mathbf{E}] = 0$ and $\mathbf{n} \cdot [\mathbf{B}] = 0$ it then follows that, just outside the metal, the following boundary conditions hold:

$$\mathbf{n} \times \mathbf{E} = 0, \qquad B_n = 0. \tag{49.2}$$

The remaining jump conditions $\mathbf{n} \cdot [\mathbf{D}] = \tau$ (where τ now denotes the surface charge density, because σ is the conductivity) and $\mathbf{n} \times [\mathbf{H}] = \mathbf{K}$ determine the surface charge and current densities on the walls:

$$\tau = -D_n, \qquad \mathbf{K} = \mathbf{H} \times \mathbf{n}. \tag{49.3}$$

In these formulae $\mathbf{n}$ points *into* the metal. The boundary conditions (49.2) and the formulae (49.3) were obtained in the limit $1/\sigma \to 0$ of a perfect conductor. We shall now seek the next-order terms, assuming that $1/\sigma$, though small, does not vanish; they will actually turn out to be of order $1/\sqrt{\sigma}$. As in Section 48, we consider the case of a monochromatic wave that impinges on a metal from a transparent medium. In the metal, eqns (49.1) become

$$i\mathbf{k} \times \mathbf{H} = \sigma\mathbf{E}, \qquad \mathbf{k} \times \mathbf{E} = \omega\mu\mathbf{H}. \tag{49.4}$$

It follows that $(\mathbf{k}' + i\mathbf{k}'')^2 = i\mu\sigma\omega$, or

$$k'^2 - k''^2 + 2i\mathbf{k}' \cdot \mathbf{k}'' = i\mu\sigma\omega. \tag{49.5}$$

Taking real and imaginary parts, we have

$$k'^2 = k''^2, \qquad \mathbf{k}' \cdot \mathbf{k}'' = \frac{\mu\sigma\omega}{2} = \frac{1}{\delta^2}, \tag{49.6}$$

where δ is the skin depth (cf. (39.6)$_3$). From the relations (48.2) we obtain, as before, $k_x = k'_x + ik''_x = k_x^{(i)}$ and $k_y = k'_y + ik''_y = 0$. Substitution in (49.5) gives

$$k_x^{(i)\,2} + k_z'^{\,2} = k_z''^{\,2}, \qquad k_z' k_z'' = \frac{1}{\delta^2}. \tag{49.7}$$

These can be solved for k_z' and k_z''. According to (49.7)$_2$, k_z' and k_z'' have the same sign, which means that the wave decays as it propagates. According to (48.3)$_1$, $k_x^{(i)}$ is at most equal to the reciprocal wavelength $(\omega/c)n_1$ in medium 1. We shall assume this to be small compared with $1/\delta$. In this approximation (of small δ), the solution of (49.7) gives $k_z' = k_z'' = 1/\delta$, so that the fields inside the metal have the dependence

$$e^{-z/\delta}e^{i[x(\omega/c)n_1 \sin\theta_i + z/\delta - \omega t]}. \tag{49.8}$$

The wave inside the metal is therefore inhomogeneous: the constant-amplitude surfaces are parallel to the metal surface, but the constant-phase surfaces are not. In the limit of small δ we may neglect this inhomogeneity and take $\mathbf{k}$ to be in the z direction $\mathbf{n}$:

$$\mathbf{k} = \frac{1+i}{\delta}\mathbf{n}. \tag{49.9}$$

Substituting this in (49.4)$_1$, we obtain

$$\mathbf{E} = \frac{i-1}{\sigma\delta}\mathbf{n} \times \mathbf{H}. \tag{49.10}$$

Since $(\sigma\delta)^{-1} \propto \delta$, this shows that $\mathbf{E}$ is small and (in the present approximation) parallel to the surface of the metal. Just inside the metal (at $z = 0$), eqn (49.10) is a relation between the tangential components of $\mathbf{E}$ and $\mathbf{H}$. But these are both continuous (in the approximation of (49.1)$_1$, with finite σ, $\mathbf{n} \times \mathbf{H}$ is continuous). Hence the fields just outside the metal must obey the same relation:

$$\mathbf{E}_s = \frac{i-1}{\sigma\delta}\mathbf{n} \times \mathbf{H}_s. \tag{49.11}$$

Thus $\mathbf{E}_s$ is of order $H_s\delta$. The normal component E_n outside the metal is *not* constrained by these considerations. Nor is it continuous. According to (48.1)$_3$ it is of the same order as H_s.

The normal component B_n is continuous. Inside the metal, according to (49.5)$_2$, it is $(\mathbf{k} \times \mathbf{E})_n/\omega$, which involves the tangential components of $\mathbf{k}$ and $\mathbf{E}$. Since the tangential component of $\mathbf{k}$ is $k_x^{(i)}$, B_n is of the order of E_s, or $H_s\delta$.

According to (49.11), the time averaged flux into the metal is

$$\Re\left(\tfrac{1}{2}\mathbf{E} \times \mathbf{H}^*\right) = \frac{|H_s|^2}{2\sigma\delta}\mathbf{n}. \tag{49.12}$$

Exercises

49.1. Show that, in the metal,

$$\int_0^\infty \Re\left(\frac{\mathbf{j} \cdot \mathbf{j}^*}{2\sigma}\right) dz = \frac{|H_s|^2}{2\sigma\delta}. \tag{49.13}$$

The time averaged flow of radiation into a thick slab of metal is therefore equal to the time averaged Joule heating. In a thin slab (of thickness $\leq \delta$) some of the radiation will leave through the other side.

49.2. Show that, in the metal,

$$\int_0^\infty \mathbf{j}\, dz = \mathbf{H} \times \mathbf{n}. \tag{49.14}$$

According to (49.8), $\mathbf{j}$ is confined to a few δ near the surface. When δ is small, the left-hand side effectively becomes a surface current density.

Since E_s/E_n and H_n/H_s are each of order δ, we obtain the ideal conditions (49.2) in the limit $\delta \to 0$. The relation (49.14) reduces in this limit to (49.3)$_2$.

---

50. Waveguides

A waveguide is a long pipe with constant cross-section. We shall assume it to be hollow, with perfectly conducting walls, and seek solutions that represent waves propagating along the waveguide. Such solutions must exist, for we can see through a metal pipe.

We choose the z axis as the axis of the waveguide. Monochromatic waves that propagate along the pipe will depend on z and t through $e^{i(k_z z - \omega t)}$. For such fields the equations

$$\mathbf{curl\ H} = -i\omega\epsilon_0\mathbf{E}, \qquad \mathbf{curl\ E} = i\omega\mu_0\mathbf{H}, \tag{50.1}$$

written out in full, are

$$\frac{\partial H_z}{\partial y} - ik_z H_y = -i\omega\epsilon_0 E_x, \qquad \frac{\partial E_z}{\partial y} - ik_z E_y = i\omega\mu_0 H_x,$$

$$ik_z H_x - \frac{\partial H_z}{\partial x} = -i\omega\epsilon_0 E_y, \qquad ik_z E_x - \frac{\partial E_z}{\partial x} = i\omega\mu_0 H_y, \tag{50.2}$$

$$\frac{\partial H_y}{\partial x} - \frac{\partial H_x}{\partial y} = -i\omega\epsilon_0 E_z, \qquad \frac{\partial E_y}{\partial x} - \frac{\partial E_x}{\partial y} = i\omega\mu_0 H_z.$$

Waves that resemble plane waves propagating in the z direction will have E_z and H_z both equal to zero; they are called *transverse electromagnetic* or, briefly, TEM waves. Strictly plane waves (with **E** and **H** independent of x or y) are, of course, impossible in a waveguide, because they will not satisfy the boundary conditions on the walls. For TEM waves, $(50.2)_{1,2,3,4}$ give

$$\omega^2 = c^2 k_z^2, \qquad \mathbf{E} \cdot \mathbf{H} = 0, \tag{50.3}$$

as in plane waves. The last two members of (50.2) (with $E_z = H_z = 0$) become

$$\frac{\partial E_x}{\partial x} + \frac{\partial E_y}{\partial y} = 0, \qquad \frac{\partial E_y}{\partial x} - \frac{\partial E_x}{\partial y} = 0; \tag{50.4}$$

these equations are also satisfied by the transverse components of **H**. From $(50.4)_2$ we deduce the existence of a potential $V(x, y)$ such that $E_x = -\partial V/\partial x$ and $E_y = -\partial V/\partial y$, or $\mathbf{E} = -\mathbf{grad}_2 V$, where $\mathbf{grad}_2$ is the two-dimensional gradient, defined over the cross-section of the waveguide. The first member of (50.4) now gives $\Delta_2 V = 0$, where Δ_2 is the two-dimensional Laplacian. Thus

$$\mathbf{E} = -\mathbf{grad}_2 V, \qquad \Delta_2 V = 0. \tag{50.5}$$

The magnetic field satisfies a similar pair of equations, with a different potential U (say); according to $(50.3)_2$, the equipotential surfaces of V must be orthogonal to those of U. The boundary condition $\mathbf{E}_s = 0$ requires V to be constant on the circumference of the cross-section. Similarly, the boundary condition $H_n = 0$ requires $\partial U/\partial n$ to vanish on the circumference of the cross-section.

TEM waves are therefore obtained by solving the two-dimensional potential problem over the cross-section, and we may use the various methods we have developed in electrostatics for such problems. It is clear that, if the cross-section is simply connected, V and U must be constant everywhere. TEM waves can therefore only propagate in waveguides with multiply connected cross-sections, for example, in the space between two pipes, one of which lies inside the other, or in the space around two parallel pipes, for only then can V be equal to different constants on different parts of the circumference. The problem of TEM waves in a waveguide is essentially the two-dimensional problem of capacitance (or resistance).

Let us now turn to a different set of solutions by assuming that only one of the fields, **H** or **E**, is transverse. Such solutions are called *transverse magnetic* (TM) and *transverse electric* (TE). In a TM wave, H_z vanishes (but E_z does not), and (50.2) become

$$ik_z H_y = i\omega\epsilon_0 E_x, \qquad \frac{\partial E_z}{\partial y} - ik_z E_y = i\omega\mu_0 H_x,$$

$$ik_z H_x = -i\omega\epsilon_0 E_y, \qquad ik_z E_x - \frac{\partial E_z}{\partial x} = i\omega\mu_0 H_y, \qquad (50.6)$$

$$\frac{\partial H_y}{\partial x} - \frac{\partial H_x}{\partial y} = -i\omega\epsilon_0 E_z, \qquad \frac{\partial E_y}{\partial x} - \frac{\partial E_x}{\partial y} = 0.$$

The first four members give

$$E_x = \frac{ik_z}{\kappa^2}\frac{\partial E_z}{\partial x}, \qquad E_y = \frac{ik_z}{\kappa^2}\frac{\partial E_z}{\partial y},$$

$$H_x = -\frac{i\omega}{\kappa^2}\epsilon_0\frac{\partial E_z}{\partial y}, \qquad H_y = \frac{i\omega}{\kappa^2}\epsilon_0\frac{\partial E_z}{\partial x}, \qquad (50.7)$$

where $\kappa^2 = \omega^2/c^2 - k_z^2$. It is seen that the x and y components of **E** and **H** are determined by the x and y derivatives of E_z. The latter are the components of the two-dimensional gradient $\mathbf{grad}_2\, E_z$ of E_z. The sixth member of (50.6) is identically satisfied by the $\mathbf{E}_2$ of (50.7)$_{1,2}$. The remaining equation (50.6)$_5$ gives

$$\Delta_2 E_z + \kappa^2 E_z = 0. \qquad (50.8)$$

The boundary condition $\mathbf{E}_s = 0$ requires E_z to vanish on the walls; in particular, on the circumference of a cross section. It is easy to check (with the aid of (50.7)) that the single requirement $E_z = 0$ ensures the vanishing of both $\mathbf{E}_s$ and H_n along the walls.

For TE waves, in which E_z vanishes (but H_z does not), we obtain, instead of (50.7), the equations

$$H_x = \frac{ik_z}{\kappa^2}\frac{\partial H_z}{\partial x}, \qquad H_y = \frac{ik_z}{\kappa^2}\frac{\partial H_z}{\partial y},$$

$$E_x = \frac{i\omega}{\kappa^2}\mu_0\frac{\partial H_z}{\partial y}, \qquad E_y = -\frac{i\omega}{\kappa^2}\mu_0\frac{\partial H_z}{\partial x}. \qquad (50.9)$$

Instead of (50.8), we have

$$\Delta_2 H_z + \kappa^2 H_z = 0. \tag{50.10}$$

The boundary condition $H_n = 0$ now requires $\partial H_z/\partial n$ to vanish on the circumference of the cross-section, and this single condition ensures that TE waves will satisfy both $\mathbf{E}_s = 0$ and $H_n = 0$ on the walls.

Both TM and TE waves are therefore obtained by solving the characteristic value problem $-\Delta_2 f = \kappa^2 f$ for the two-dimensional Laplacian operator on the cross-section of the waveguide. The TM waves are given by modes that satisfy the boundary condition $f = 0$ on the circumference; the TE waves, by modes that satisfy $\partial f/\partial n = 0$. Unlike TEM waves, TM and TE waves can be propagated in waveguides with simply connected cross-sections.

In a waveguide which is filled with a transparent material, (50.1) must be replaced by

$$\mathbf{curl\ H} = -i\omega\epsilon\mathbf{E}, \qquad \mathbf{curl\ E} = i\omega\mu\mathbf{H}. \tag{50.11}$$

It is easily seen that solutions of this system are obtained from the solutions of (50.1) by the replacements

$$\mathbf{E} \to \sqrt{\frac{\epsilon}{\epsilon_0}}\mathbf{E}, \qquad \mathbf{H} \to \sqrt{\frac{\mu}{\mu_0}}\mathbf{H}, \qquad \omega \to \omega\sqrt{\frac{\epsilon\mu}{\epsilon_0\mu_0}}. \tag{50.12}$$

Exercise

50.1. By using the identity

$$\mathrm{div}_2(f\,\mathbf{grad}_2\,g) = f\Delta_2 g + \mathbf{grad}_2\,f \cdot \mathbf{grad}_2\,g,$$

prove that the TM characteristic modes, or the TE characteristic modes, which satisfy $-\Delta f_n = \kappa_n^2 f_n$, all have positive characteristic values. In fact

$$\kappa_n^2 = \frac{\int |\mathbf{grad}_2\,f_n|^2\,dA}{\int |f_n|^2\,dA}, \tag{50.13}$$

the integration being over the area of the cross-section. Show further that characteristic modes belonging to different characteristic values κ_m^2 and κ_n^2 are orthogonal in the sense that

$$\int \mathbf{grad}_2\,f_m \cdot \mathbf{grad}_2\,f_n\,dA = 0. \tag{50.14}$$

Equation (50.13) shows that the κ_n^2 are all positive, the smallest one being of the order of ℓ^{-2}, where ℓ is the diameter of the cross-section. For each of the κ_n^2, there is a relation

$$\omega^2 = c^2(k_z^2 + \kappa^2) \tag{50.15}$$

between the frequency ω and the wave number k_z of the wave. These waves have the group velocity

$$\frac{\partial \omega}{\partial k_z} = \frac{ck_z}{\sqrt{k_z^2 + \kappa^2}} = \frac{c^2 k_z}{\omega}, \tag{50.16}$$

which varies from 0 (for $k_z = 0$) to c (for $k_z \to \infty$).

Equation (50.15) shows that, for each cross-section, there is a frequency $\omega_{min} = c\kappa_{min}$ below which no waves can be propagated in a waveguide with a simply connected cross-section. This *cut-off frequency* is of the order of c/ℓ, where ℓ is the diameter of the cross-section. It explains, for example, why low-frequency radio stations cannot be received while driving through a tunnel (metal-reinforced concrete walls provide an approximation to metallic walls).

Exercise

50.2. A waveguide with a rectangular cross-section has sides a_x and a_y. Show that the characteristic modes of $-\Delta_2$ for such a rectangle are

$$\begin{aligned}
\text{TM}_{n_x n_y} : \quad & \sin \frac{n_x \pi x}{a_x} \sin \frac{n_y \pi y}{a_y}, \\
\text{TE}_{n_x n_y} : \quad & \cos \frac{n_x \pi x}{a_x} \cos \frac{n_y \pi y}{a_y},
\end{aligned} \tag{50.17}$$

where n_x and n_y are integers; both of them non-zero for the TM modes, and at least one of them non-zero for the TE modes. Show that, for either type,

$$\kappa^2 = \pi^2 \left(\frac{n_x^2}{a_x^2} + \frac{n_y^2}{a_y^2} \right). \tag{50.18}$$

Deduce that the smallest characteristic value (corresponding to a TE mode) is π^2, divided by the larger of a_x^2 and a_y^2.

It is a straightforward matter to calculate the z component of the time averaged energy flux $\Re(\frac{1}{2}\mathbf{E} \times \mathbf{H}^*)$ for the various modes. Integration over the cross-section then yields the time averaged rates of energy flow through the cross-section. The results are:

$$\text{TEM:} \quad \frac{c}{2} \int \epsilon_0 |\,\mathbf{grad}_2\, V|^2 \, dA,$$

$$\text{TM:} \quad \frac{\omega k_z}{2\kappa^2} \int \epsilon_0 |E_z|^2 \, dA, \qquad (50.19)$$

$$\text{TE:} \quad \frac{\omega k_z}{2\kappa^2} \int \mu_0 |H_z|^2 \, dA.$$

Our formulae were obtained on the assumption that the walls were perfectly conducting. Actual metallic walls have finite conductivity and will absorb some of the radiation. To the first order in the skin depth of the walls, we may calculate the rate of absorption by substituting the foregoing ideal solutions in (49.12). In this approximation, the walls will absorb energy, per unit length of the waveguide, at the rate $(2\sigma\delta)^{-1} \oint |H_s|^2 \, dl$, where the line integral is taken along the circumference of the cross-section.

Exercise

50.3. Show that, per unit length of the waveguide, the rates of absorption are

$$\text{TEM:} \quad \frac{1}{2\mu_0\sigma\delta} \oint \epsilon_0 |\,\mathbf{grad}_2\, V|^2 \, dl,$$

$$\text{TM:} \quad \frac{\omega^2}{2\mu_0\sigma\delta c^2 \kappa^4} \oint \epsilon_0 |\,\mathbf{grad}_2\, E_z|^2 \, dl, \qquad (50.20)$$

$$\text{TE:} \quad \frac{1}{2\mu_0\sigma\delta} \oint \mu_0 \left(|H_z|^2 + \frac{k_z^2}{\kappa^4} |\,\mathbf{grad}_2\, H_z|^2 \right) dl.$$

Absorption by the walls will cause the energy flow through the cross-section of the waveguide to decay as $e^{-\alpha z}$, where—for each mode—the *attenuation coefficient α* is the ratio of the appropriate expression in (50.20) to the corresponding expression in (50.19). The field amplitudes, too, will decay exponentially (with a decay factor per unit length equal to $\alpha/2$).

Waveguides are used for transmitting electromagnetic radiation; for example, from a device that generates the radiation to an antenna. Despite the exponential decay of the energy flow, transmission of radiation along waveguides may—over not too long distances—be far more efficient than the alternative transmission through free space: radiation emanating from a source propagates as a spherical wave, and its intensity decreases as the square of the distance; moreover, the collecting device may be a target that subtends a minute solid angle at the source.

A waveguide need not always have the form of a pipe. There may be openings in the walls. Consider, for example, a pair of parallel, perfectly conducting planes. These provide a waveguide that allows the propagation of a TEM wave in the form of a plane wave, polarized normally to the planes. Practical designs may take the form of two parallel strips, supported by a transparent dielectric slab between them; in this case we must apply the transformations (50.12). An arrangement that allows several such strip lines uses a common ground plane and is shown in Fig. 50.1.

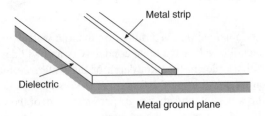

Fig. 50.1

If we close the ends of a waveguide, we get a *resonator*. The added boundary conditions at the closed ends will then select a discrete set of wave numbers k_z (and a corresponding discrete set of frequencies ω). It is easily seen that the allowed k_z's are determined by the condition that the distance between the closed ends be an integral number of half-wavelengths. The effect is similar to that of clamping a taut string: instead of travelling along a waveguide, the waves in a resonator are *standing electromagnetic waves*.

We may also join the two ends of a waveguide, rather than close each one of them; that would result in a ring-shaped resonator. Again, a resonator need not be closed on all sides. Figure 50.2 shows a circular strip-line resonator.

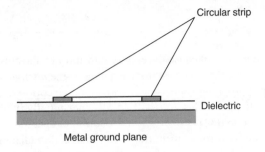

Fig. 50.2

14

A review of continuum mechanics

In the first seven chapters of this book we have laid down three principles which have led us to various relations between the electromagnetic fields, culminating in Maxwell's equations for polarizable and magnetizable materials. But these equations did not tell us what the electromagnetic fields were supposed to do.

In the case of charged particles we introduced a further axiom—the Lorentz force—which provided the link between electromagnetism and the mechanics of material points. But in dealing with macroscopic bodies we had to resort to some *ad hoc* assumptions about the dynamical and thermal effects which the electromagnetic field exerts on such bodies. The formulae for the stress on a charged conductor, the force on a current-carrying wire, the Joule heating and the radiative energy flux were all assumptions of this kind. Each one of them was introduced with a promise of later justification. We must now fulfil those promises, and we shall do so by creating a coherent link between electromagnetism and the mechanics of macroscopic, or space filling, bodies.

The motion of macroscopic bodies—or continuous media—is subject to the laws of mechanics, but their known irreversible behaviour is governed by thermodynamic laws, which cannot be deduced from mechanics. Although we are used to thinking of bodies as made up of particles, the fact is that they do *not* behave as mere assemblies of particles that are governed by an action principle.

Thus, in the case of macroscopic bodies, we must set up a link between three disciplines: electromagnetism, mechanics and thermodynamics. The precise way in which this link is set up is far from trivial. It must be done with great care and therefore requires some patience.

We begin with a review of continuum mechanics. In this classical theory all tensors will be three-dimensional Cartesian tensors. These are tensors with respect to orthogonal space transformations only, and for them the difference between covariance and contravariance disappears (cf. the discussion in last paragraph of Section 3). Hence subscripts suffice.

51. The laws of continuum mechanics

In the mechanics of mass points the trajectory of the ith mass point is given by specifying its position as a function of the time, viz. $\mathbf{x}_i(t)$. In a continuous body each material point has a trajectory, too, but the material points cannot be labelled by an integer. A method of labelling the material points of a continuous body was devised by Euler (Fig. 51.1). Let $\mathbf{X} = (X_1, X_2, X_3)$ be the position of a material point at $t = 0$. Then the triad (X_1, X_2, X_3) can serve as a label for this material point, and its trajectory given by $\mathbf{x}(\mathbf{X}, t)$.

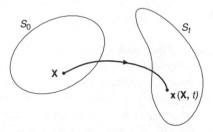

Fig. 51.1

It is not difficult to suggest other methods of labelling: there is nothing to prevent us from choosing a time different from $t = 0$; we can use any other triad (Y_1, Y_2, Y_3) related to (X_1, X_2, X_3) by a one-to-one mapping; and so on. But for our purposes, Euler's method will do. Given the vector function $\mathbf{x}(\mathbf{X}, t)$, we obtain the trajectory of a material point $\mathbf{X}$ by keeping $\mathbf{X}$ fixed and letting t vary. Its velocity and acceleration will be given by the partial time derivatives

$$\dot{\mathbf{x}} = \frac{\partial \mathbf{x}(\mathbf{X}, t)}{\partial t}, \qquad \ddot{\mathbf{x}} = \frac{\partial^2 \mathbf{x}(\mathbf{X}, t)}{\partial t^2}. \tag{51.1}$$

We can also fix t and vary $\mathbf{X}$ in $\mathbf{x}(\mathbf{X}, t)$. This gives us the positions of the various material points $\mathbf{X}$ at time t—a *picture* of the body at time t. The nine partial derivatives,

$$F_{ij}(\mathbf{X}, t) = \frac{\partial x_i(\mathbf{X}, t)}{\partial X_j}, \tag{51.2}$$

form the *deformation tensor*, or *matrix*. Its determinant is the Jacobian of the transformation from $\mathbf{X}$ to $\mathbf{x}$, and we shall assume that it never vanishes. This ensures the invertibility of $\mathbf{x}(\mathbf{X}, t)$: given $\mathbf{x}$ in the body, there exists one material point $\mathbf{X}$ which is there at time t. In Euler's method $\det F = 1$ at $t = 0$.

The mass m of a body is assumed to be a continuous function of its volume (cf. (1.1)):

$$m = \int \rho \, dV. \tag{51.3}$$

Of course, the density ρ is positive[†].

The first principle of continuum mechanics is the conservation of mass. Thus the integral in (51.3) is conserved, but since the body may be moving and deforming, the limits of the integral may be changing, and it does not follow that ρ is constant. We can, however, change to an integral over the X_i by writing $\int \rho \, d^3x = \int \rho \det F \, d^3X$. Now the limits are constant, and the law of mass conservation, in local form, is

$$\rho(\mathbf{X}, t) \det F(\mathbf{X}, t) = \rho_0(\mathbf{X}). \tag{51.4}$$

This is Euler's equation of continuity.

Exercises

51.1. Let $A(t)$ be a non-singular matrix with elements that are functions of a variable t, and let $\dot{A}(t)$ denote the matrix with elements that are the derivatives, with respect to t, of the elements of A. Prove the formula for the derivative of the determinant:

$$\frac{d}{dt}(\det A) = (\det A) \operatorname{tr} A^{-1}\dot{A},$$

where $\operatorname{tr} A^{-1}\dot{A}$ is the trace of the matrix product $A^{-1}\dot{A}$.

51.2. By differentiating (51.4) and using the result of the preceding exercise, prove that

$$\dot{\rho} + \rho \operatorname{tr} \dot{F} F^{-1} = 0. \tag{51.5}$$

Besides the *material description* of any quantity f as a function $A(\mathbf{X}, t)$ of $\mathbf{X}$ and t, there is the *spatial description*, in which the same quantity is regarded as a function $a(\mathbf{x}, t)$ of position $\mathbf{x}$ and t.[‡] The connection between them is, of course, that at time t the position $\mathbf{x}$ is one occupied by a material point $\mathbf{X}$ such that $\mathbf{x} = \mathbf{x}(\mathbf{X}, t)$. The material time derivative

$$\dot{f} = \frac{\partial A(\mathbf{X}, t)}{\partial t} \tag{51.6}$$

[†] As in the case of electric charge, the assumption (51.3) is easily relaxed. It is only for reasons of economy that we do not mention distributions of mass along lines, surfaces, etc.

[‡] The material description, introduced by Euler, is usually called 'Lagrangian'; the spatial description, due to d'Alembert, is usually called 'Eulerian'.

is connected with the partial derivatives of the spatial function $a(\mathbf{x}, t)$: if we use the chain rule in $A(\mathbf{X}, t) = a(\mathbf{x}(\mathbf{X}, t), t)$ we obtain $\dot{f} = \dot{\mathbf{x}} \cdot \mathbf{grad}\, a + a_t$. In Euler's notation,

$$\dot{f} = f_t + \dot{\mathbf{x}} \cdot \mathbf{grad}\, f. \tag{51.7}$$

Of particular interest are the nine space derivatives of the velocity in the spatial description:

$$(\mathrm{grad}\,\dot{\mathbf{x}})_{ij} = \frac{\partial \dot{x}_i(\mathbf{x}, t)}{\partial x_j}; \tag{51.8}$$

this tensor, or matrix, is called the *velocity gradient*. Its trace is div $\dot{\mathbf{x}}$.

Exercises

51.3. Prove that

$$\ddot{\mathbf{x}} = \dot{\mathbf{x}}_t + (\mathrm{grad}\,\dot{\mathbf{x}})\dot{\mathbf{x}}, \tag{51.9}$$

where the last term is the vector obtained by multiplying the matrix grad $\dot{\mathbf{x}}$ by the vector $\dot{\mathbf{x}}$.

51.4. Prove that

$$\mathrm{grad}\,\dot{\mathbf{x}} = \dot{F} F^{-1}. \tag{51.10}$$

From (51.5) and (51.10) we obtain

$$\dot{\rho} + \rho\, \mathrm{div}\,\dot{\mathbf{x}} = 0, \tag{51.11}$$

another Euler equation of continuity expressing mass conservation. Equation (51.4) is of course a first integral of (51.11).

Exercise

51.5. Deduce the equation of continuity in the spatial description:

$$\rho_t + \mathrm{div}\,\rho\dot{\mathbf{x}} = 0. \tag{51.12}$$

We have expressed the global mass conservation law in various local forms. These are of course valid only where ρ is smooth. On a surface across which ρ is discontinuous, mass conservation requires the jump condition

$$[\![\rho(\dot{\mathbf{x}} - \mathbf{v})]\!] \cdot \mathbf{n} = 0, \tag{51.13}$$

where **v** is the velocity of the surface. This jump condition expresses the inability of a surface of discontinuity to store mass. It equates the mass flux arriving at the surface with the mass flux leaving it.

The laws of mechanics require for their formulation the concepts of momentum, force, angular momentum (or moment of momentum) and torque. These are defined, respectively, by

$$\mathbf{G} = \int \dot{\mathbf{x}} \, dm,$$

$$\mathbf{F} = \oint \mathbf{t} \, dS + \int \mathbf{b} \, dm,$$

$$\mathbf{L}_O = \int (\mathbf{x} - \mathbf{x}_O) \times \dot{\mathbf{x}} \, dm,$$

$$\mathbf{M}_O = \oint (\mathbf{x} - \mathbf{x}_O) \times \mathbf{t} \, dS + \int (\mathbf{x} - \mathbf{x}_O) \times \mathbf{b} \, dm.$$

(51.14)

Note that the force is assumed to consist of two kinds: a *surface*, or *contact*, force **t** per unit area of the surface of the body; and a *body* force **b** per unit mass, which is the same as $\rho\mathbf{b}$ per unit volume. As an example of a surface force, we cite the inward pointing pressure force $-p\mathbf{n} \, dS$ on a surface element dS, where p (the pressure) is the force per unit area. Of course $\mathbf{t} \, dS$ need not in general be normal to the surface. A common example of a body force is provided by gravity; then **b** is simply the acceleration of gravity. The angular momentum and torque are defined with respect to a point O with position $\mathbf{x}_O$.

We now assume the existence of a frame, called inertial, in which for a fixed point O

$$\dot{\mathbf{G}} = \mathbf{F}, \qquad \dot{\mathbf{L}}_O = \mathbf{M}_O.$$

(51.15)

These are Euler's laws of mechanics[†].

Exercises

51.6. Use $(51.15)_1$ to show that the forces exerted by any two bodies on each other are equal and opposite (whether or not there are any other forces acting on these bodies).

51.7. Use both of Euler's laws to show that the line of action of the force that a *mass point* exerts on *another* is the line joining them.

[†]These were announced by him in the 1770s, towards the end of his life. The law $m\ddot{\mathbf{x}} = \mathbf{F}$ was also announced by Euler in 1750, in a work entitled *Découverte d'un nouveau principe de mécanique*, but he later realized that it was insufficient. In those days Newton's writings were still read, and it occurred to no one that Euler's $m\ddot{\mathbf{x}} = \mathbf{F}$ was contained in, or anticipated by, Newton's second law. Indeed this equation is nowhere to be found in Newton's writings.

If we substitute (51.14) in Euler's laws, we obtain

$$\frac{d}{dt} \int \dot{\mathbf{x}} \, dm = \oint \mathbf{t} \, dS + \int \mathbf{b} \, dm,$$

$$\frac{d}{dt} \int (\mathbf{x} - \mathbf{x}_O) \times \dot{\mathbf{x}} \, dm = \oint (\mathbf{x} - \mathbf{x}_O) \times \mathbf{t} \, dS + \int (\mathbf{x} - \mathbf{x}_O) \times \mathbf{b} \, dm. \tag{51.16}$$

These are vector equations. The ith Cartesian component of $(51.16)_1$,

$$\frac{d}{dt} \int \dot{x}_i \, dm = \oint t_i \, dS + \int b_i \, dm,$$

is a conservation law for the ith component of the momentum. By Cauchy's theorem (cf. (1.3)) it follows that, whenever t_i is smooth, it must be a linear, homogeneous function (with three coefficients) of the Cartesian components of the normal $\mathbf{n}$ to the surface. Since this holds for each one of the components t_i, we conclude that there exist nine functions T_{ij}—the components of the *stress tensor*—such that, in matrix notation,

$$\mathbf{t} = T\mathbf{n}. \tag{51.17}$$

As in Section 7, we shall assume all quantities to be piecewise smooth. It then follows that, wherever the quantities are smooth, the integral law $(51.16)_1$ is equivalent to Cauchy's first equation,

$$\rho\ddot{\mathbf{x}} = \mathbf{div}\, T + \rho\mathbf{b}, \tag{51.18}$$

where $\mathbf{div}\, T$ denotes the vector whose ith component is $(\mathbf{div}\, T)_i = \partial T_{ij}/\partial x_j$ (cf. (4.20)). Across a surface of discontinuity whose velocity is $\mathbf{v}$, the integral law leads to the jump conditions

$$[\![\rho\dot{\mathbf{x}} \otimes (\dot{\mathbf{x}} - \mathbf{v}) - T]\!]\mathbf{n} = 0. \tag{51.19}$$

Equation (51.19) is also written in matrix notation and has three components. In addition, we have used the notation for the direct, or outer, product:

$$A = \mathbf{b} \otimes \mathbf{c} \qquad \text{means} \qquad A_{ij} = b_i c_j. \tag{51.20}$$

It should perhaps be emphasized that (51.19) is not a condition for equilibrium. It is a consequence of the inability of a massless surface of discontinuity to store momentum, and it involves the normal velocity v_n of the surface, but not its acceleration. If the material does not cross the surface, that is whenever $\dot{x}_n = v_n$, the stress $\mathbf{t} = T\mathbf{n}$ must be continuous.

Euler's second law $(51.16)_2$ can now be shown to lead to Cauchy's second equation,

$$T^T = T, \tag{51.21}$$

where T^T is the transpose of T. Thus T is symmetric.

Exercise

51.8. Derive eqn (51.21).

---

52. The laws of thermodynamics

We now turn to the principles of thermodynamics. These require some additional primitives. The first is the *energy E*, which we assume, as usual, to be a continuous function of the mass, or of the volume, of a body[†]:

$$E = \int \varepsilon \, dm = \int \rho \varepsilon \, dV. \tag{52.1}$$

We warn against confusing this primitive with a first integral of the laws of mechanics. Such a first integral, even if called energy, would be a derived quantity, not a primitive. (We trust that the energy E will not easily be confused with the electric field **E**.)

Another primitive is the *heating Q* of a body, which we shall presently assume to contribute to the rate of change of its energy. We refrain from saying that Q is the rate at which a body receives heat, because Q is not assumed to be the time derivative of anything. Heating may generally be taken to be a combination of *surface*, or *contact*, heating β per unit area and *body* heating h per unit mass (or ρh per unit volume):

$$Q = \oint \beta \, dS + \int h \, dm. \tag{52.2}$$

In interpreting the theory, we might think of the first term as describing conduction of heat through the boundary of the body, and of the second as representing a distribution of heat sources inside the body, perhaps through the absorption of some kind of radiation. Heating may be negative, in which case it is called *cooling*. It may also vanish, and if it does so over a time interval, the body is said to undergo an *adiabatic* process.

The third concept we need—not a primitive—is the power of the forces acting on the body:

$$\Pi = \oint \dot{\mathbf{x}} \cdot T\mathbf{n} \, dS + \int \dot{\mathbf{x}} \cdot \mathbf{b} \, dm. \tag{52.3}$$

The first principle of thermodynamics states that

$$\dot{E} = \Pi + Q. \tag{52.4}$$

[†]This assumption may need to be relaxed. Even for a volume-filling body, a surface integral should be added to (52.1) if surface tension is important.

If we substitute (52.1)–(52.3), this becomes

$$\frac{d}{dt} \int \varepsilon \, dm = \oint \dot{\mathbf{x}} \cdot T\mathbf{n} \, dS + \int \dot{\mathbf{x}} \cdot \mathbf{b} \, dm + \oint \beta \, dS + \int h \, dm. \qquad (52.5)$$

By Cauchy's theorem β must be of the form $\beta = -\mathbf{q} \cdot \mathbf{n} = -q_n$, where $\mathbf{q}$ is called the *heat flux*; the sign is conventional. (The use of the same letter for the heat flux vector $\mathbf{q}$ and for the electric charge density q is unfortunate, but we regard the danger of confusing the two as unlikely.) Thus

$$\frac{d}{dt} \int \varepsilon \, dm = \oint \dot{\mathbf{x}} \cdot T\mathbf{n} \, dS + \int \dot{\mathbf{x}} \cdot \mathbf{b} \, dm - \oint q_n \, dS + \int h \, dm. \qquad (52.6)$$

This statement of the first principle, or axiom, of thermodynamics is also (appropriately) called *the law of energy balance*. It leads, as usual, to the local law

$$\rho \dot{\varepsilon} = \text{div } T^T \dot{\mathbf{x}} + \rho \dot{\mathbf{x}} \cdot \mathbf{b} - \text{div } \mathbf{q} + \rho h, \qquad (52.7)$$

and to the jump condition

$$[\![\rho\varepsilon(\dot{\mathbf{x}} - \mathbf{v}) - T^T\dot{\mathbf{x}} + \mathbf{q}]\!] \cdot \mathbf{n} = 0. \qquad (52.8)$$

Two further primitives are needed for the formulation of the second principle of thermodynamics. The first is *temperature* ϑ, a property we associate with the material points of a body. Like density, it has a material description $\vartheta = A(\mathbf{X}, t)$ or a spatial description $\vartheta = a(\mathbf{x}, t)$, the latter being equal to the former for the material point $\mathbf{X}$ passing through $\mathbf{x}$ at time t. It is meant to describe the hotness (or coldness) of material points. We shall further assume that temperature has a universal lower bound below which no body can ever be cooled. Of course, the assumption of a universal lower bound is really an axiom. It is analogous to the axiom, which we seldom bother to mention, that mass is always positive, in all frames and in every system of units. If we measure ϑ from this lower bound, as we shall, it becomes positive[†].

The second primitive is the *entropy S* of a body. We assume that, like energy, it is a continuous function of mass[‡]:

$$S = \int s \, dm, \qquad (52.9)$$

where s is the entropy per unit mass, or the *specific entropy*.

The second principle of thermodynamics states that the entropy increases at least as rapidly as the sum of heatings, each divided by the temperature at which it takes

[†]In the classical kinetic theory of gases the foregoing analogy becomes a theorem: temperature is there defined to be the average kinetic energy of a molecule; it is then positive *because* mass is positive.

[‡]Again, this may need to be relaxed. Cf. the first footnote of this section.

place:

$$\dot{S} = \frac{d}{dt} \int s \, dm \geq \int \frac{h}{\vartheta} \, dm - \oint \frac{q_n}{\vartheta} \, dA. \qquad (52.10)$$

(In the last integral on the right-hand side the element of area is denoted by dA, because dS might now be regarded as an increment of entropy.) Of course, $\dot{S}$ need not be positive. If the inequality in (52.10) becomes an equality throughout a time interval, we say that the body is undergoing a *reversible process* during that time interval. A process that is not reversible is *irreversible*.

For piecewise smooth functions the integral inequality (52.10) reduces to the local inequality

$$\rho \dot{s} \geq \rho h/\vartheta - \operatorname{div}(\mathbf{q}/\vartheta), \qquad (52.11)$$

and to the jump inequality

$$[\![\rho s (\dot{\mathbf{x}} - \mathbf{v}) + \mathbf{q}/\vartheta]\!] \cdot \mathbf{n} \geq 0, \qquad (52.12)$$

which is—like all the other jump conditions—subject to the prescription following (7.1): having chosen the unit normal $\mathbf{n}$ to the surface (in one of the two possible ways), the side to which $\mathbf{n}$ points is the positive one, and the jump of any quantity f is $[\![f]\!] = f_+ - f_-$. The product of $[\![f]\!]$ and $\mathbf{n}$ is then independent of the choice of $\mathbf{n}$. The energy jump condition (52.8) and the entropy jump condition (52.12), respectively, express the inability of a surface of discontinuity to store energy or entropy.

While the law of energy balance, like the laws of mechanics, is assumed to hold everywhere and at all times in all bodies, the second principle of thermodynamics does not enjoy the same status. Bodies may exist that have no unique temperature, or even no temperature at all. They lie outside the scope of thermodynamics. From now on, when we speak of ϑ, the qualifying phrase 'if it exists' will be understood.

When we say that a body has a temperature ϑ, rather than a temperature *field* $\vartheta(\mathbf{x}, t)$, we imply that its temperature is uniform, and is at most a function of time. For such bodies, which almost exhaust the subject matter of treatises on classical thermodynamics, the temperature can be taken outside the integrals of (52.10), and the second principle reduces to the Clausius–Planck inequality $\dot{S} \geq Q/\vartheta$.[†]

According to the second principle of thermodynamics (52.10) the dimensions of the product ϑS are those of energy. If temperature is measured in units of energy, entropy is dimensionless. If ϑ is measured in *degrees*, the dimensions of entropy are those of energy per degree, for example joule K^{-1}.

[†]The history of the mathematical statement of the second principle of thermodynamics spans an entire century. Clausius (1862) stated the inequality $0 \geq \oint (Q/\vartheta) \, dt$ for cyclic process. Planck (1887) gave it essentialy in the form $\Delta S \geq \int (Q/\vartheta) \, dt$. Duhem (1901) extended the Clausius–Planck inequality to non-uniform temperature, and wrote it in the form $\dot{S} \geq - \oint (q_n/\vartheta) \, dA$. The inequality (52.10), which includes body heating, is due to Truesdell and Toupin (1960).

53. The method of Coleman and Noll

We can eliminate the body force **b** from the energy balance equation (52.7) by substituting for it from Cauchy's first equation (51.18). This gives

$$\rho\dot{\varepsilon} = \operatorname{div} T^T \dot{\mathbf{x}} + \dot{\mathbf{x}} \cdot (\rho\ddot{\mathbf{x}} - \operatorname{\mathbf{div}} T) + \rho h - \operatorname{div} \mathbf{q}$$
$$= T_{ij}(\operatorname{grad}\dot{\mathbf{x}})_{ij} + \rho(\tfrac{1}{2}\dot{x}^2)^{\cdot} + \rho h - \operatorname{div} \mathbf{q}. \tag{53.1}$$

It is convenient to define, for any two matrices A and B, a *scalar product* by

$$A \cdot B = \operatorname{tr} A^T B = A_{ij} B_{ij}. \tag{53.2}$$

It can be shown that the name is justified, because this product has all the properties of a scalar product. For example, $B \cdot A = A \cdot B$; if $A \cdot A = 0$ then $A = 0$; if $A \cdot X = 0$ for all X, then $A = 0$. If I denotes the unit matrix, with $I_{ij} = \delta_{ij}$, then $A \cdot I$ is the trace of A. For example, $I \cdot \operatorname{grad}\dot{\mathbf{x}} = \operatorname{tr} \operatorname{grad}\dot{\mathbf{x}} = \operatorname{div}\dot{\mathbf{x}}$.

The difference $\varepsilon - \tfrac{1}{2}\dot{x}^2$, in which the specific kinetic energy is subtracted from the total specific energy, is called the *specific internal energy*, and denoted by u. With this notation, and with the notation we have just introduced for the scalar product of matrices, we can write (53.1) in the form

$$\rho\dot{u} = T \cdot \operatorname{grad}\dot{\mathbf{x}} + \rho h - \operatorname{div} \mathbf{q}. \tag{53.3}$$

Next, we substitute h from this equation into the second law of thermodynamics (52.11):

$$\rho\dot{s} \geq [\rho\dot{u} - T \cdot \operatorname{grad}\dot{\mathbf{x}} + \operatorname{div}\mathbf{q}]/\vartheta - \operatorname{div}(\mathbf{q}/\vartheta). \tag{53.4}$$

This may be multiplied by ϑ (since ϑ is positive) and written as

$$\rho(\vartheta\dot{s} - \dot{u}) + T \cdot \operatorname{grad}\dot{\mathbf{x}} - (\mathbf{q} \cdot \operatorname{\mathbf{grad}}\vartheta)/\vartheta \geq 0. \tag{53.5}$$

We shall refer to (53.5) as the *entropy inequality*.

Let us now assemble the laws that govern the behaviour of continuous bodies. For reasons that will become clear later, we list only the local equations, leaving out the jump conditions for the present:

$$\dot{\rho} + \rho\operatorname{div}\dot{\mathbf{x}} = 0, \tag{53.6}$$
$$\rho\ddot{\mathbf{x}} = \operatorname{\mathbf{div}} T + \rho\mathbf{b}, \tag{53.7}$$
$$T_{[ij]} = 0, \tag{53.8}$$
$$\rho\dot{u} = T \cdot \operatorname{grad}\dot{\mathbf{x}} + \rho h - \operatorname{div}\mathbf{q}, \tag{53.9}$$
$$\rho(\vartheta\dot{s} - \dot{u}) + T \cdot \operatorname{grad}\dot{\mathbf{x}} - (\mathbf{q} \cdot \operatorname{\mathbf{grad}}\vartheta)/\vartheta \geq 0. \tag{53.10}$$

These laws are obviously under-determined and must be supplemented by further information. This information takes the form of *constitutive relations* for the various material properties. We have, in fact, already met examples of constitutive relations in the response functions of linear dielectrics we have introduced in Sections 22–23. We shall now have to consider constitutive relations for various other quantities that, in the present context of the system (53.6)–(53.10), we regard as determined by material properties, namely the eleven quantities

$$T, \quad u, \quad \mathbf{q} \quad \text{and} \quad s. \tag{53.11}$$

(According to (53.8), T has only six independent components.)

Not every quantity appearing in (53.6)–(53.10) is determined by a constitutive relation. Ocean tides, for example, are the result of an imbalance between **div** T and a body force $\rho\mathbf{b}$ arising from the gravitational pull of the earth, the sun and the moon. But, whereas the stress tensor T is a material property—depending, presumably, on the material constitution and on its state (density, temperature, etc.)—the proximity of the moon is not a property of sea water. Thus **b** is *not* to be determined by a constitutive relation. Similarly, we regard the heat flux vector $\mathbf{q}$ as a material property, but the body heating h is, like **b**, an extraneous agent which is *not* subject to a constitutive relation.

Let us now consider a class of materials with constitutive relations of the form

$$f = \mathcal{F}(\rho, \vartheta, \mathbf{grad}\,\vartheta) \tag{53.12}$$

for each of the eleven quantities (53.11). We shall assume each of the eleven $\mathcal{F}$'s to be differentiable. The dependence on the temperature gradient is natural enough when one thinks of the heat flux vector $\mathbf{q}$, but in the case of u or s it may appear strange, if not outright perverse. We are indeed trying to overcome such 'physical' prejudice by consciously adopting an attitude of extreme impartiality. But then, one might ask, why should we restrict the arguments in (53.12) to the five variables $(\rho, \vartheta, \mathbf{grad}\,\vartheta)$? There is, indeed, no theoretical reason for such a restriction. That is why we regard the constitutive relations (53.12) as describing 'a class of materials'.

Actually, we call these materials *perfect* or *inviscid fluids*, because (53.12) corresponds to some of our prejudices regarding such fluids. In a viscous fluid we expect the constitutive relations to depend also on gradients of the velocity components, in fact on the nine components of grad $\dot{\mathbf{x}}$. In a solid we expect these relations to depend on the distortion of the material—perhaps through the various components of the deformation tensor F—rather than merely on its density.

The question of whether (53.12) is capable of describing all perfect fluids, or even any particular kind of fluid, has no meaning, because we have not really defined 'fluid' (or 'solid', for that matter). Such definitions are quite complicated—and even controversial—and they certainly lie outside the scope of this book. We shall, however, discuss several other classes of materials in the next chapter.

In mechanics, it is customary to think of the initial value problem as follows: for a given assembly of masses, assume any given initial place and any given initial velocity for each mass and, starting from these initial values, proceed to calculate the trajectories by integrating the equations of motion. The initial instant may, of course, be chosen at any moment. The modification required by the laws (53.6)–(53.10) is obvious: assume initial *distributions* of mass, place and velocity; that is, $\rho(\mathbf{X}, 0)$, $\mathbf{x}(\mathbf{X}, 0)$ and $\dot{\mathbf{x}}(\mathbf{X}, 0)$. But in the system (53.6)–(53.10) we have all the other quantities (53.11), with a constitutive relation of the form (53.12) for each one. Clearly, then, we ought to assume some arbitrary initial temperature distribution $\vartheta(\mathbf{X}, 0)$ as well. This, in turn, will determine the initial temperature gradient at each point.

We have in the three equations (53.6), (53.7) and (53.9) a system which we may expect to be integrable, provided that the extraneous body force $\mathbf{b}$ and body heating h be given and the constitutive relation for the specific internal energy be solvable for the temperature; this is a condition on the specific heat—the partial derivative u_ϑ.

In the foregoing discussion of the initial value problem we have entirely ignored the entropy inequality (53.10). Will it be satisfied at each stage, and in particular at the initial instant? It may be satisfied, but more importantly, it may not. Clearly, unless it is identically satisfied, it restricts either the initial distributions of temperature, density, place or velocity, or the extraneous $\mathbf{b}$ or h, or the constitutive relations.

Since restrictions on initial values or extraneous agents violate the whole of our physical experience, we consider them unthinkable. We are then forced, by the entropy inequality, to consider the last alternative, of accepting restrictions on the constitutive relations. In other words, initial conditions and extraneous body forces and heatings are entirely at our disposal, but we cannot expect every imaginable material to actually exist.

In order to investigate the way in which the entropy inequality restricts the constitutive relations, it is convenient to use, instead of the specific internal energy, the function

$$\varphi = u - \vartheta s, \tag{53.13}$$

which is called the *specific free energy*. Its significance will emerge presently. In terms of φ the entropy inequality (53.10) is

$$-\rho(\dot{\varphi} + s\dot{\vartheta}) + T \cdot \operatorname{grad} \dot{\mathbf{x}} - (\mathbf{q} \cdot \mathbf{grad}\, \vartheta)/\vartheta \geq 0. \tag{53.14}$$

Since both u and s are given by constitutive relations of the form (53.12), the same is true of φ. By the chain rule its material derivative will be given by

$$\dot{\varphi} = \varphi_\rho \dot{\rho} + \varphi_\vartheta \dot{\vartheta} + \varphi_{\mathbf{grad}\,\vartheta} \cdot (\mathbf{grad}\, \vartheta)^{\cdot}, \tag{53.15}$$

where, for any function $f(\mathbf{a})$ of a vector $\mathbf{a}$, $f_{\mathbf{a}}$ denotes the vector with components $(f_{\mathbf{a}})_i = \partial f / \partial a_i$. We substitute this in (53.14) and obtain

$$
\begin{aligned}
&- \rho(\varphi_\vartheta + s)\dot{\vartheta} - \rho\varphi_{\mathbf{grad}\,\vartheta} \cdot (\mathbf{grad}\, \vartheta)^{\cdot} \\
&+ (T + \rho^2 \varphi_\rho I) \cdot \operatorname{grad} \dot{\mathbf{x}} - (\mathbf{q} \cdot \mathbf{grad}\, \vartheta)/\vartheta \geq 0,
\end{aligned}
\tag{53.16}
$$

where use has been made of $\dot{\rho} = -\rho \operatorname{div} \dot{\mathbf{x}}$ and $\operatorname{div} \dot{\mathbf{x}} = I \cdot \operatorname{grad} \dot{\mathbf{x}}$.

If we choose any initial distribution of ρ and ϑ, the constitutive relations determine each of the coefficients of $\dot{\vartheta}$, $(\mathbf{grad}\ \vartheta)\dot{}$ and grad $\dot{\mathbf{x}}$, as well as the last term of the inequality (53.16). We can then still choose the initial velocity distribution so as to give the nine components of grad $\dot{\mathbf{x}}$ any arbitrarily chosen values. Next, the derivatives $\dot{\vartheta}$ and $(\mathbf{grad}\ \vartheta)\dot{}$ can be given arbitrary values by an appropriate choice of h at the point considered and in its neighbourhood. Thus (53.16) is of the form $\sum a_r(\mathbf{x})y_r + b(\mathbf{x}) \geq 0$, with arbitrary y_r, which has the consequences $a_r(\mathbf{x}) = 0$ and $b(\mathbf{x}) \geq 0$. Hence

$$s = -\varphi_\vartheta, \tag{53.17}$$

$$\varphi_{\mathbf{grad}\ \vartheta} = 0, \tag{53.18}$$

$$T = -\rho^2\varphi_\rho I, \tag{53.19}$$

$$-(\mathbf{q}\cdot\mathbf{grad}\ \vartheta)/\vartheta \geq 0. \tag{53.20}$$

Since any instant may be regarded as initial, this system of ten equations and one inequality must hold everywhere at all times.

Equation (53.18) shows that φ cannot depend on the temperature gradient. Hence it is (at most) a function $\varphi(\rho, \vartheta)$ of the two variables ρ and ϑ. According to (53.17), the specific entropy cannot be given by an arbitrary constitutive relation, as we have assumed above: on the contrary, it must be equal to $-\varphi_\vartheta$; in particular it, too, must be independent of $\mathbf{grad}\ \vartheta$. Similar remarks hold for u, which is now given by $\varphi + \vartheta s$, and for T. Moreover, the latter is manifestly symmetric, and thus satisfies Cauchy's second equation (53.8). We now have

$$s = -\varphi_\vartheta(\rho, \vartheta), \qquad u = \varphi(\rho, \vartheta) - \vartheta\varphi_\vartheta(\rho, \vartheta), \tag{53.21}$$

$$T = -pI, \qquad p = \rho^2\varphi_\rho(\rho, \vartheta), \tag{53.22}$$

and the remnant (53.20) of the entropy inequality, with which we shall deal presently. The quantity p, defined by $(53.22)_2$, is called the *pressure*. Equation $(53.22)_2$ can also be written as $p = -\partial\varphi/\partial\rho^{-1}$, which is minus the partial derivative of the specific free energy with respect to the specific volume.

In this way, eleven constitutive relations, each involving five variables, have been reduced to a single function of two variables, plus the three components of $\mathbf{q}(\rho, \vartheta, \mathbf{grad}\ \vartheta)$, which are still subject to an inequality. Since the entropy, the stress (or the pressure) and the internal energy are all determined by derivatives of the free energy, the latter deserves to be called a *thermodynamic potential*.

We now turn to the inequality (53.20), which is all that survives of the original entropy inequality. For an *insulator* we have $\mathbf{q} = 0$ by definition, and the inequality (53.20) reduces to the *equality* $0 = 0$. We conclude that insulating perfect fluids are *incapable of undergoing irreversible processes*. It is this property that makes them 'perfect'. For *conducting* perfect fluids, all processes throughout which $\mathbf{grad}\ \vartheta = 0$ are reversible. There may, of course, be other processes for which (53.20) becomes an equality.

According to (53.20), the heat flux vector cannot have a component in the direction of the temperature gradient. But unlike the other quantities, which all derive from the free energy and depend only on ρ and ϑ, there is no reason why $\mathbf{q}$ should not depend on $\mathbf{grad}\,\vartheta$ as well. It can be shown that (53.20) requires $\mathbf{q}$ to vanish whenever the temperature gradient does, that is, whenever the temperature is uniform throughout the body:

$$\mathbf{q}(\rho, \vartheta, 0) = 0. \tag{53.23}$$

This is a generalization to several variables of the statement that, if $x f(x) \geq 0$, then $f(0) = 0$. We shall not pause to prove (53.23): it will follow as a special case from a more general theorem that we shall state and prove in Section 57.

The set of values which the arguments of the constitutive relations assume throughout a body at a given instant is called a *state* of the body. We shall refer to the states with $\mathbf{grad}\,\vartheta = 0$ as *special*, and to the processes that are sequences of special states as *special processes*. Although the special states do not imply any balance of forces or torques, it is clear that they, together with the special processes—which need not be 'slow'—have the essential properties that classical thermodynamics associates with 'equilibrium' and 'quasi-static process'. For example, along a special process, $\Delta S = \int (Q/\vartheta)\,dt$. *Cyclic* special processes can be shown to have maximal efficiency, and can be used (after Kelvin) to construct an absolute temperature scale that is independent of the particular perfect fluid substance undergoing these processes.

It is to Coleman and Noll (1963) that we owe the clear separation between initial values, extraneous body forces and heatings, and constitutive relations, as well as the simple, logical and straightforward method by which the results (53.17)–(53.20) have been obtained.

We may substitute $T = -pI$ in (53.7), which leads to

$$\rho\ddot{\mathbf{x}} = -\mathbf{grad}\,p + \rho\mathbf{b}. \tag{53.24}$$

In hydrodynamics, this equation is called Euler's equation of motion. Similarly, we can use (53.21)–(53.22) in the internal energy balance equation (53.9). The result is

$$\rho\vartheta\dot{s} = \rho h - \mathrm{div}\,\mathbf{q}. \tag{53.25}$$

Despite its appearance, eqn (53.25) is the first principle of thermodynamics (the energy balance) for materials in the class under consideration. It is *not* the second principle for reversible processes, as has sometimes been claimed. In hydrodynamics (53.25) is also called the equation of heat conduction[†].

[†] If, beside ρ and ϑ, φ depends on further variables, which we collectively denote by $\mathbf{y}$, there will be an added term $+\rho\varphi_\mathbf{y} \cdot \dot{\mathbf{y}}$ on the left-hand side of (53.25). This will be the case when the perfect fluid is inhomogeneous: $\mathbf{y}$ will then stand for the concentrations of the various constituents; and additional equations will then be needed for the components of $\dot{\mathbf{y}}(\rho, \vartheta, \mathbf{y})$ (for example, chemical reaction rates).

To sum up the results of this lengthy discussion, a perfect fluid, that is, a material with constitutive relations depending on $(\rho, \vartheta, \mathbf{grad}\,\vartheta)$, is determined by a vector $\mathbf{q}$, which must satisfy the inequality (53.20), and by a function $\varphi(\rho, \vartheta)$. The entropy, stress, pressure and internal energy are determined by (53.21)–(53.22). The behaviour of such materials is governed by the law of mass conservation (53.6), the momentum balance law (53.24) and the energy balance law (53.25). Solutions of these equations need not be checked for compliance with Euler's second law of mechanics or with the second principle of thermodynamics, because these laws have already 'been taken care of'.

Since the whole analysis concerned the local laws, the results we have found apply on either side of any surface of discontinuity, irrespective of whether or not the surface is a boundary between two materials with different constitutive relations. Across any such surface, *all* the jump conditions must be satisfied. In particular, the entropy jump inequality (52.12) must not be forgotten: it has *not* 'been taken care of' by the foregoing treatment. The jump conditions may affect the speed at which the surface is moving (as was the case when we applied the electromagnetic jump conditions across a shock), its location and shape, and even the solutions far away from the surface. The entropy jump inequality determines the *direction* in which a shock will advance through matter.

One may think of the free energy of a continuous body in much the same way as one thinks of the Lagrangian of a dynamical system in analytical mechanics. Any given free energy $\varphi(\rho, \vartheta)$ can be used in (53.21)–(53.22): it may describe an ideal gas or an aqueous solution of some salt—just as a Lagrangian may describe a harmonic oscillator or a planet moving around the sun—but it may just as well stand for a fantastic material that cannot be found, or has not yet been found, in 'real nature'.

The analogy between L and φ extends to changes of variables: just as one may solve $p = L_{\dot{q}}(q, \dot{q}, t)$ for $\dot{q}$ and then replace L by the Hamiltonian $H(p, q, t) = p\dot{q} - L$, so one may solve $p = \rho^2 \varphi_\rho(\rho, \vartheta)$ for ρ and then replace φ by the specific Gibbs potential $\gamma(p, \vartheta) = \varphi + p/\rho$.

Problem 53.1 The monatomic ideal gas is defined by the equations

$$p = R\rho\vartheta, \qquad u = \tfrac{3}{2}R\vartheta + c_1,$$

where R is the gas constant and c_1 is a reference constant for the specific internal energy. Determine the specific free energy on the assumption that the gas is a perfect fluid, that is, $\varphi = \varphi(\rho, \vartheta)$.

Solution The equation $p = \rho^2\varphi_\rho = R\rho\vartheta$ has the integral

$$\varphi = R\vartheta \ln \rho + f(\vartheta),$$

where f is arbitrary. Now

$$u = \varphi + \vartheta s = \varphi - \vartheta \varphi_\vartheta = f(\vartheta) - \vartheta f'(\vartheta)$$
$$= -\vartheta^2 (f/\vartheta)' = \tfrac{3}{2} R\vartheta + c_1.$$

Thus

$$f(\vartheta) = -\tfrac{3}{2} R\vartheta \ln \vartheta + c_1 + c_2 \vartheta,$$

where c_2 is an integration constant. The specific free energy

$$\varphi(\rho, \vartheta) = R\vartheta \ln \frac{\rho}{\vartheta^{\frac{3}{2}}} + c_1 + c_2 \vartheta$$

is therefore determined up to an arbitrary linear function of ϑ. The specific entropy is

$$s = -\varphi_\vartheta = R \ln \frac{\vartheta^{\frac{3}{2}}}{\rho} + \tfrac{3}{2} R - c_2.$$

Thus c_2 may be regarded as an 'entropy constant'.

Problem 53.2   Use Euler's equation of motion (53.24) to derive the formula for a steady horizontal wind in the earth's atmosphere.

Solution   For a steady wind in the earth's atmosphere

$$2\rho \, \Omega \times \dot{\mathbf{x}} = - \operatorname{\mathbf{grad}} p + \rho \mathbf{g},$$

where Ω is the earth's angular spin velocity, $2\Omega \times \dot{\mathbf{x}}$ is the Coriolis acceleration, and $\mathbf{g}$ is the acceleration of gravity. We have neglected the centrifugal acceleration, which is at most 0.3% of g. Taking the vector product with the upward vertical unit vector $\mathbf{k}$, and noting that $\mathbf{k} \cdot \dot{\mathbf{x}} = \mathbf{k} \times \mathbf{g} = 0$—since $\dot{\mathbf{x}}$ is horizontal and $\mathbf{g}$ vertical—we obtain

$$2\rho (\mathbf{k} \cdot \Omega) \dot{\mathbf{x}} = \mathbf{k} \times \operatorname{\mathbf{grad}} p.$$

From this follows the *geostrophic wind formula* [†],

$$\pm (2\rho \Omega \sin \lambda) \dot{\mathbf{x}} = \mathbf{k} \times \operatorname{\mathbf{grad}} p,$$

where λ denotes the latitude; the plus sign refers to the northern hemisphere ($\mathbf{k} \cdot \Omega > 0$), and the minus to the southern ($\mathbf{k} \cdot \Omega < 0$). Although the pressure force density is $- \operatorname{\mathbf{grad}} p$, the wind blows along the lines of constant pressure (the *isobars*). For a given pressure gradient the wind increases with height, because the air density ρ decreases.

[†] Sailors and aviators sort out the directions by using (the empirical) *Buys Ballot's law*: 'If, in the northern hemisphere, you stand with your back to the wind, the low pressure is on your left and the high pressure on your right; in the southern hemisphere, the low pressure area is on your right'.

Problem 53.3 A star is a spherically symmetric, self-gravitating configuration, in which all possible motions are radial, as is the heat flux vector **q**. The spherical symmetry has the effect that material points are replaced by material spheres. For any such sphere the mass $m = \int_0^r 4\pi r^2 \rho\, dr$ inside the sphere, which must remain constant, can serve as a material label. All quantities (including r) then become functions of m and t. Write the equations of continuity (mass conservation), of motion and of energy balance in a form appropriate to a star.

Solution If $V(m, t) = 4\pi r^3(m, t)/3$ is the volume at time t of a sphere of mass m, then at any instant $dm = \rho\, dV$. This determines the partial derivative $\partial V(m, t)/\partial m$:

$$\rho \frac{\partial V}{\partial m} = 4\pi r^2 \rho \frac{\partial r}{\partial m} = 1,$$

which is the form that Euler's equation of continuity (51.4) assumes in the present case.

Euler's equation of motion (53.24), which has only a radial non-vanishing component, is

$$\ddot{r} = -\frac{1}{\rho} \frac{\partial p}{\partial r} + b_r.$$

Since **q** is radial, the energy balance equation (53.25) is

$$\vartheta \dot{s} = h - \frac{1}{\rho r^2} \frac{\partial}{\partial r}(r^2 q).$$

Using the previous equation of continuity in order to transform the derivatives with respect to r into derivatives with respect to m, and substituting for b_r the gravitational acceleration $-Gm/r^2$, which is only affected by the mass m inside the sphere of radius $r(m, t)$, we obtain the equations

$$\frac{\partial}{\partial m} r(m, t) = \frac{1}{4\pi r^2 \rho},$$

$$\frac{\partial^2}{\partial t^2} r(m, t) = -4\pi r^2 \frac{\partial p}{\partial m} - \frac{Gm}{r^2},$$

$$\vartheta \frac{\partial}{\partial t} s(m, t) = h - \frac{\partial}{\partial m} 4\pi r^2 q.$$

15

The fusion of electromagnetism with mechanics and thermodynamics

54. Continuum mechanics and electromagnetism

In Sections 51–52 we formulated five principles of continuum mechanics: conservation of mass, Euler's two laws, the law of energy balance and the entropy inequality. A theory capable of describing the dynamic and thermal effects of electromagnetic fields requires a certain generalization of these principles. Finally, of course, the principles of electromagnetism must be added.

We generalize, or modify, the laws of continuum mechanics in two places. First, we employ a more general expression than the velocity $\dot{\mathbf{x}}$ for the momentum per unit mass, and denote it by $\mathbf{g}$. Thus we take $\mathbf{G}$ and $\mathbf{L}_O$ to be

$$\mathbf{G} = \int \mathbf{g}\,dm, \qquad \mathbf{L}_O = \int (\mathbf{x} - \mathbf{x}_O) \times \mathbf{g}\,dm. \qquad (54.1)$$

With this modification, Euler's first law, $\dot{\mathbf{G}} = \mathbf{F}$, leads to the equation

$$\rho\dot{\mathbf{g}} = \operatorname{div} T + \rho\mathbf{b} \qquad (54.2)$$

and to the three jump conditions

$$[\![\rho\mathbf{g} \otimes (\dot{\mathbf{x}} - \mathbf{v}) - T]\!]\mathbf{n} = 0. \qquad (54.3)$$

His second law, $\dot{\mathbf{L}}_O = \mathbf{M}_O$, leads to the three equations

$$\rho\dot{x}_{[i}g_{j]} + T_{[ij]} = 0. \qquad (54.4)$$

Note that, unless $\dot{x}_i g_j = \dot{x}_j g_i$, the stress tensor is no longer symmetric. Equations (54.2)–(54.4) replace (51.18), (51.19) and (51.21).

We shall keep an open mind as to what $\mathbf{g}$ is, but if the modified theory is to be a generalization of the original one, we must require that $\mathbf{g}$ reduce to $\dot{\mathbf{x}}$ in the absence of an electromagnetic field.

The second modification is in the balance of energy: we introduce (cf. Section 45), alongside the heat flux vector $\mathbf{q}$, an extra energy flux vector equal to $\mathcal{E} \times \mathcal{H}$, the vector product of the electromotive intensity $\mathcal{E} = \mathbf{E} + \dot{\mathbf{x}} \times \mathbf{B}$ and the magnetomotive intensity $\mathcal{H} = \mathbf{H} - \dot{\mathbf{x}} \times \mathbf{D}$.[†] With this extra term, (52.7)–(52.8) are replaced by

$$\rho \dot{\varepsilon} = \operatorname{div} T^T \dot{\mathbf{x}} + \rho \dot{\mathbf{x}} \cdot \mathbf{b} - \operatorname{div}(\mathbf{q} + \mathcal{E} \times \mathcal{H}) + \rho h, \tag{54.5}$$

$$[\![\rho \varepsilon (\dot{\mathbf{x}} - \mathbf{v}) - T^T \dot{\mathbf{x}} + \mathbf{q} + \mathcal{E} \times \mathcal{H}]\!] \cdot \mathbf{n} = 0. \tag{54.6}$$

Two features of this modification of the law of energy balance should be noted. First, it is obviously an electromagnetic generalization, because $\mathcal{E} = \mathbf{E} + \dot{\mathbf{x}} \times \mathbf{B}$ vanishes whenever both $\mathbf{E}$ and $\mathbf{B}$ do. Secondly, this modification replaces the Lorentz force axiom we have laid down for charged particles. We must therefore redefine the electromagnetic SI units, which in the case of charged particles were based on the Lorentz force axiom. This is easy: by asserting that $\mathcal{E} \times \mathcal{H}$, which has the dimensions $\Phi C/(LT)^2$, is an energy flux (energy per unit area per unit time), we conclude that ΦC is dimensionally the same as NLT, where N is the dimension of force. Hence the constant μ_0, with dimensions $(\Phi/C)/(L/T)$, has the dimensions of $N/(C/T)^2$, that is, force divided by squared current. We can therefore define the SI unit of electric current, the *ampère*, by the same formula as in Section 15:

$$\mu_0 = 4\pi \times 10^{-7} \text{ newton amp}^{-2}, \tag{54.7}$$

(cf. (15.2)), and use this definition to deduce all the other electromagnetic SI units.

It is important to realize that the modifications we have made in the concept of momentum and in the law of energy balance will force us to give up some of our pre-electromagnetic notions regarding motion and energy. While, according to $\dot{\mathbf{G}} = \mathbf{F}$, imbalance of applied forces is still the same as change of momentum, the latter is *not* the same as acceleration (even in a rigid body), for $\mathbf{G}$ is no longer $\int \dot{\mathbf{x}} \, dm$. Similarly, $\dot{\mathbf{L}}_O = \mathbf{M}_O$; hence imbalance of torques is still the same as change of angular momentum, but the latter is no longer directly related to change in angular velocity (even in a rigid body). Finally, change of energy is no longer the same as imbalance between mechanical power and heating, because there is now an extra rate of energy increase, namely $\oint -\mathcal{E} \times \mathcal{H} \cdot \mathbf{n} \, dS$. Thus an electric wire's energy may change even when it is thermally insulated (no heating) and at rest (no mechanical power).

We make no change in the law of mass conservation or in the entropy inequality.

It is time to make a list of all the classical laws that govern the behaviour of continuous bodies. Since we intend to apply the method of Coleman and Noll, we

---

[†]This assumption is motivated by applications of Poynting's theorem (which we shall derive below; cf. (55.4)) to special situations, *and* on identifications of various expressions as energy, mechanical power or heating. From such considerations it appears plausible that $\mathcal{E} \times \mathcal{H}$, plus the curl of an unspecified vector, represents an energy flux. A final assumption is then made to the effect that $\mathcal{E} \times \mathcal{H}$ is itself an energy flux. We prefer an unambiguous and frank statement of this hypothesis.

list only the local equations:

$$\dot{\rho} + \rho \operatorname{div} \dot{\mathbf{x}} = 0, \tag{54.8}$$

$$\rho \dot{\mathbf{g}} = \operatorname{\mathbf{div}} T + \rho \mathbf{b}, \tag{54.9}$$

$$\rho \dot{x}_{[i} g_{j]} + T_{[ij]} = 0, \tag{54.10}$$

$$\rho \dot{\varepsilon} = \operatorname{div} T^T \dot{\mathbf{x}} + \rho \dot{\mathbf{x}} \cdot \mathbf{b} + \rho h - \operatorname{div}(\mathbf{q} + \mathcal{E} \times \mathcal{H}), \tag{54.11}$$

$$\rho \dot{s} \geq \rho h / \vartheta - \operatorname{div}(\mathbf{q}/\vartheta), \tag{54.12}$$

$$\operatorname{div} \mathbf{D} = q, \tag{54.13}$$

$$\operatorname{\mathbf{curl}} \mathcal{H} = \mathcal{J} + \overset{*}{\mathbf{D}}, \tag{54.14}$$

$$\operatorname{div} \mathbf{B} = 0, \tag{54.15}$$

$$\operatorname{\mathbf{curl}} \mathcal{E} = -\overset{*}{\mathbf{B}}, \tag{54.16}$$

$$\mathbf{D} = \epsilon_0 \mathbf{E} + \mathbf{P}, \tag{54.17}$$

$$\mathcal{H} = \mathbf{B}/\mu_0 - \dot{\mathbf{x}} \times \epsilon_0 \mathbf{E} - \mathcal{M}. \tag{54.18}$$

Maxwell's equations (54.14) and (54.16) are written in the forms (25.9) and (25.11), which involve only Galilean invariants. The aether relations (54.17)–(54.18) hold in any aether frame. The laws (54.8)–(54.12) derive from classical continuum mechanics, and are restricted to the inertial frames of that theory. It is possible to formulate relativistic analogues of continuum mechanics, but they are not definitive, because an accepted relativistic thermodynamics does not exist. Several relativistic theories of mechanics and thermodynamics have indeed been proposed, but the absence—so far—of any clear experimental evidence has made it impossible to decide which one of them, if any, should be preferred. The one feature they all share—beside Lorentz invariance—is that they reduce to the classical set of mechanical and thermodynamic laws when $\dot{x}^2$ is everywhere small compared with c^2.

So long as we limit ourselves to systems of bodies in which the relative velocities are all small compared with c, we shall always be able to choose an aether (or Lorentz) frame relative to which $\dot{x}^2 \ll c^2$ everywhere. In this aether frame every relativistic theory will reduce to the laws (54.8)–(54.18). It is in this sense, and under these limited circumstances, that we shall henceforth apply this system of laws.

So far as terrestrial experiments are concerned, the foregoing limitation is not serious, because macroscopic bodies cannot be accelerated to relativistic speeds. It is only in astronomy that we find, in accordance with some interpretations, evidence for relative velocities approaching the velocity of light, and the circumstances surrounding these astronomical phenomena seem to call for a relativistic theory, not only of mechanics and thermodynamics, but also of gravitation. Thus special-relativistic, as opposed to general-relativistic, thermodynamic theories appear to be neither here nor there.

The system (54.8)–(54.18) will, in any case, suffice for macroscopic bodies moving at speeds that are less than thirty million kilometres per hour. Such bodies can, at the same time, be described by classical (Galilean) constitutive relations.

For charged particles, on the other hand, a relativistic theory is both necessary and available, but it is not the same as the one that we are setting up for macroscopic (often called *ponderable*) bodies. After all, a particle has no energy (in the thermodynamic sense), temperature or entropy. It does not even have a surface.

Would it simplify matters if we regarded the sum $\mathbf{q} + \mathcal{E} \times \mathcal{H}$, which appears in the energy balance equation (54.11), as a generalized heat flux $\mathbf{p}$ (say)? Could we then simply forget about the extra, electromagnetic energy flux $\mathcal{E} \times \mathcal{H}$? No, we could not, for we would then have to substitute $\mathbf{q} = \mathbf{p} - \mathcal{E} \times \mathcal{H}$ in the inequality (54.12).

The laws (54.8)–(54.12) are written in terms of material derivatives. Maxwell's equations (54.14) and (54.16) involve flux derivatives (cf. (25.7)), but these can be related to material derivatives with the help of Euler's formula (51.7). The relation, for any vector field $\mathbf{A}(\mathbf{x}, t)$, is

$$\overset{*}{\mathbf{A}} = \dot{\mathbf{A}} + \mathbf{A} \operatorname{div} \dot{\mathbf{x}} - (\mathbf{A} \cdot \mathbf{grad})\dot{\mathbf{x}}. \tag{54.19}$$

If we leave out the law of angular momentum balance (54.10) and the second law (54.12), the remaining equations of the system (54.8)–(54.18) constitute an initial value problem. In order to solve it we shall have to assume constitutive relations for the 26 quantities

$$\mathbf{g}, \quad T, \quad \varepsilon, \quad \mathbf{q}, \quad s, \quad \mathcal{J}, \quad \mathbf{P} \text{ and } \mathcal{M}. \tag{54.20}$$

But the two laws we have left out must still be satisfied, everywhere and at all times. This means that we shall now have to impose the second law of thermodynamics, *as well as* the three components of Euler's second law of mechanics, as restrictions on the constitutive relations. In the following sections we shall do this for several classes of materials.

55. The electromagnetic entropy inequality

In order to determine the restrictions on the constitutive relations that follow from the second law of thermodynamics, we must derive the entropy inequality for our augmented system of laws of continuum mechanics by eliminating the body force $\mathbf{b}$ and the body heating h. This can be done once and for all, before deciding on any specific class of constitutive relations.

We first substitute $\mathbf{b}$ from (54.9) into (54.11). This gives

$$\begin{aligned} \rho\dot{\varepsilon} &= \operatorname{div} T^T \dot{\mathbf{x}} + \dot{\mathbf{x}} \cdot (\rho\dot{\mathbf{g}} - \mathbf{div}\, T) + \rho h - \operatorname{div}(\mathbf{q} + \mathcal{E} \times \mathcal{H}) \\ &= T \cdot \operatorname{grad} \dot{\mathbf{x}} + \rho(\dot{\mathbf{g}} \cdot \dot{\mathbf{x}} + h) - \operatorname{div}(\mathbf{q} + \mathcal{E} \times \mathcal{H}), \end{aligned} \tag{55.1}$$

where the matrix scalar product $T \cdot \mathrm{grad}\, \dot{\mathbf{x}}$ is defined as in Section 53. Next, we substitute h from this equation into the entropy inequality (54.12):

$$\rho \dot{s} \geq [\rho \dot{\varepsilon} - T \cdot \mathrm{grad}\, \dot{\mathbf{x}} - \rho \dot{\mathbf{g}} \cdot \dot{\mathbf{x}} + \mathrm{div}(\mathbf{q} + \mathcal{E} \times \mathcal{H})]/\vartheta - \mathrm{div}(\mathbf{q}/\vartheta). \qquad (55.2)$$

We may multiply this by ϑ (since ϑ is positive) and write it as

$$\rho(\vartheta \dot{s} - \dot{\varepsilon} + \dot{\mathbf{g}} \cdot \dot{\mathbf{x}}) + T \cdot \mathrm{grad}\, \dot{\mathbf{x}} - \mathrm{div}\,\mathcal{E} \times \mathcal{H} - (\mathbf{q} \cdot \mathrm{grad}\, \vartheta)/\vartheta \geq 0. \qquad (55.3)$$

Now, we have the vector identity

$$-\mathrm{div}\,\mathcal{E} \times \mathcal{H} = \mathcal{E} \cdot \mathbf{curl}\, \mathcal{H} - \mathcal{H} \cdot \mathbf{curl}\, \mathcal{E}.$$

If we substitute the curls from Maxwell's equations (54.14) and (54.16), we obtain

$$-\mathrm{div}\,\mathcal{E} \times \mathcal{H} = \mathcal{J} \cdot \mathcal{E} + \mathcal{E} \cdot \overset{*}{\mathbf{D}} + \mathcal{H} \cdot \overset{*}{\mathbf{B}}. \qquad (55.4)$$

This relation (or, more especially, the one to which it reduces when $\dot{\mathbf{x}} = 0$) is called Poynting's theorem. It is often, and erroneously, regarded as a balance law for 'electromagnetic energy'. Of course, it is nothing of the kind, because it is just an *identity* satisfied by all fields that are solutions of Maxwell's equations.

Exercises

55.1. Prove the identity

$$\mathbf{b} \cdot \overset{*}{\mathbf{a}} = \mathbf{b} \cdot \dot{\mathbf{a}} + [(\mathbf{b} \cdot \mathbf{a})I - \mathbf{b} \otimes \mathbf{a}] \cdot \mathrm{grad}\, \dot{\mathbf{x}},$$

where I is the unit matrix.

55.2. Prove the identity

$$\mathbf{a} \times \mathbf{b} \otimes \mathbf{c} + \mathbf{b} \times \mathbf{c} \otimes \mathbf{a} + \mathbf{c} \times \mathbf{a} \otimes \mathbf{b} = (\mathbf{a} \times \mathbf{b} \cdot \mathbf{c})I,$$

where (cf. (51.20)) $(\mathbf{a} \times \mathbf{b} \otimes \mathbf{c})_{ij} = (\mathbf{a} \times \mathbf{b})_i c_j$.

55.3. Use the preceding two identities to cast (55.4) in the form

$$\begin{aligned}
-\mathrm{div}\,\mathcal{E} \times \mathcal{H} = {} & \mathcal{J} \cdot \mathcal{E} - \mathbf{P} \cdot \dot{\mathcal{E}} - \mathcal{M} \cdot \dot{\mathbf{B}} + \epsilon_0 \mathbf{E} \times \mathbf{B} \cdot \ddot{\mathbf{x}} \\
& + \left[(\epsilon_0 E^2 + B^2/\mu_0 - \epsilon_0 \mathbf{E} \times \mathbf{B} \cdot \dot{\mathbf{x}} + \mathcal{E} \cdot \mathbf{P} - \mathcal{M} \cdot \mathbf{B})I \right. \\
& \left. - \epsilon_0 \mathbf{E} \otimes \mathbf{E} - \mathbf{B} \otimes \mathbf{B}/\mu_0 - \mathcal{E} \otimes \mathbf{P} + \mathcal{M} \otimes \mathbf{B} - \epsilon_0 \mathbf{E} \times \mathbf{B} \otimes \dot{\mathbf{x}} \right] \cdot \mathrm{grad}\, \dot{\mathbf{x}} \\
& + \left[\tfrac{1}{2}\epsilon_0 E^2 + \tfrac{1}{2}B^2/\mu_0 + \mathcal{E} \cdot \mathbf{P} - \epsilon_0 \mathbf{E} \times \mathbf{B} \cdot \dot{\mathbf{x}} \right]^{\cdot},
\end{aligned}$$

where $[\ldots]^{\cdot}$ denotes the material derivative of $[\ldots]$.

By using of the result of Exercise 55.3, the entropy inequality (55.3) can be cast in the form

$$-\rho\dot{\varphi} - \rho s\dot{\vartheta} - (\rho\mathbf{g} - \rho\dot{\mathbf{x}} - \epsilon_0\mathbf{E} \times \mathbf{B}) \cdot \ddot{\mathbf{x}} - \mathbf{P} \cdot \dot{\mathcal{E}} - \mathcal{M} \cdot \dot{\mathbf{B}}$$
$$+ \tau \cdot \operatorname{grad} \dot{\mathbf{x}} + \mathcal{J} \cdot \mathcal{E} - (\mathbf{q} \cdot \operatorname{grad} \vartheta)/\vartheta \geq 0, \qquad (55.5)$$

where

$$\tau = T + \left[\tfrac{1}{2}\epsilon_0 E^2 + \tfrac{1}{2}B^2/\mu_0 - \mathcal{M} \cdot \mathbf{B}\right] I$$
$$- \epsilon_0\mathbf{E} \otimes \mathbf{E} - \mathbf{B} \otimes \mathbf{B}/\mu_0 - \mathcal{E} \otimes \mathbf{P} + \mathcal{M} \otimes \mathbf{B} - \epsilon_0\mathbf{E} \times \mathbf{B} \otimes \dot{\mathbf{x}}, \qquad (55.6)$$

and the function φ, which we call *the (electromagnetic) specific free energy*, is defined by

$$\varphi = \varepsilon - \vartheta s - \mathbf{g} \cdot \dot{\mathbf{x}} + \tfrac{1}{2}\dot{x}^2$$
$$- \left[\tfrac{1}{2}\epsilon_0 E^2 + \tfrac{1}{2}B^2/\mu_0 + \mathcal{E} \cdot \mathbf{P} - \epsilon_0\mathbf{E} \times \mathbf{B} \cdot \dot{\mathbf{x}}\right]/\rho. \qquad (55.7)$$

In the absence of an electromagnetic field, when $\mathbf{g} = \dot{\mathbf{x}}$, φ reduces to $\varepsilon - \tfrac{1}{2}\dot{x}^2 - \vartheta s$, which agrees with (53.13).

The inequality (55.5) is the desired electromagnetic entropy inequality, of which we shall make extensive use. We shall regard it as a restriction on the constitutive relations for the 26 quantities

$$\mathbf{g}, \quad \tau, \quad \varphi, \quad \mathbf{q}, \quad s, \quad \mathcal{J}, \quad \mathbf{P} \text{ and } \mathcal{M}. \qquad (55.8)$$

This list is the same as (54.20), except that T has been replaced by τ, and ε by φ.

We conclude this section with a corollary that follows immediately from the entropy inequality (55.5). If ϑ, $\dot{\mathbf{x}}$, $\mathcal{E}$ and $\mathbf{B}$ are all constant, then

$$\rho\dot{\varphi} \leq \tau \cdot \operatorname{grad} \dot{\mathbf{x}} + \mathcal{J} \cdot \mathcal{E} - (\mathbf{q} \cdot \operatorname{grad} \vartheta)/\vartheta. \qquad (55.9)$$

This sets an upper limit to the rate of increase of φ. With so many variables held constant, this statement has content only when φ depends on other variables, in addition (perhaps) to ϑ, $\dot{\mathbf{x}}$, $\mathcal{E}$ and $\mathbf{B}$. For example, φ may depend on the concentrations of substances undergoing a chemical reaction. The corollary therefore refers to a *partial rate of increase* (in the sense of a partial derivative). If the right-hand side of (55.9) vanishes—as it will for a uniformly moving insulator ($\mathcal{J} = \mathbf{q} = 0$), or whenever grad $\dot{\mathbf{x}}$, $\mathcal{E}$ and $\operatorname{grad} \vartheta$ all vanish—then $\dot{\varphi} \leq 0$. Under these circumstances the free energy cannot increase; again, this usually refers to a partial time derivative. In this special form the corollary furnishes the basis for considerations of thermodynamic stability: let α be a variable on which φ depends, in addition to ϑ, $\dot{\mathbf{x}}$, $\mathcal{E}$ and $\mathbf{B}$, and let α_m be such that

$$\varphi_\alpha(\vartheta, \dot{\mathbf{x}}, \mathcal{E}, \mathbf{B}, \alpha_m) = 0, \qquad \varphi_{\alpha\alpha}(\vartheta, \dot{\mathbf{x}}, \mathcal{E}, \mathbf{B}, \alpha_m) > 0, \qquad (55.10)$$

so that φ has a minimum at $\alpha = \alpha_m$. Then α_m—which according to $(55.10)_1$ is a function of the other variables—is a *stable* value of α with respect to every change throughout which the other variables have fixed values and the right-hand side of (55.9) vanishes, for φ cannot increase during any such change; therefore the value of α at the end cannot be different from α_m.

56. Perfect electromagnetic fluids

Consider a class of materials for which the constitutive relations for the quantities (55.8) are all of the form

$$f = \mathcal{F}(\rho, \vartheta, \dot{\mathbf{x}}, \mathcal{E}, \mathbf{B}, \mathbf{grad}\,\vartheta). \tag{56.1}$$

If we compare (56.1) with (53.12) we see that nine arguments—the components of the velocity $\dot{\mathbf{x}}$, the electromotive intensity $\mathcal{E}$ and the magnetic field $\mathbf{B}$—have been added to the original five. We shall call the class of materials characterized by (56.1) *perfect (or inviscid) electromagnetic fluids*. The velocity appears in (56.1) because $\mathbf{g}$ must reduce to $\dot{\mathbf{x}}$ in the absence of an electromagnetic field, and because the relations (55.5)–(55.7) involve $\dot{\mathbf{x}}$ itself, rather than just grad $\dot{\mathbf{x}}$. Since $\mathcal{E} = \mathbf{E} + \dot{\mathbf{x}} \times \mathbf{B}$, it does not matter whether we choose the nine arguments $\{\dot{\mathbf{x}}, \mathbf{E}, \mathbf{B}\}$ or $\{\dot{\mathbf{x}}, \mathcal{E}, \mathbf{B}\}$.

According to (56.1), φ is now a function of the fourteen arguments $\{\rho, \vartheta, \dot{\mathbf{x}}, \mathcal{E}, \mathbf{B}, \mathbf{grad}\,\vartheta\}$. By the chain rule its material derivative will be given by

$$\dot{\varphi} = \varphi_\rho \dot{\rho} + \varphi_\vartheta \dot{\vartheta} + \varphi_{\dot{\mathbf{x}}} \cdot \ddot{\mathbf{x}}$$
$$+ \varphi_{\mathcal{E}} \cdot \dot{\mathcal{E}} + \varphi_{\mathbf{B}} \cdot \dot{\mathbf{B}} + \varphi_{\mathbf{grad}\,\vartheta} \cdot (\mathbf{grad}\,\vartheta)^{\cdot}, \tag{56.2}$$

(cf. (53.15)). We substitute this in (55.5) and obtain

$$-\rho(\varphi_\vartheta + s)\dot{\vartheta} - (\rho\varphi_{\dot{\mathbf{x}}} + \rho\mathbf{g} - \rho\dot{\mathbf{x}} - \epsilon_0\mathbf{E} \times \mathbf{B}) \cdot \ddot{\mathbf{x}}$$
$$-\rho\varphi_{\mathbf{grad}\,\vartheta} \cdot (\mathbf{grad}\,\vartheta)^{\cdot} - (\rho\varphi_{\mathcal{E}} + \mathbf{P}) \cdot \dot{\mathcal{E}} - (\rho\varphi_{\mathbf{B}} + \mathcal{M}) \cdot \dot{\mathbf{B}}$$
$$+(\tau + \rho^2\varphi_\rho I) \cdot \mathrm{grad}\,\dot{\mathbf{x}} + \mathcal{J} \cdot \mathcal{E} - (\mathbf{q} \cdot \mathrm{grad}\,\vartheta)/\vartheta \geq 0. \tag{56.3}$$

By an argument which is completely analogous to the one we have followed in Section 53 we conclude that the factors of $\dot{\vartheta}$, $\ddot{\mathbf{x}}$, $(\mathbf{grad}\,\vartheta)^{\cdot}$, $\dot{\mathcal{E}}$, $\dot{\mathbf{B}}$ and grad $\dot{\mathbf{x}}$ in (56.3) must all vanish. In particular, we must have $\varphi_{\mathbf{grad}\,\vartheta} = 0$, which means that φ cannot depend on the temperature gradient: it can only be a function $\varphi(\rho, \vartheta, \dot{\mathbf{x}}, \mathcal{E}, \mathbf{B})$ of the eleven variables $\rho, \vartheta, \dot{\mathbf{x}}, \mathcal{E}$ and $\mathbf{B}$. Thus

$$s = -\varphi_\vartheta, \tag{56.4}$$

$$\mathbf{g} = \dot{\mathbf{x}} + \epsilon_0\mathbf{E} \times \mathbf{B}/\rho - \varphi_{\dot{\mathbf{x}}}, \tag{56.5}$$

$$\mathbf{P} = -\rho\varphi_{\mathcal{E}}, \tag{56.6}$$

$$\mathcal{M} = -\rho\varphi_{\mathbf{B}}, \tag{56.7}$$

$$T = -\left[p + \tfrac{1}{2}\epsilon_0 E^2 + \tfrac{1}{2}B^2/\mu_0 - \mathcal{M} \cdot \mathbf{B}\right]I$$
$$+ \epsilon_0\mathbf{E} \otimes \mathbf{E} + \mathbf{B} \otimes \mathbf{B}/\mu_0 + \mathcal{E} \otimes \mathbf{P} - \mathcal{M} \otimes \mathbf{B} + \epsilon_0\mathbf{E} \times \mathbf{B} \otimes \dot{\mathbf{x}}, \qquad (56.8)$$

$$p = \rho^2\varphi_\rho, \qquad (56.9)$$

$$\mathcal{J} \cdot \mathcal{E} - (\mathbf{q} \cdot \mathbf{grad}\,\vartheta)/\vartheta \geq 0, \qquad (56.10)$$

where we have used (55.6) to write the equation $\tau = -\rho^2\varphi_\rho I$ in the form (56.8), with the (electromagnetic) pressure p defined by (56.9). According to (56.4), the specific entropy (the fifth on the list (55.8) of 26 quantities), cannot be given by an arbitrary constitutive relation: it must be equal to $-\varphi_\vartheta$; in particular it, too, must be independent of $\mathbf{grad}\,\vartheta$. Similar remarks hold for $\mathbf{P}$, $\mathcal{M}$ and T.

The specific momentum $\mathbf{g}$ is fixed by (56.5). If $\mathbf{g}$ is to reduce to $\dot{\mathbf{x}}$ in the absence of an electromagnetic field, we must have $\varphi_{\dot{\mathbf{x}}}(\rho, \vartheta, \dot{\mathbf{x}}, 0, 0) = 0$. According to (56.6)–(56.7), $\mathbf{P}$ and $\mathcal{M}$ are functions of $\rho, \vartheta, \dot{\mathbf{x}}, \mathcal{E}$ and $\mathbf{B}$. But, according to Section 20, $\mathbf{P}$ and $\mathcal{M}$ are Galilean invariants, as are their arguments $\rho, \vartheta, \mathcal{E}$ and $\mathbf{B}$, but *not* the argument $\dot{\mathbf{x}}$. Thus $\varphi_\mathcal{E}$ and $\varphi_\mathbf{B}$ must both be independent of $\dot{\mathbf{x}}$, which is the same as saying that the mixed partial derivatives $\varphi_{\mathcal{E}\dot{\mathbf{x}}}$ and $\varphi_{\mathbf{B}\dot{\mathbf{x}}}$ must vanish. Hence $\varphi_{\dot{\mathbf{x}}}$ is independent of either $\mathcal{E}$ or $\mathbf{B}$, and the foregoing requirement that $\varphi_{\dot{\mathbf{x}}}(\rho, \vartheta, \dot{\mathbf{x}}, 0, 0) = 0$ leads to the conclusion that $\varphi_{\dot{\mathbf{x}}} = 0$ *always*. The variables on which φ may depend are thus reduced by another three: the free energy is a function $\varphi(\rho, \vartheta, \mathcal{E}, \mathbf{B})$ of eight arguments, and so are $s = -\varphi_\vartheta(\rho, \vartheta, \mathcal{E}, \mathbf{B})$, $p = \rho^2\varphi_\rho(\rho, \vartheta, \mathcal{E}, \mathbf{B})$, $\mathbf{P} = -\rho\varphi_\mathcal{E}(\rho, \vartheta, \mathcal{E}, \mathbf{B})$ and $\mathbf{M} = -\rho\varphi_\mathbf{B}(\rho, \vartheta, \mathcal{E}, \mathbf{B})$.

According to (55.7) and (56.5), the specific energy is given by

$$\varepsilon = \varphi + \vartheta s + \tfrac{1}{2}\dot{x}^2 + \left[\tfrac{1}{2}\epsilon_0 E^2 + \tfrac{1}{2}B^2/\mu_0 + \mathcal{E} \cdot \mathbf{P}\right]/\rho. \qquad (56.11)$$

Since φ, s and $\mathbf{P}$ are independent of $\mathbf{grad}\,\vartheta$, so is ε. But since it involves $\tfrac{1}{2}\dot{x}^2$ and $\tfrac{1}{2}\epsilon_0 E^2$, it is not a Galilean invariant. Similarly, the momentum density $\rho\mathbf{g} = \rho\dot{\mathbf{x}} + \epsilon_0\mathbf{E} \times \mathbf{B}$ and the stress T of (56.8) are not Galilean invariants. We shall, however, presently obtain an equation of motion, and an energy equation, that are both invariant under Galilean transformations.

The inequality (56.10) is the remnant of the entropy inequality. It restricts the constitutive relations for $\mathcal{J}$ and $\mathbf{q}$. Since the entropy, the specific energy, the stress, the polarization and the magnetization are all determined by derivatives of the (electromagnetic) free energy, the latter again deserves to be called a thermodynamic potential (actually it is an electromagnetothermodynamic potential).

In the present electromagnetic case we are not yet done, for we must still impose Euler's second law of mechanics (54.10) as a *second* constraint on the constitutive relations. If we substitute (56.5)–(56.8) in (54.10), we obtain the conditions

$$\mathcal{E}_i\varphi_{\mathcal{E}_j} + B_i\varphi_{B_j} = \mathcal{E}_j\varphi_{\mathcal{E}_i} + B_j\varphi_{B_i}. \qquad (56.12)$$

Exercise

56.1. If $f(\mathbf{a})$ is differentiable, prove that $f(Q\mathbf{a}) = f(\mathbf{a})$ for every orthogonal matrix Q if, and only if,

$$a_i \frac{\partial f}{\partial a_j} = a_j \frac{\partial f}{\partial a_i} \qquad (56.13)$$

for all i and j.

The conditions (56.12) restrict the way in which φ depends on the vectors $\mathcal{E}$ and $\mathbf{B}$: it must be invariant with respect to orthogonal transformations (including reflections). For example, φ cannot be a function of the sum of the Cartesian components of $\mathbf{B}$. It can only depend on such invariants as $\mathcal{E}^2$, B^2 or $(\mathcal{E} \cdot \mathbf{B})^2$ (note that $(\mathcal{E} \times \mathbf{B})^2 = (\mathcal{E} \cdot \mathbf{B})^2 - \mathcal{E}^2 B^2$).

The original list (55.8) of 26 functions of 14 variables has been narrowed down to six ($\mathcal{J}$ and $\mathbf{q}$), plus one function φ of the five arguments $\rho, \vartheta, \mathcal{E}^2, B^2$ and $(\mathcal{E} \cdot \mathbf{B})^2$. This is obviously a considerable reduction, but it is not complete, because the constitutive relations for the conduction current density $\mathcal{J}$ and the heat flux $\mathbf{q}$ must still satisfy the inequality (56.10), which is all that survives of the original entropy inequality. For an *insulator* (with respect to *both* electricity *and* heat) these constitutive relations are $\mathcal{J} = \mathbf{q} = 0$ by definition, and the inequality (56.10) reduces to the *equality* $0 = 0$. We conclude that, among the materials in the class under consideration, *the insulators are incapable of undergoing irreversible processes*. Even for conductors, all processes throughout which $\mathcal{E} = \mathbf{grad}\,\vartheta = 0$ are reversible. There may, of course, be other processes for which (56.10) becomes an equality.

As a generalization of the special states and processes we have introduced in Section 53, we shall now refer to the states with $\mathcal{E} = \mathbf{grad}\,\vartheta = 0$ as *special*, and to the processes that are sequences of special states as *special processes*. Again, these special states and processes have the essential properties that classical thermodynamics associates with 'equilibrium' and 'quasi-static process'.

As in Section 53, we can now substitute $\mathbf{g}$ and T in Euler's first law (54.9). But the stress (56.8) is rather complicated, and we must brace ourselves for some computational effort.

Exercises

56.2. Prove the identity

$$\mathbf{div}\left(\mathbf{E} \otimes \mathbf{E} - \tfrac{1}{2}E^2 I\right) = \mathbf{E}\,\mathrm{div}\,\mathbf{E} - \mathbf{E} \times \mathbf{curl}\,\mathbf{E}. \qquad (56.14)$$

56.3. Use Maxwell's equations to prove that, for the T of (56.8),

$$\text{div } T = - \text{grad } p + (\mathcal{M} \cdot \text{grad})B + \mathcal{M} \times \text{curl } B$$
$$+ (P \cdot \text{grad})\mathcal{E} + qE + (j + \overset{*}{P}) \times B + \rho(\epsilon_0 E \times B/\rho)^{\cdot}. \tag{56.15}$$

56.4. Verify that

$$qE + j \times B = q\mathcal{E} + \mathcal{J} \times B. \tag{56.16}$$

We can now substitute (56.5) and (56.14)–(56.16) in (54.9), and obtain Euler's first law in the form

$$\rho\ddot{x} = - \text{grad } p + q\mathcal{E} + (\mathcal{J} + \overset{*}{P}) \times B + (P \cdot \text{grad})\mathcal{E}$$
$$+ (\mathcal{M} \cdot \text{grad})B + \mathcal{M} \times \text{curl } B + \rho b. \tag{56.17}$$

This is the generalization of Euler's equation (53.24). We note that the electromagnetic part $\epsilon_0 E \times B/\rho$ of g has dropped out. It should perhaps be emphasized that, unless P and $\mathcal{M}$ both vanish, in which case φ is independent of $\mathcal{E}$ or B, the pressure $p = \rho^2 \varphi_\rho(\rho, \vartheta, \mathcal{E}, B)$ depends on the electromagnetic field, and the term $- \text{grad } p$ on the right-hand side of (56.17) is as 'electromagnetic' as the others. Another fact worth noting is that (56.17) involves only Galilean invariants. It is therefore invariant under Galilean transformations.

Except for the terms involving the polarization or magnetization, there are two terms, $q\mathcal{E}$ and $\mathcal{J} \times B$, that represent forces independent of material properties. The first, which is simply the Lorentz force density, acts on any space charge present. It may provide motivation for the Lorentz force axiom for a classical charged particle. But the derivation of (56.17) was based on assumptions, like surface terms and thermodynamics, that have no meaning for a particle. Moreover, the term $q\mathcal{E} = q(E + \dot{x} \times B)$ involves the entire electromagnetic field, *including* the self-fields that derive from q through $\text{div } D = q$ and the aether relations. Thus the term $q\mathcal{E}$ provides motivation, not justification, for the Lorentz force axiom.

The force density $\mathcal{J} \times B$ acts on any conduction current which crosses a magnetic field. We have already discussed this force in Section 40, but there it was assumed, not derived. Similarly, we assumed in the electrostatic discussion of Section 26 that a conductor in vacuum was acted upon by an electric surface force $t = \frac{1}{2}\sigma E$ per unit area. Indeed, for a conductor at rest, in the absence of magnetic fields, the electric part of the stress tensor (56.8) is $T = \epsilon_0 E \otimes E - \frac{1}{2}\epsilon_0 E^2 I$. Therefore $t = Tn = \epsilon_0(E_n E - \frac{1}{2}E^2 n)$. But the field at the surface of a conductor is normal to the surface and related to the surface charge density by $\epsilon_0 E_n = \sigma$. Hence $\epsilon_0 E_n E = \sigma E$, $\epsilon_0 E^2 n = \sigma E$ and $t = \frac{1}{2}\sigma E$.

Since **P** and $\mathcal{M}$ are the rest frame electric and magnetic moments per unit volume (cf. the remarks following (20.7) and (20.8)), and $\mathcal{E}$ is the rest frame electric field, the terms $(\mathbf{P} \cdot \mathbf{grad})\mathcal{E}$ and $(\mathcal{M} \cdot \mathbf{grad})\mathbf{B} + \mathcal{M} \times \mathbf{curl}\ \mathbf{B}$ in (56.17) are consistent with the forces $(\mathbf{d} \cdot \mathbf{grad})\mathbf{E}$ and $(\mathbf{m} \cdot \mathbf{grad})\mathbf{B} + \mathbf{m} \times \mathbf{curl}\ \mathbf{B}$ on single electric and magnetic dipoles at rest (Problem 15.1 and (40.21)). But **P** and $\mathcal{M}$ are now continuous distributions, determined by (56.6)–(56.7), and there is no stipulation that self-fields be left out.

The only term in (56.17) that we have not met before is $\overset{*}{\mathbf{P}} \times \mathbf{B}$. It is analogous to $\mathcal{J} \times \mathbf{B}$, since $\overset{*}{\mathbf{P}}$ is a 'polarization conduction current density': according to (20.5) and (25.7) $\mathcal{J}_R = \mathbf{curl}\ \mathcal{M} + \overset{*}{\mathbf{P}}$.

Exercises

56.5. Prove that an electromagnetic wave impinging on a perfectly conducting wall exerts on the latter an inward stress

$$T\mathbf{n} = \frac{1}{2}(\tau\mathbf{E} + \mathbf{K} \times \mathbf{B}),$$

where τ, as in Section 49, denotes the surface charge density. This is a normal stress. The first term is a tension, and the second a pressure (called *radiation pressure*).

56.6. Show that the energy balance law (54.11) can be written in the form

$$\rho\vartheta\dot{s} = \mathcal{J}\cdot\mathcal{E} + \rho h - \mathrm{div}\ \mathbf{q}. \tag{56.18}$$

Equation (56.18) is the generalization of (53.25). Like (56.17), it is invariant under Galilean transformations. The product $\mathcal{J} \cdot \mathcal{E}$ appears alongside the heating term $\rho h - \mathrm{div}\ \mathbf{q}$. So far as its effect on the entropy rate $\dot{s}$ is concerned, it acts as an additional heating term. It is of course due to the extra energy flux $\mathcal{E} \times \mathcal{H}$ (cf. Poynting's formula (55.4)), and is therefore not a true heating source. Its position in (56.18) is the excuse, rather than the justification, for the name *Joule heating*.

As in Section 53, we emphasize that (56.18) is the first principle of thermodynamics for materials in the class under consideration. It is *not* 'the second principle for reversible processes': if $\mathcal{J}\cdot\mathcal{E} \neq 0$ then, according to (56.10), reversibility requires that $\mathcal{J} \cdot \mathcal{E}$ be exactly cancelled by $(\mathbf{q} \cdot \mathbf{grad}\ \vartheta)/\vartheta$, everywhere and over an interval of time. In an adiabatic process the last two terms on the right-hand side of (56.18) vanish, but the product $\mathcal{J} \cdot \mathcal{E}$ remains. Thus 'adiabatic' (that is, $\rho h - \mathrm{div}\ \mathbf{q} = 0$) is generally not the same as 'isentropic'.

Although the specific energy ε of (56.11) does not appear in the energy balance equation (56.18), it may still be needed if there is a discontinuity somewhere, because it appears in the jump condition (54.6). Similarly, the momentum density $\rho\mathbf{g} = \rho\dot{\mathbf{x}} + \epsilon_0\mathbf{E} \times \mathbf{B}$ does not appear in the equation of motion (56.17), but it may be needed in applying the jump condition (54.3).

Of course, Coleman and Noll's method can, and indeed must, be applied separately to each class of materials, that is, to each class of constitutive relations. We must not expect the terms on the right-hand side of the equation of motion (56.17) to be the same for all bodies: there is no *universal* or *standard* expression for the electromagnetic force on a body (except in the case of charged particles, where a universal expression is laid down as an axiom). This becomes clear if we remember that in (56.17) p, $\mathbf{P}$ and $\mathcal{M}$ are not just any pressure, polarization or magnetization: they are $\rho^2\varphi_\rho$, $-\rho\varphi_\mathcal{E}$ and $-\rho\varphi_\mathbf{B}$. Otherwise, the entropy inequality is not satisfied. Similarly, s in (56.18) stands for $-\varphi_\vartheta$; it is not 'any entropy'. In this sense, electromagnetism is really a theory of materials. We shall treat other classes of materials in Sections 58–59.

57. The conduction current and the heat flux

Like $\mathbf{P}$ and $\mathcal{M}$, the Galilean invariants $\mathcal{J}$ and $\mathbf{q}$ cannot depend on $\dot{\mathbf{x}}$. But unlike $\mathbf{P}$ and $\mathcal{M}$, which are related by (56.6)–(56.7) to the derivatives of the free energy, and must therefore be independent of the temperature gradient, $\mathcal{J}$ and $\mathbf{q}$ may well depend on $\mathbf{grad}\,\vartheta$. Thus they are functions of the arguments $\{\rho, \vartheta, \mathcal{E}, \mathbf{B}, \mathbf{grad}\,\vartheta\}$.

We can establish an important property of the vector fields $\mathcal{J}$ and $\mathbf{q}$ by setting $\mathcal{E} = \lambda\mathbf{a}$ and $\mathbf{grad}\,\vartheta = \lambda\mathbf{b}$ and considering, for arbitrary ρ, ϑ, $\mathbf{B}$, $\mathbf{a}$ and $\mathbf{b}$, the function

$$\gamma(\lambda) = \mathcal{J}(\rho, \vartheta, \lambda\mathbf{a}, \mathbf{B}, \lambda\mathbf{b}) \cdot \lambda\mathbf{a} - \mathbf{q}(\rho, \vartheta, \lambda\mathbf{a}, \mathbf{B}, \lambda\mathbf{b}) \cdot \lambda\mathbf{b}/\vartheta. \tag{57.1}$$

According to (56.10), $\gamma \geq 0$ always. Evidently $\gamma(0) = 0$, which means that the lower bound of $\gamma(\lambda)$ is actually a *minimum* at $\lambda = 0$. If $\mathcal{J}$ and $\mathbf{q}$ are differentiable, it follows that $\gamma'(0) = 0$ (a necessary, but not a sufficient, condition for a minimum). According to (57.1), this condition requires

$$\mathcal{J}(\rho, \vartheta, 0, \mathbf{B}, 0) \cdot \mathbf{a} - \mathbf{q}(\rho, \vartheta, 0, \mathbf{B}, 0) \cdot \mathbf{b}/\vartheta = 0 \tag{57.2}$$

for arbitrary ρ, ϑ, $\mathbf{B}$, $\mathbf{a}$ and $\mathbf{b}$. Hence

$$\mathcal{J}(\rho, \vartheta, 0, \mathbf{B}, 0) = \mathbf{q}(\rho, \vartheta, 0, \mathbf{B}, 0) = 0, \tag{57.3}$$

which means that $\mathcal{J}$ and $\mathbf{q}$ must both vanish whenever the electromotive intensity $\mathcal{E}$ and the temperature gradient both vanish, that is, whenever the state is a special one. We emphasize that this is a corollary, arrived at by Coleman and Noll's method. We have certainly not assumed anything like Ohm's law $\mathcal{J} = C\mathcal{E}$ or Fourier's law $\mathbf{q} = -\kappa\,\mathbf{grad}\,\vartheta$. (In classical thermodynamics 'equilibrium' is defined as a state in which, among other things, $\mathbf{grad}\,\vartheta$ and $\mathbf{q}$ *both* vanish.)

We have just shown that the vectors $\mathcal{J}$ and $\mathbf{q}$ must both vanish whenever $\mathcal{E} = \mathbf{grad}\,\vartheta = 0$, but this corollary does not exhaust the implications of the inequality (56.10), because it is a necessary, not a sufficient, condition for the inequality to hold. For the subclass of *linearly conducting perfect fluids*, $\mathcal{J}$ and $\mathbf{q}$ are assumed to be linear in $\mathcal{E}$ and $\mathbf{grad}\,\vartheta$. By the corollary, these linear relations must be homogeneous:

$$\mathcal{J} = a\mathcal{E} + b\,\mathbf{grad}\,\vartheta,$$
$$\mathbf{q} = c\mathcal{E} + d\,\mathbf{grad}\,\vartheta, \tag{57.4}$$

where a, b, c and d are matrices with elements depending on ρ, ϑ and $\mathbf{B}$. These matrices are still subject to conditions that follow from (56.10). Writing $a = \frac{1}{2}(a + a^T) + \frac{1}{2}(a - a^T)$ and introducing a vector $\mathbf{a}$ through the equation

$$\epsilon_{ijk}a_k = \tfrac{1}{2}(a_{ij} - a_{ji}),$$

we have

$$a\mathcal{E} = \tfrac{1}{2}(a + a^T)\mathcal{E} + \mathcal{E} \times \mathbf{a}.$$

Instead of (57.4), we can therefore write for linearly conducting materials the relations

$$\mathcal{J} = a\mathcal{E} + \mathcal{E} \times \mathbf{a} + b\,\mathbf{grad}\,\vartheta + \mathbf{grad}\,\vartheta \times \mathbf{b},$$
$$\mathbf{q} = c\mathcal{E} + \mathcal{E} \times \mathbf{c} + d\,\mathbf{grad}\,\vartheta + \mathbf{grad}\,\vartheta \times \mathbf{d}, \tag{57.5}$$

where a, b, c and d are now *symmetric* matrices. These symmetric matrices, and the vectors $\mathbf{a}$, $\mathbf{b}$, $\mathbf{c}$ and $\mathbf{d}$, express material properties. They are not all independent, because the inequality (56.10) imposes restrictions on them. For example, the symmetric matrices a and $(-d)$ must be positive definite. For highly isotropic materials each of the matrices reduces to a multiple of the identity, and each of the vectors reduces to a multiple of $\mathbf{B}$. For such materials, the relations (57.5) are commonly written in the form

$$\mathcal{E} = \mathcal{J}/\sigma + \alpha\,\mathbf{grad}\,\vartheta + \mathcal{R}\mathbf{B} \times \mathcal{J} + \mathcal{N}\mathbf{B} \times \mathbf{grad}\,\vartheta,$$
$$\mathbf{q} = \Pi\mathcal{J} - \kappa\,\mathbf{grad}\,\vartheta + \mathcal{S}\mathbf{B} \times \mathcal{J} + \mathcal{L}\mathbf{B} \times \mathbf{grad}\,\vartheta. \tag{57.6}$$

Each term in (57.6) is responsible for a well-known effect. For example, the term $\mathcal{R}\mathbf{B} \times \mathcal{J}$ in (57.6)₁ gives rise to an electromotive intensity which is perpendicular to the conduction current and to the magnetic field; this is called the *Hall effect*. The term with the coefficient $\mathcal{N}$ is connected with an $\mathcal{E}$ which is normal to both the magnetic field and the temperature gradient; this is the *Nernst effect*. The third term on the right-hand side of (57.6)₂ gives rise to a heat flux perpendicular to the magnetic field and to the conduction current; this is the *Ettingshausen effect*. The last term, with the coefficient $\mathcal{L}$, gives a heat flux normal to the temperature gradient and to the magnetic field; this is called the *Leduc–Righi effect*.

If the transverse terms are absent, as they must be whenever the magnetic field is negligible, we have

$$\mathcal{E} = \mathcal{J}/\sigma + \alpha \, \mathbf{grad} \, \vartheta,$$
$$\mathbf{q} = \Pi \mathcal{J} - \kappa \, \mathbf{grad} \, \vartheta. \tag{57.7}$$

If α, called *the Thomson coefficient*, vanishes, $(57.7)_1$ becomes Ohm's law $\mathcal{J} = \sigma \mathcal{E}$. The coefficient σ is called *the electric conductivity*. If Π, called *the Peltier coefficient*, vanishes, $(57.7)_2$ reduces to Fourier's law $\mathbf{q} = -\kappa \, \mathbf{grad} \, \vartheta$. The coefficient κ is called *the thermal conductivity*. It is easy to obtain the conditions that the coefficients σ, α, Π and κ must satisfy, because, with (57.7), (56.10) becomes an inequality for a quadratic form. The conditions[†] are:

$$\sigma \geq 0, \qquad \kappa \geq 0, \qquad \vartheta \kappa / \sigma \geq (\Pi - \vartheta \alpha)^2. \tag{57.8}$$

58. Elastic materials

Materials with constitutive relations of the form

$$f = \mathcal{F}(F, \vartheta, \dot{\mathbf{x}}, \mathcal{E}, \mathbf{B}, \mathbf{grad} \, \vartheta), \tag{58.1}$$

where F is the deformation tensor (51.2), are called *elastic*. They are often used as prototypes of solid materials, but this must not be taken too literally. Indeed the remarks following (53.12) apply here as well.

The perfect fluids defined by (56.1) are a subclass of the elastic materials; they depend on the components of F through $\rho \propto (\det F)^{-1}$. Frankly, the reason for calling the materials defined by (58.1) elastic is that $f = \mathcal{F}(F, \vartheta, \mathbf{grad} \, \vartheta)$ is the constitutive relation used in most studies devoted to the theory of elasticity. We simply extend this to electromagnetism by adding the electromagnetic vectors $\dot{\mathbf{x}}$, $\mathcal{E}$ and $\mathbf{B}$.

If we now try to follow the analysis of the previous section with the new kind of constitutive relation, we immediately see that the term $\varphi_\rho \dot{\rho}$ in (56.2) must be replaced by the double sum $(\partial \varphi / \partial F_{ij}) \dot{F}_{ij}$. It is convenient to define a *matrix* φ_F by $(\varphi_F)_{ij} = \partial \varphi / \partial F_{ij}$, so that this sum becomes the matrix scalar product $\varphi_F \cdot \dot{F}$. This causes the replacement, in the entropy inequality (56.3), of the term $(\tau + \rho^2 \varphi_\rho I) \cdot \mathbf{grad} \, \dot{\mathbf{x}}$ by

$$\tau \cdot \mathbf{grad} \, \dot{\mathbf{x}} - \rho \varphi_F \cdot \dot{F}.$$

[†]Onsager proposed in 1931 a theory of irreversible processes, according to which, among other relations, $\Pi = \vartheta \alpha$. The theory was flawed, and Onsager's relations are violated by many substances.

Exercise

58.1. Use (51.10) and tr $BA = $ tr AB to show that

$$\tau \cdot \text{grad}\,\dot{\mathbf{x}} - \rho\varphi_F \cdot \dot{F} = [\tau(F^T)^{-1} - \rho\varphi_F] \cdot \dot{F}.$$

From the last result it follows that (56.8)–(56.9) are replaced by $\tau = \rho\varphi_F F^T$. Indeed the whole argument leading up to (56.4)–(56.10) goes through with this single change, and we obtain the relations

$$s = -\varphi_\vartheta (F, \vartheta, \mathcal{E}, \mathbf{B}), \tag{58.2}$$

$$\mathbf{g} = \dot{\mathbf{x}} + \epsilon_0 \mathbf{E} \times \mathbf{B}/\rho, \tag{58.3}$$

$$\mathbf{P} = -\rho\varphi_\mathcal{E}, \tag{58.4}$$

$$\mathcal{M} = -\rho\varphi_\mathbf{B}, \tag{58.5}$$

$$T = \rho\varphi_F F^T - \left[\tfrac{1}{2}\epsilon_0 E^2 + \tfrac{1}{2} B^2/\mu_0 - \mathcal{M} \cdot \mathbf{B}\right] I$$
$$+ \epsilon_0 \mathbf{E} \otimes \mathbf{E} + \mathbf{B} \otimes \mathbf{B}/\mu_0 + \mathcal{E} \otimes \mathbf{P} - \mathcal{M} \otimes \mathbf{B} + \epsilon_0 \mathbf{E} \times \mathbf{B} \otimes \dot{\mathbf{x}}, \tag{58.6}$$

$$\mathcal{J} \cdot \mathcal{E} - (\mathbf{q} \cdot \text{grad}\,\vartheta)/\vartheta \geq 0. \tag{58.7}$$

It should, perhaps, be noted that the partial derivative φ_ϑ in (58.2) is with F held constant, whereas in the corresponding equation (56.4) it was taken with ρ (or det F) held constant. The same remark applies to $\varphi_\mathcal{E}$ and $\varphi_\mathbf{B}$.

From the remnant (58.7) of the entropy inequality we conclude that, in elastic materials, as in perfect fluids, all special processes throughout which $\mathcal{E} = \text{grad}\,\vartheta = 0$ are reversible. Of course, (58.7) may also become an equality for other kinds of processes. The discussion of the properties of $\mathcal{J}$ and $\mathbf{q}$, and in particular the corollary (57.3) and the relations (57.4) for linearly conducting perfect fluids, can be applied to elastic materials by replacing the argument ρ by F.

In the equation of motion (56.17) we need only replace $-\,\text{grad}\,p = \text{div}\,(-pI)$ by $\text{div}\,\rho\varphi_F F^T$. The energy balance equation (56.18) is unchanged. It must, of course, be understood that, in applying these equations to elastic materials, we must also use the relations (58.2)–(58.5), rather than (56.4)–(56.7): the partial derivatives of φ must be at constant F, rather than at constant ρ.

When we now impose Euler's second law of mechanics as a further restriction on the constitutive relations, we find that the free energy must be such that

$$\varphi(QF, \vartheta, Q\mathcal{E}, Q\mathbf{B}) = \varphi(F, \vartheta, \mathcal{E}, \mathbf{B}) \tag{58.8}$$

hold for every orthogonal matrix Q. This requirement can be satisfied by making use of an algebraic theorem of Cauchy: any non-singular matrix F has a unique decomposition of the form

$$F = RU, \tag{58.9}$$

where R is orthogonal and U is a symmetric, positive definite matrix. We use this decomposition for the deformation F. In (58.8) we may choose the orthogonal matrix R^T for Q. Then $QF = R^T RU = U$, $Q\mathcal{E} = R^T \mathcal{E} = U^{-1}UR^T\mathcal{E} = U^{-1}F^T\mathcal{E}$ and, similarly, $Q\mathbf{B} = U^{-1}F^T\mathbf{B}$. Hence

$$\varphi(F, \vartheta, \mathcal{E}, \mathbf{B}) = \varphi(U, \vartheta, U^{-1}F^T\mathcal{E}, U^{-1}F^T\mathbf{B}).$$

But $F^T F = (RU)^T RU = U^T U = U^2$. It follows that φ is a function of $F^T F$, ϑ, $F^T\mathcal{E}$ and $F^T\mathbf{B}$:

$$\varphi = \Phi(F^T F, \vartheta, F^T\mathcal{E}, F^T\mathbf{B}). \tag{58.10}$$

Conversely, if φ is *any* function of the form (58.10), it automatically satisfies the requirement (58.8) for *any* orthogonal Q. We shall therefore refer to (58.10) as the free energy in *reduced* form. Whereas $\varphi(F, \vartheta, \mathcal{E}, \mathbf{B})$ depends on the nine components of the deformation matrix F, the reduced Φ depends only on the six independent components of the *symmetric* matrix $F^T F$. This reduction in the number of arguments is, of course, a direct result of imposing the requirements expressed by the three components of Euler's second law of mechanics.

59. Viscous fluids

Viscous fluids are those in which velocity gradients matter, so that the nine components of the velocity gradient grad $\dot{\mathbf{x}}$ must be added as further arguments in the constitutive relations. Instead of (56.1), they have the form

$$f = \mathcal{F}(\rho, \vartheta, \dot{\mathbf{x}}, \mathcal{E}, \mathbf{B}, \mathbf{grad}\,\vartheta, \text{grad}\,\dot{\mathbf{x}}). \tag{59.1}$$

Starting from the entropy inequality (55.5), it can easily be shown that φ does not depend on *either* $\mathbf{grad}\,\vartheta$, *or* grad $\dot{\mathbf{x}}$, or $\dot{\mathbf{x}}$. The relations (56.4)–(56.7) still hold. Moreover, like the φ of Section 56, and for the same reason, φ in a viscous fluid can only depend on the five arguments ρ, ϑ, $\mathcal{E}^2$, B^2 and $(\mathcal{E} \cdot \mathbf{B})^2$. But since the τ of (55.6) now depends on grad $\dot{\mathbf{x}}$, $\tau + pI$ no longer vanishes. In fact, it now joins $\mathcal{J}$ and $\mathbf{q}$ in the inequality

$$(\tau + pI) \cdot \text{grad}\,\dot{\mathbf{x}} + \mathcal{J} \cdot \mathcal{E} - (\mathbf{q} \cdot \mathbf{grad}\,\vartheta)/\vartheta \geq 0, \tag{59.2}$$

which replaces (56.8)–(56.10). It follows that, in viscous fluids, all processes throughout which $\mathcal{E}$, $\mathbf{grad}\,\vartheta$ *and* grad $\dot{\mathbf{x}}$ all vanish are reversible. Of course, there may be other processes for which (59.2) becomes an equality.

By a direct generalization of the argument that has led to (57.3) we arrive at the following result: whenever grad $\dot{\mathbf{x}}$, $\mathcal{E}$ and **grad** ϑ all vanish, so do $\tau + pI$, $\mathcal{J}$ and $\mathbf{q}$; in particular, the stress T will then be given by (56.8)–(56.9). For viscous fluids, the special states are obviously those in which the temperature and the velocity are both uniform; moreover, the motion and the electromagnetic field are everywhere constrained by $\mathbf{E} + \dot{\mathbf{x}} \times \mathbf{B} = 0$. The special processes that are sequences of these special states have the essential properties that classical thermodynamics associates with 'quasi-static processes'. For example, along a special process, $\Delta S = \int (\mathcal{Q}/\vartheta)\, dt$.

In *linearly viscous* fluids, $\tau + pI$, $\mathcal{J}$ and $\mathbf{q}$ are assumed to be linear in grad $\dot{\mathbf{x}}$, $\mathcal{E}$ and **grad** ϑ. By the arguments of Section 57 these linear relations must be homogeneous. In isotropic materials, they become especially simple: the matrix $\tau + pI$ cannot depend on either of the vectors $\mathcal{E}$ or **grad** ϑ, and the vectors $\mathcal{J}$ and $\mathbf{q}$ cannot depend on the matrix grad $\dot{\mathbf{x}}$. For $\mathcal{J}$ and $\mathbf{q}$ we are, again, led to (57.4). For $\tau + pI$, we obtain a linear relation of the form

$$\tau + pI = 2\lambda d + \zeta (\text{div } \dot{\mathbf{x}}) I, \tag{59.3}$$

where

$$2d = \text{grad } \dot{\mathbf{x}} + (\text{grad } \dot{\mathbf{x}})^T, \tag{59.4}$$

and λ and ζ are scalar coefficients. Finally, it follows from (59.2) that λ and ζ are restricted by the *Stokes–Duhem inequalities*

$$\lambda \geq 0, \qquad 3\zeta + 2\lambda \geq 0. \tag{59.5}$$

Exercises

59.1. Show that, for uniform λ and ζ,

$$\mathbf{div}\, [2\lambda d + \zeta (\text{div } \dot{\mathbf{x}}) I] = (2\lambda + \zeta)\, \mathbf{grad}\, \text{div } \dot{\mathbf{x}} - \lambda\, \mathbf{curl}^2\, \dot{\mathbf{x}}. \tag{59.6}$$

In a viscous fluid, these are the the terms that should be added on the right-hand side of (56.17).

59.2. Show that, for linearly viscous fluids, the law of energy balance (cf. (56.18)) becomes

$$\rho \vartheta \dot{s} = \mathcal{J} \cdot \mathcal{E} + 2\lambda (d \cdot d) + \zeta (\text{tr } d)^2 + \rho h - \text{div } \mathbf{q}. \tag{59.7}$$

Because of the first three terms on the right-hand side of (59.7), an adiabatic process, in which the last two terms vanish, is generally not an isentropic one.

16

Magnetohydrodynamics

60. Purely conducting fluids

Among the electromagnetic fluids, the subclass of those that are neither polarizable nor magnetizable is especially simple: according to (56.6) and (56.7), they are characterized by a free energy that depends only on ρ and ϑ. Such materials, sometimes called *pure conductors*, furnish the subject matter of *magnetohydrodynamics*. For them eqn (56.17) becomes

$$\rho \ddot{\mathbf{x}} = -\operatorname{\mathbf{grad}} p(\rho, \vartheta) + q\boldsymbol{\mathcal{E}} + \boldsymbol{\mathcal{J}} \times \mathbf{B} + \rho \mathbf{b}$$
$$= -\operatorname{\mathbf{grad}} p + q\mathbf{E} + \mathbf{j} \times \mathbf{B} + \rho \mathbf{b}. \tag{60.1}$$

For this subclass of materials, the $\operatorname{\mathbf{grad}} p(\rho, \vartheta)$ term is purely 'thermodynamic', because it does not involve $\mathbf{E}$ or $\mathbf{B}$. Similarly (again, only for this subclass), the specific energy (cf. (56.11)),

$$\varepsilon = \varphi(\rho, \vartheta) - \vartheta \varphi_\vartheta(\rho, \vartheta) + \tfrac{1}{2}\dot{x}^2 + \left[\tfrac{1}{2}\epsilon_0 E^2 + \tfrac{1}{2}B^2/\mu_0 \right]/\rho, \tag{60.2}$$

is the sum of parts which may be called thermodynamic, kinetic, electric and magnetic. There is no harm in using these names for the various terms in (60.2). But there is the real danger that use may lead to abuse: for example, it is easy to forget that the statement '$\tfrac{1}{2}\epsilon_0 E^2$ is the electric energy density' is just the consequence of a particularly simple constitutive relation, and then proceed to apply it to polarizable materials—with disastrous effects. Indeed the relation for general fluids is (56.11). Thus, for example, $\int \tfrac{1}{2}D^2/\epsilon d^3x$, which was the subject of Thomson's theorem (Section 31), is not the energy of a linear dielectric. Nor is it the free energy (which we shall derive in Section 66).

For pure conductors, the stress tensor is given by

$$T = -\left[p(\rho, \vartheta) + \tfrac{1}{2}\epsilon_0 E^2 + \tfrac{1}{2}B^2/\mu_0 \right] I$$
$$+ \epsilon_0 \mathbf{E} \otimes \mathbf{E} + \mathbf{B} \otimes \mathbf{B}/\mu_0 + \epsilon_0 \mathbf{E} \times \mathbf{B} \otimes \dot{\mathbf{x}}. \tag{60.3}$$

The *incompressible* fluids form another useful subclass. These fluids have a constant density, which may be omitted from the constitutive relations. By the law of mass conservation their velocity fields are constrained to be volume-preserving: div $\dot{\mathbf{x}} = 0$. This has the following consequences: first, the terms involving φ_ρ drop out of (56.2) and (56.3), and the sixth term in (56.3) becomes simply $\tau \cdot$ grad $\dot{\mathbf{x}}$. Secondly, this scalar product of the matrices τ and grad $\dot{\mathbf{x}}$ has to satisfy the entropy inequality, but now grad $\dot{\mathbf{x}}$ is not arbitrary, because div $\dot{\mathbf{x}}$ must vanish. We note, however, that div $\dot{\mathbf{x}} = I \cdot$ grad $\dot{\mathbf{x}}$. The requirement that $\tau \cdot$ grad $\dot{\mathbf{x}} = 0$ for all grad $\dot{\mathbf{x}}$ such that $I \cdot$ grad $\dot{\mathbf{x}} = 0$ therefore means that τ must be perpendicular to every matrix that is perpendicular to the unit matrix. It follows that τ must be parallel to I:

$$\tau = -pI. \tag{60.4}$$

This relation, in which the pressure p is now an arbitrary number (at each event in space-time), replaces (56.9). All subsequent relations are unchanged if we replace $\rho^2 \varphi_\rho$ everywhere by $p(\mathbf{x}, t)$ and remember that ρ is a given constant.

It should, perhaps, be noted that the pressure $p(\mathbf{x}, t)$ in an incompressible fluid will generally not be the same as in a *compressible* fluid that happens to undergo a volume-preserving motion, even if it is the *same* motion. In the latter case p must equal $\rho^2 \varphi_\rho$, and in the resulting solution ρ will generally fail to be a constant.

61. The magnetohydrodynamic approximation

Our use of classical, non-relativistic mechanics already requires the assumption $\dot{x}^2 \ll c^2$. We shall now seek to determine the circumstances under which the displacement term $\mathbf{D}_t$ and the convection current density $q\dot{\mathbf{x}}$ can be neglected.

Let ω^{-1} and ℓ be the characteristic time and length scales over which the electromagnetic fields vary. Then **curl E** $= -\mathbf{B}_t$ requires $E/\ell \sim \omega B$ or $E \sim \omega \ell B$. The displacement term $\mathbf{D}_t$ is thus of order $\omega \epsilon_0 E \sim \epsilon_0 \omega^2 \ell B$. We compare it with **curl H**, which is of order $(\mu_0 \ell)^{-1} B$. The requirement is therefore that $\epsilon_0 \mu_0 (\omega \ell)^2 \ll 1$:

$$\epsilon_0 \mu_0 (\omega \ell)^2 = \left(\frac{\omega \ell}{c}\right)^2 \ll 1 \quad \Longrightarrow \quad \mathbf{curl\ H} = \mathbf{j}. \tag{61.1}$$

The convection current density $q\dot{\mathbf{x}} = (\text{div } \mathbf{D})\dot{\mathbf{x}}$ is of order $\ell^{-1} \epsilon_0 E \dot{x} \sim \epsilon_0 \omega \dot{x} B$. We compare it with $\mathbf{j} = \mathbf{curl\ H}$, which is of order $(\mu_0 \ell)^{-1} B$. Thus we can neglect $q\dot{\mathbf{x}}$ in Ohm's law $\mathbf{j} - q\dot{\mathbf{x}} = \sigma \mathcal{E}$ for a moving body if $\epsilon_0 \mu_0 \omega \ell \dot{x} \ll 1$:

$$\epsilon_0 \mu_0 \omega \ell \dot{x} = \frac{\omega \ell \dot{x}}{c^2} \ll 1 \quad \Longrightarrow \quad \mathbf{j} = \sigma(\mathbf{E} + \dot{\mathbf{x}} \times \mathbf{B}). \tag{61.2}$$

The two inequalities in (61.1)–(61.2) constitute the *magnetohydrodynamic approximation*. They have further consequences. First, the electric energy density $\frac{1}{2}\epsilon_0 E^2 \sim \epsilon_0 (\omega \ell)^2 B^2$ becomes small, compared with the magnetic energy density $\frac{1}{2} B^2/\mu_0$, because their ratio is $\epsilon_0\mu_0(\omega\ell)^2 \ll 1$. The same relation holds between the terms $\epsilon_0\mathbf{E} \otimes \mathbf{E}$ and $\mathbf{B} \otimes \mathbf{B}/\mu_0$ in the stress tensor T. Finally, in the equation of motion the term $q\mathbf{E}$, which is of order $\epsilon_0 E^2/\ell$, can be neglected in comparison with $\mathbf{j} \times \mathbf{B}$, which is of order $B^2/(\mu_0\ell)$.

In the magnetohydrodynamic (MHD) approximation a linearly viscous, linearly conducting fluid is therefore governed by the following equations:

$$\mathbf{curl\,B} = \mu_0\mathbf{j}, \qquad \mathbf{j} = \sigma(\mathbf{E} + \dot{\mathbf{x}} \times \mathbf{B}),$$
$$\mathrm{div\,}\mathbf{B} = 0, \qquad \mathbf{curl\,E} = -\mathbf{B}_t,$$
$$T = -\left[p + \tfrac{1}{2}B^2/\mu_0 + \zeta\,\mathrm{div}\,\dot{\mathbf{x}}\right]I + \mathbf{B} \otimes \mathbf{B}/\mu_0 + 2\lambda d,$$
$$\varepsilon = \varphi - \vartheta\varphi_\vartheta + \tfrac{1}{2}\dot{x}^2 + \tfrac{1}{2}B^2/(\mu_0\rho), \tag{61.3}$$
$$\rho\ddot{\mathbf{x}} = -\mathbf{grad\,}p + \mathbf{j} \times \mathbf{B} + (2\lambda + \zeta)\,\mathbf{grad}\,\mathrm{div}\,\dot{\mathbf{x}} - \lambda\,\mathbf{curl}^2\,\dot{\mathbf{x}} + \rho\mathbf{b},$$
$$\rho\vartheta\dot{s} = j^2/\sigma + 2\lambda d \cdot d + \zeta(\mathrm{tr}\,d)^2 + \rho h - \mathrm{div}\,\mathbf{q},$$

where λ and ζ are the coefficients of viscosity (Section 59). The system (61.3) constitutes the laws of magnetohydrodynamics. The name signifies the preponderance of the magnetic terms, but it would be a mistake to conclude that the electric field is unimportant, let alone negligible. The equations $\mathrm{div}\,\epsilon_0\mathbf{E} = q$ and $\mathbf{n} \cdot [\![\epsilon_0\mathbf{E}]\!] = \sigma$ still hold, of course; they are omitted from (61.3) only because they are read from right to left (cf. Section 24), with $\mathbf{E} = \mathbf{j}/\sigma - \dot{\mathbf{x}} \times \mathbf{B}$ given by Ohm's law $(61.3)_2$. But wherever there is a vacuum, as there must be around any finite system of conductors, Ohm's law is indeed *replaced* by $\mathrm{div}\,\mathbf{D} = \mathrm{div}\,\epsilon_0\mathbf{E} = 0$, and $(61.3)_1$ by $\mathbf{curl\,B} = 0$.

The magnetic part $\mathbf{j} \times \mathbf{B}$ of the force density is the divergence of the magnetic part $T_m = \mu_0^{-1}[-(\tfrac{1}{2}B^2)I + \mathbf{B} \otimes \mathbf{B}]$ of the stress tensor. The first term of this stress, which is proportional to the unit matrix, is equivalent to a *magnetic pressure*

$$p_m = \frac{B^2}{2\mu_0}. \tag{61.4}$$

The second term $\mu_0^{-1}\mathbf{B} \otimes \mathbf{B}$ represents a *tension* B^2/μ_0 along the direction of $\mathbf{B}$. The magnetic field lines therefore act like taut strings which repel one another. One may expect to find transverse waves travelling along the field lines with a velocity u_A, called *the Alfvén velocity*, determined by the rule 'squared phase speed equals tension divided by density', that is,

$$u_A^2 = \frac{B^2}{\mu_0\rho}. \tag{61.5}$$

Exercise

61.1. Prove the identity

$$(\text{grad } \dot{\mathbf{x}})\dot{\mathbf{x}} = (\textbf{curl } \dot{\mathbf{x}}) \times \dot{\mathbf{x}} + \textbf{grad } \tfrac{1}{2}\dot{x}^2. \tag{61.6}$$

Hence deduce that $(61.3)_7$ can be written in the form

$$\rho \left[\dot{\mathbf{x}}_t + (\textbf{curl } \dot{\mathbf{x}}) \times \dot{\mathbf{x}} + \textbf{grad } \tfrac{1}{2}\dot{x}^2 \right]$$
$$= - \textbf{grad } p + (\textbf{curl } \mathbf{B}) \times \mathbf{B}/\mu_0 + (2\lambda + \zeta) \textbf{grad } \text{div } \dot{\mathbf{x}}$$
$$- \lambda \textbf{curl}^2 \dot{\mathbf{x}} + \rho \mathbf{b}. \tag{61.7}$$

---

62. Hartmann's flow

As a non-trivial example of (61.3), we consider the steady, rectilinear flow (in the x direction) of an incompressible, conducting, viscous fluid between two parallel horizontal planes (at $z = \pm a$) in the presence of an imposed magnetic field in the direction normal to the planes[†].

We shall assume all quantities, except for the pressure, to depend on z only; a pressure gradient in the flow direction x may be necessary for maintaining the viscous flow. The condition div $\dot{\mathbf{x}} = 0$ for incompressible flow is satisfied by any velocity field $v_x(z)$ in the x direction.

Consider first the flow in the absence of a magnetic field. For the external body force we substitute gravity, $\mathbf{b} = \mathbf{g}$. For incompressible flow $\textbf{curl}^2 \mathbf{v} = \textbf{grad } \text{div } \mathbf{v} - \Delta \mathbf{v} = -\Delta \mathbf{v}$. Since the flow is steady (unaccelerated), the equation of motion is

$$\textbf{grad } p = \lambda \Delta \mathbf{v} + \rho \mathbf{g}. \tag{62.1}$$

Since $\mathbf{v}$ depends only on z it follows that $\textbf{grad}(\partial p/\partial x) = 0$. Thus $\partial p/\partial x$ is a constant. From the x component of (62.1),

$$\frac{\partial p}{\partial x} = \lambda \frac{d^2 v_x}{dz^2}, \tag{62.2}$$

it then follows that $v_x(z)$ is a polynomial of second degree. We impose the boundary condition that v_x is to vanish at $z = \pm a$, where the viscous fluid is assumed to adhere to the walls. Then

$$v_x = -\frac{1}{\lambda} \frac{\partial p}{\partial x} (a^2 - z^2). \tag{62.3}$$

[†]Hartmann (1937).

In terms of the velocity $v_x(0)$ on the mid-plane $z = 0$,

$$v_x = v_x(0)(1 - z^2/a^2). \tag{62.4}$$

This simple parabolic law for incompressible, viscous flow between parallel planes is called *Poiseuille flow*.

Now we turn to the MHD case. From div $\mathbf{B} = dB_z/dz = 0$ we see that an imposed, uniform B_z is indeed a consistent assumption. From Ohm's law we have $j_y = \sigma(E_y - v_x B_z)$, where E_y is constant because **curl E** $= 0$. The equation of motion becomes

$$\mathbf{grad}\, p = \mathbf{j} \times \mathbf{B} + \lambda \Delta \mathbf{v} + \rho \mathbf{g}. \tag{62.5}$$

Since p is the only quantity which may depend on x we find, again, that $\mathbf{grad}(\partial p/\partial x) = 0$. Thus, again, $\partial p/\partial x$ is a constant. The x component of (62.5) is

$$\frac{\partial p}{\partial x} = j_y B_z + \lambda \frac{d^2 v_x}{dz^2}. \tag{62.6}$$

Substituting Ohm's law yields

$$\lambda \frac{d^2 v_x}{dz^2} - \sigma B_z^2 v_x = \frac{\partial p}{\partial x} - \sigma E_y B_z. \tag{62.7}$$

This is an ordinary differential equation with constant coefficients (and a constant inhomogeneous term). It is easily solved in accordance with the no-slip boundary conditions on the planes. In terms of the mid-plane velocity, the solution is

$$v_x = v_x(0) \frac{\cosh(a/h) - \cosh(z/h)}{\cosh(a/h) - 1}, \tag{62.8}$$

where $h = (\lambda/\sigma)^{1/2}/B_z$. The velocity profile (62.8) depends on B_z through the *Hartmann number* a/h. When $a/h \to 0$ (or $B_z \to 0$), the velocity profile becomes parabolic:

$$v_x = v_x(0)(1 - z^2/a^2), \tag{62.9}$$

as it should, because in this limit we recover the Poiseuille flow. When $a/h \to \infty$ (or $B_z \to \infty$), the profile becomes flattened, except for boundary layers of thickness $\approx h$ along the planes:

$$v_x = v_x(0)(1 - e^{(a-|z|)/h}). \tag{62.10}$$

The foregoing solution provides an approximation to the flow in a rectangular channel if the width in the y direction is large compared with $2a$. If we integrate $j_y = \sigma(E_y - v_x B_z)$ with respect to z, we obtain

$$i_y = 2a\sigma(E_y - B_z v_m), \tag{62.11}$$

where $i_y = \int j_y \, dz$ is the total current per unit lateral length of the channel and $v_m = (2a)^{-1} \int v_x \, dz$ is the mean velocity. The current i_y will depend on the external return path between the lateral planes ($y = \pm \text{const.}$). If this path is open, $i_y = 0$. The electric field component $E_y = B_z v_m$ can be measured, and the arrangement provides an electromagnetic flow meter. If the path is short-circuited, $E_y = 0$ and $i_y = -2a\sigma B_z v_m$. In other cases, depending on the external resistance and the pressure gradient $-\partial p / \partial x$, the device may function as an electromagnetic pump, brake or generator.

63. Magnetic diffusion and flux freezing

From $\text{div } \mathbf{B} = 0$ and Ohm's law $\mathbf{j} = \sigma(\mathbf{E} + \dot{\mathbf{x}} \times \mathbf{B})$ we obtain

$$\begin{aligned}
\text{curl}^2 \mathbf{B} &= \text{grad div } \mathbf{B} - \Delta \mathbf{B} \\
&= -\Delta \mathbf{B} \\
&= \mu_0 \, \text{curl } \mathbf{j} \\
&= \mu_0 \sigma [\text{curl } \mathbf{E} + \text{curl } (\dot{\mathbf{x}} \times \mathbf{B})] \\
&= \mu_0 \sigma [-\mathbf{B}_t + \text{curl } (\dot{\mathbf{x}} \times \mathbf{B})],
\end{aligned}$$

or

$$\mathbf{B}_t = \frac{1}{\mu_0 \sigma} \Delta \mathbf{B} + \text{curl } (\dot{\mathbf{x}} \times \mathbf{B}). \tag{63.1}$$

In stationary matter, we have

$$\mathbf{B}_t = \frac{1}{\mu_0 \sigma} \Delta \mathbf{B}. \tag{63.2}$$

This is, mathematically speaking, a diffusion equation. For an isolated body it leads to a decay, or relaxation, of the magnetic field on a time scale of $\mu_0 \sigma \ell^2 / (2\pi)^2$. For example, if the earth's magnetic field has its sources in electric currents that flow in its conducting, molten core, the magnetic field should decay after a few times 10^4 years (based on $\sigma \approx 10^7$ siemen m^{-1} for the molten iron in the earth's core, and $\ell \approx 10^3$ km).

Since magnetically aligned sediments have been found that are *millions* of years old, we infer that the earth's interior must have been stirred, at least occasionally, because only the last term of (63.1) can halt the decay of the magnetic field. Such stirring could be provided by thermal convection, but the proof that convection can actually maintain the magnetic field against the inexorable diffusive decay is difficult, and constitutes the *dynamo problem*.

In some situations, mainly astronomical, the magnetic diffusion term—the first on the right-hand side of (63.1)—may become small. This will happen whenever

$$\mu_0 \sigma \omega \ell^2 \gg 1 \qquad \text{or} \qquad \mu_0 \sigma \dot{x} \ell \gg 1. \tag{63.3}$$

Under such circumstances (loosely called 'infinite conductivity') the flux derivative

$$\overset{*}{\mathbf{B}} = \mathbf{B}_t + \dot{\mathbf{x}} \operatorname{div} \mathbf{B} - \mathbf{curl}\,(\dot{\mathbf{x}} \times \mathbf{B})$$

will vanish, and the magnetic flux $\int B_n\, dS$ through any open surface, which is carried along with the material, will be constant (cf. Section 25)—no matter what the motion is. This is called *(magnetic) flux freezing*. It can of course be directly derived from Faraday's law of induction for a moving surface (cf. (25.12)),

$$\oint_c \mathcal{E} \cdot \mathbf{ds} = -\frac{d}{dt} \int B_n\, dS$$

by taking the limit $\mathcal{E} = \mathbf{j}/\sigma \to 0$ as $\sigma \to \infty$.

Thus, whenever the conductivity satisfies one of the inequalities (63.3), the material behaves like a perfect conductor: Ohm's law $\mathcal{E} = \mathbf{j}/\sigma$ reduces to $\mathcal{E} = 0$, the Joule heating term j^2/σ in (61.3)$_8$ becomes negligible, and the magnetic field is 'frozen into the matter', which means that the number of magnetic field lines through any closed curve (cf. Section 8) that is swept along with the fluid motion is a constant.

Exercises

In the following exercises the fluid is assumed to be non-viscous.

63.1. Let $\mathbf{B}$ be any solenoidal field. Show that, if $\mathbf{b} = -\mathbf{grad}\,U$, solutions of the MHD equations (61.3) for a perfectly conducting, incompressible fluid are obtained by choosing

$$\dot{\mathbf{x}} = \pm \frac{\mathbf{B}}{\sqrt{\mu_0 \rho}} \qquad \text{and} \qquad p + \tfrac{1}{2}\rho\dot{x}^2 + \rho U = \text{const.}$$

63.2. Let $\mathbf{B}$ be a solenoidal field which becomes uniform, $\mathbf{B} \to \mathbf{B}_\infty$, at infinity. For a perfectly conducting, incompressible fluid with a conservative body force, the simple solutions of Exercise 61.1 tend to a constant flow, $\mathbf{u} = \pm \mathbf{B}_\infty/\sqrt{\mu_0 \rho}$, at infinity. By applying a Galilean transformation $\mathbf{x}' = \mathbf{x} - \mathbf{u}t$, show that the solutions in the (x') frame become

$$\mathbf{B}' = \mathbf{B}(\mathbf{x}' \pm \mathbf{u}t),$$

$$\dot{\mathbf{x}}' = \pm \frac{\mathbf{B}(\mathbf{x}' \pm \mathbf{u}t) - \mathbf{B}_\infty}{\sqrt{\mu_0 \rho}}.$$

These are finite-amplitude waves (called *Alfvén waves*) travelling at the constant velocity $-\mathbf{u} = \mp \mathbf{B}_\infty/\sqrt{\mu_0 \rho}$.

---

64. Magnetohydrodynamic equilibrium

In the absence of motion we have

$$0 = -\operatorname{grad} p + \mathbf{j} \times \mathbf{B} + \rho \mathbf{b}. \qquad (64.1)$$

In practical applications the external body force is due to gravity, $\mathbf{b} = \mathbf{g}$. The boldest assumption one can make about the magnetic force $\mathbf{j} \times \mathbf{B}$ is that it vanishes. Such *force-free* fields seem to be required in the outer regions of stars, where $\operatorname{div} \mathbf{B} = 0$ requires $\int B_n \, dS$ to remain constant along any magnetic flux tube emerging from the interior, so that the magnetic field cannot diminish very quickly. The temperature gradient is constrained by the requirement that the star radiates the heat produced in its interior, and this constraint forces the density to diminish along with the pressure. Ultimately, when $p \ll B^2/(2\mu_0)$, the terms $-\operatorname{grad} p + \rho \mathbf{g}$ can no longer provide a balance against a non-zero magnetic force.

For a magnetic field to be force-free, $\mathbf{j} \times \mathbf{B} = 0$, the current $\mathbf{j} = \mu_0^{-1} \operatorname{curl} \mathbf{B}$ must either vanish, in which case $\mathbf{B}$ becomes a potential field; or $\mathbf{j}$ must be parallel to $\mathbf{B}$,

$$\operatorname{curl} \mathbf{B} = \alpha \mathbf{B}. \qquad (64.2)$$

The function α is not quite arbitrary: taking the divergence of the last equation, we obtain $0 = \mathbf{B} \cdot \operatorname{grad} \alpha$, which requires α to be constant along each field line. If the constant is the same along all field lines, α is constant everywhere. Applying another **curl** to (64.2) we then obtain

$$\operatorname{curl}^2 \mathbf{B} = \alpha^2 \mathbf{B}. \qquad (64.3)$$

Mathematically speaking, the solutions of this equation are the same as the monochromatic fields we have encountered in connection with the skin effect (cf. (39.5)). In axisymmetric situations they become Bessel functions (cf. (39.6)). Still, although mathematically tractable, the assumption that α is constant everywhere, not just along each field line, is hard to justify.

If the magnetic field is not force-free, equilibrium in the absence of an external body force $\mathbf{b}$ is governed by the equations

$$\begin{aligned}
\operatorname{curl} \mathbf{B} &= \mu_0 \mathbf{j}, \\
\operatorname{div} \mathbf{B} &= 0, \\
T &= -\left[p + \tfrac{1}{2} B^2/\mu_0 \right] I + \mathbf{B} \otimes \mathbf{B}/\mu_0, \\
\operatorname{grad} p &= \mathbf{j} \times \mathbf{B}.
\end{aligned} \qquad (64.4)$$

According to the last line, the current and the field are both perpendicular to the pressure gradient. Thus the vectors $\mathbf{B}$ and $\mathbf{j}$ must both lie on the surfaces $p = \text{const.}$ Any such surface may be regarded as the boundary of the fluid: eqn $(64.4)_4$ only determines p to within an arbitrary constant, so that any isobaric surface may be taken as $p = 0$.

Equation (64.4)$_4$ is of course the same as **div** $T = 0$. We integrate the scalar product $\mathbf{x} \cdot$ **div** $T = x_i \partial_j T_{ij}$ over a volume. Noting that $x_i \partial_j T_{ij} = \partial_j (x_i T_{ij}) - T_{ii}$ and using Gauss's theorem $\int dV \partial_j = \oint dS n_j$, we obtain

$$\int \left(3p + \frac{B^2}{2\mu_0} \right) dV + \oint \mathbf{x} \cdot T\mathbf{n} \, dS = 0. \tag{64.5}$$

At a surface of discontinuity T must satisfy the jump condition (54.3), which in the present case is simply $[\![T]\!]\mathbf{n} = 0$. Thus $T\mathbf{n}$ is continuous, and (64.5) holds for *any* region. If the fluid and the field sources both occupy a finite region, we can extend the region of integration in (64.5) to infinity. The surface integral will vanish because $\mathbf{B} \to 1/r^3$, but the volume integral is positive, which means that the assumptions are inconsistent. Thus the fluid may occupy a finite region, but the field sources then cannot (and the surface integral will then extend over *their* surfaces). Or else the whole configuration must be unbounded.

As an example of a simple unbounded configuration, we consider an infinite cylinder of fluid. Using cylindrical coordinates (r, ϕ, z), with all quantities dependent on r only, we find, as usual, that B_r must vanish; otherwise div $\mathbf{B} = r^{-1}d(r B_r)/dr = 0$ would cause it to become singular on the axis. Since div $\mathbf{j} = $ div μ_0^{-1} **curl** $\mathbf{B} = 0$, j_r must vanish too. The non-zero components of (64.4)$_1$ are

$$j_\phi = -\frac{1}{\mu_0}\frac{d B_z}{dr}, \qquad j_z = \frac{1}{\mu_0}\frac{d(r B_\phi)}{dr}. \tag{64.6}$$

From the second of these we obtain

$$B_\phi = \frac{\mu_0 i(r)}{2\pi r}, \qquad i(r) = \int_0^r 2\pi r j_z \, dr. \tag{64.7}$$

Thus the equation of equilibrium (64.4)$_4$ becomes

$$\frac{dp}{dr} + \frac{1}{2\mu_0}\frac{d B_z^2}{dr} + \frac{1}{8\pi^2 r^2}\frac{d i^2(r)}{dr} = 0, \tag{64.8}$$

which is the *pinch* equation. The name refers to the action of the magnetic force, which balances the non-zero pressure gradient.

Exercises

64.1. Solve (64.8) for the *linear pinch*, when only $j_z \neq 0$.

64.2. Solve (64.8) for the *theta pinch*, when only $j_\phi \neq 0$.

65. Magnetohydrodynamic waves

We shall consider small, adiabatic perturbations in an inviscid, perfectly conducting fluid at rest. The unperturbed fluid is supposed to be uniform, and we assume that a uniform magnetic field B in the x direction is present. Let ρ', p' and $\mathbf{B}'$ be the perturbations in the density, pressure and magnetic field, and let $\mathbf{v}$ and $\mathbf{E}$ (themselves small quantities) be the velocity and electric field in the perturbed state. The linearized equation of continuity is

$$\frac{\partial \rho'}{\partial t} + \rho \operatorname{div} \mathbf{v} = 0. \tag{65.1}$$

With $\mathbf{E} = -\mathbf{v} \times \mathbf{B}$, Maxwell's equation **curl E** $= -\mathbf{B}_t$ becomes

$$\frac{\partial \mathbf{B}'}{\partial t} = \mathbf{curl}\,(\mathbf{v} \times \mathbf{B}). \tag{65.2}$$

We shall not need div $\mathbf{B}' = 0$, because we shall consider oscillatory perturbations, and for those div $\mathbf{B}' = 0$ is a consequence of (65.2). The linearized equation of motion is

$$\rho \frac{\partial \mathbf{v}}{\partial t} = -\operatorname{\mathbf{grad}} p' + \frac{1}{\mu_0}(\mathbf{curl}\,\mathbf{B}') \times \mathbf{B}. \tag{65.3}$$

Since we are considering adiabatic perturbations, and the fluid is inviscid and perfectly conducting, the energy equation is $\rho\vartheta\dot{s} = 0$. Thus the specific entropy s remains the same as in the uniform state. Regarding p as a function of ρ and s (which can be thought of as the result of eliminating ϑ between $p(\rho, \vartheta)$ and $s(\rho, \vartheta)$), we thus have the following relation between the pressure and density perturbations:

$$p' = \frac{\partial p(\rho, s)}{\partial \rho} \rho' = u_0^2 \rho', \tag{65.4}$$

where, as we shall see presently,

$$u_0^2 = \frac{\partial p(\rho, s)}{\partial \rho} \tag{65.5}$$

is the (uniform) speed of sound in the fluid. Taking the time derivative of (65.3) and using (65.4), we obtain

$$\rho \frac{\partial^2 \mathbf{v}}{\partial t^2} = -\operatorname{\mathbf{grad}}\left(u_0^2 \frac{\partial \rho'}{\partial t}\right) + \frac{1}{\mu_0}\left(\mathbf{curl}\,\frac{\partial \mathbf{B}'}{\partial t}\right) \times \mathbf{B}. \tag{65.6}$$

If we now substitute (65.1)–(65.2), we have

$$\frac{\partial^2 \mathbf{v}}{\partial t^2} = u_0^2 \operatorname{\mathbf{grad}} \operatorname{div} \mathbf{v} - \frac{1}{\mu_0 \rho}\mathbf{B} \times \mathbf{curl}^2(\mathbf{v} \times \mathbf{B}). \tag{65.7}$$

If $\mathbf{B} = 0$, the last term in this equation disappears. For plane wave perturbations, which have the behaviour $e^{i(\mathbf{k}\cdot\mathbf{x}-\omega t)}$, we have div $\mathbf{v} = i\mathbf{k}\cdot\mathbf{v}$, **curl** $\mathbf{v} = i\mathbf{k}\times\mathbf{v}$ and **grad** $f = i\mathbf{k}f$. In the absence of $\mathbf{B}$ the velocity plane wave perturbations then satisfy

$$\omega^2\mathbf{v} = u_0^2\mathbf{k}(\mathbf{k}\cdot\mathbf{v}), \tag{65.8}$$

which shows that the waves are longitudinal ($\mathbf{v}$ in the direction of the wave propagation vector $\mathbf{k}$) and have the phase velocity $u = \omega/k = u_0$. Such waves are sound waves. We can therefore identify u_0 with the speed of sound.

Now let $\mathbf{B} \neq 0$. We may choose coordinates such that $\mathbf{B}$ is in the x direction and the plane formed by $\mathbf{k}$ and $\mathbf{B}$ is the xy plane. With this choice of coordinates there is no variation in the z direction (since $k_z = 0$). After some algebra the three components of (65.7) take the form

$$\frac{\partial^2 v_z}{\partial t^2} = u_A^2 \frac{\partial^2 v_z}{\partial x^2}. \tag{65.9}$$

and

$$\frac{\partial^2 v_x}{\partial t^2} = u_0^2 \frac{\partial}{\partial x} \text{ div } \mathbf{v}, \quad \frac{\partial^2 v_y}{\partial t^2} = u_0^2 \frac{\partial}{\partial y} \text{ div } \mathbf{v} + u_A^2 \Delta v_y, \tag{65.10}$$

where

$$u_A = \frac{B}{\sqrt{\mu_0 \rho}} \tag{65.11}$$

is the Alfvén speed (cf. (61.5)). The waves fall into two groups. The first, governed by (65.9), is transverse ($v_z \neq 0$, whereas $\mathbf{k}$ is in the xy plane) and has the phase velocity u_A. There is no density perturbation, since div $\mathbf{v} = 0$ in (65.1). These *Alfvén waves* propagate parallel, or antiparallel, to $\mathbf{B}$.

The second group is governed by (65.10). In these waves both v_x and v_y are non-zero. Substitution of $\partial/\partial t = -i\omega$, and $\partial/\partial x = ik_x$, and so on, leads to a pair of linear, homogeneous equations for v_x and v_y:

$$\begin{aligned}
(u_0^2 k_x^2 - \omega^2)v_x + u_0^2 k_x k_y v_y &= 0, \\
u_0^2 k_x k_y v_x + (u_0^2 k_y^2 + u_A^2 k^2 - \omega^2)v_y &= 0.
\end{aligned} \tag{65.12}$$

The condition for a non-trivial solution is that $u = \omega/k$ satisfy the quartic equation

$$u^4 - (u_0^2 + u_A^2)u^2 + u_0^2 u_A^2 (k_x/k)^2 = 0. \tag{65.13}$$

The roots of this equation are $\pm u_f$, $\pm u_s$ (say), u_f denoting the larger. The choice of signs for these *fast* and *slow magnetoacoustic waves* merely indicates that they may travel in either direction.

Because of the appearance of the ratio k_x/k in (65.13), the two speeds depend on the direction of propagation. For magnetoacoustic waves propagating in the direction of **B**, or in the opposite direction, u_f is equal to the larger of u_0 and u_A, and u_s to the smaller one. For magnetoacoustic waves propagating in a direction normal to **B**, $u_f = \sqrt{(u_0^2 + u_A^2)}$ and $u_s = 0$. It is easily shown that, for any direction of propagation,

$$u_f > u_0 > u_s \quad \text{and} \quad u_f > u_A > u_s. \tag{65.14}$$

Besides the small-amplitude waves, the equations of MHD allow various kinds of finite amplitude, or non-linear waves. One such class is the Alfvén waves in an incompressible fluid (Exercise 63.2), but the equations also allow finite-amplitude, adiabatic waves in a compressible fluid. We shall not pursue the subject any further, however. Nor shall we discuss MHD discontinuities, or shocks, which are governed by the jump conditions (51.13), (54.3), (54.6) and (52.12), with the T and ε of (61.3)$_{5,6}$. For these matters the reader is referred to treatises on magneto-hydrodynamics.

Exercise

65.1. A perfectly conducting, plane plate is inclined at $45°$ to a uniform magnetic field **B**. The plate is immersed in a stationary, perfectly conducting, inviscid adiabatic fluid, for which $u_0\sqrt{\mu_0\rho} = 2B$. Find how the plate must be caused to vibrate if it is to excite either slow or fast magnetoacoustic plane waves alone. Discuss what waves are generated if the plate is non-conducting.

17

Electric materials

66. Forces on linear dielectrics

A dielectric fluid which is not magnetizable is characterized by $\mathcal{M} = -\rho\varphi_\mathbf{B} = 0$. Thus the free energy density is independent of the magnetic field $\mathbf{B}$. So is

$$\mathbf{P} = -\rho\varphi_\mathcal{E}. \tag{66.1}$$

If the dielectric is *linear* $\mathbf{P}$ and $\mathcal{E}$ are linearly related:

$$\mathbf{P} = (\epsilon - \epsilon_0)\mathcal{E}, \tag{66.2}$$

where ϵ may depend on the temperature and the density. Integration of (66.1) gives

$$\rho\varphi = \rho\psi - \tfrac{1}{2}\mathbf{P} \cdot \mathcal{E} = \rho\psi - \tfrac{1}{2}(\epsilon - \epsilon_0)\mathcal{E}^2, \tag{66.3}$$

where $\psi(\rho, \vartheta)$ is independent of $\mathcal{E}$; it is the specific free energy in the absence of an electromagnetic field, and it can, in principle, be determined from measurements carried out with $\mathbf{E} = \mathbf{B} = 0$. We may, of course, substitute for ψ any of the standard forms from classical thermodynamics, such as the specific free energy of an ideal gas or a solution.

The pressure in a linear dielectric fluid is

$$p = \rho^2\varphi_\rho = \rho^2\psi_\rho + \tfrac{1}{2}(\epsilon - \epsilon_0 - \rho\epsilon_\rho)\mathcal{E}^2. \tag{66.4}$$

The first term, $\rho^2\psi_\rho = \pi(\rho, \vartheta)$ (say), is the pressure in the absence of an electromotive intensity $\mathcal{E}$. For a fluid dielectric at rest the stress tensor (56.8) in the absence of a magnetic field reduces to

$$
\begin{aligned}
T &= -(p + \tfrac{1}{2}\epsilon_0 E^2)I + \epsilon_0\mathbf{E} \otimes \mathbf{E} + \mathbf{E} \otimes \mathbf{P} \\
&= -\left[\pi(\rho, \vartheta) + \tfrac{1}{2}\mathbf{E} \cdot \mathbf{D} - \tfrac{1}{2}\rho\epsilon_\rho E^2\right]I + \mathbf{E} \otimes \mathbf{D}.
\end{aligned}
$$

If ϵ is constant the stress is

$$\mathbf{t} = T\mathbf{n} = -\left[\pi(\rho, \vartheta) + \tfrac{1}{2}\mathbf{E} \cdot \mathbf{D}\right]\mathbf{n} + E D_n. \tag{66.5}$$

Exercise

66.1. A conducting sphere of radius a in an infinite medium of constant, uniform permittivity ϵ carries a charge Q. It is divided in two by a diametral plane. Neglecting the pressure of the dielectric, prove that a force of magnitude $Q^2/(32\pi\epsilon a^2)$ is required in order to separate the two halves.

At the boundary of a dielectric at rest the jump condition (51.19) requires the stress $\mathbf{t} = T\mathbf{n}$ to be continuous. The tangential component of (66.5) is $\mathbf{n} \times \mathbf{t} = \mathbf{n} \times E D_n$. The factor $\mathbf{n} \times \mathbf{E}$ is continuous at any stationary boundary. At a dielectric–dielectric boundary D_n, too, is continuous. At a conductor–dielectric boundary D_n is discontinuous, but $\mathbf{n} \times \mathbf{E} = 0$. Hence we need only worry about the normal component,

$$\mathbf{t} \cdot \mathbf{n} = -\pi(\rho, \vartheta) + \tfrac{1}{2}(\epsilon^{-1} D_n^2 - \epsilon E_s^2) \tag{66.6}$$

where E_s is the continuous tangential component of $\mathbf{E}$. Thus, at a boundary between a fluid dielectric and air (for which we may set $\epsilon = \epsilon_0$),

$$-\pi(\rho, \vartheta) + \tfrac{1}{2}(\epsilon^{-1} D_n^2 - \epsilon E_s^2) = -\pi_0 + \tfrac{1}{2}(\epsilon_0^{-1} D_n^2 - \epsilon_0 E_s^2)$$

or

$$\pi_0 - \pi(\rho, \vartheta) = \tfrac{1}{2}(\epsilon_0^{-1} - \epsilon^{-1}) D_n^2 + \tfrac{1}{2}(\epsilon - \epsilon_0) E_s^2. \tag{66.7}$$

Since $\epsilon > \epsilon_0$, the right-hand side is positive. Thus the air pressure π_0 must exceed the pressure π of the dielectric, and the right-hand side of (66.7) may be regarded as the electric tension pulling the dielectric fluid outward. For given air pressure, temperature and fields, (66.7) determines the density of the dielectric at the boundary.

For an insulating dielectric, in the absence of a magnetic field, the equation of motion (56.17) becomes

$$\rho\ddot{\mathbf{x}} = -\operatorname{grad} p + (\mathbf{P} \cdot \operatorname{grad})\mathcal{E} + \rho\mathbf{b}. \tag{66.8}$$

As an example, we consider a dielectric liquid at rest ($\mathcal{E} = \mathbf{E}$) under gravity ($\mathbf{b} = -\operatorname{grad} gz$). The surviving terms of (66.8) are then

$$0 = -\operatorname{grad} p + (\mathbf{P} \cdot \operatorname{grad})\mathbf{E} - \rho g \operatorname{grad} z. \tag{66.9}$$

The ith component of $(\mathbf{P} \cdot \operatorname{grad})\mathbf{E}$ is $P_j \partial_j E_i = (\epsilon - \epsilon_0) E_j \partial_j E_i$. Under the present circumstances $\operatorname{curl} \mathbf{E} = -\mathbf{B}_t = 0$, so that $\partial_j E_i = \partial_i E_j$. Hence

$E_j \partial_j E_i = E_j \partial_i E_j = \partial_i \frac{1}{2} E^2$, and (66.9) becomes

$$0 = - \mathbf{grad}\, p + (\epsilon - \epsilon_0)\, \mathbf{grad}\, \tfrac{1}{2} E^2 - \rho\, \mathbf{grad}\, gz. \tag{66.10}$$

We regard the liquid as incompressible, with constant ρ and ϵ. Then, to within an additive constant,

$$p = \tfrac{1}{2}(\epsilon - \epsilon_0) E^2 - \rho g z. \tag{66.11}$$

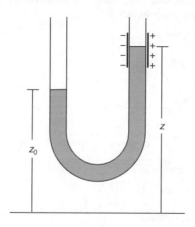

Fig. 66.1

If the liquid is in a U-shaped laboratory tube (Fig. 66.1), with one of the arms placed inside a capacitor (electric field), the tops of the liquid columns in either arm are at the same atmospheric pressure. Thus

$$\tfrac{1}{2}(\epsilon - \epsilon_0) E^2 - \rho g z = -\rho g z_0, \tag{66.12}$$

where z is the height of the liquid in the electrified arm, and z_0 in the other one. The difference in heights is therefore

$$z - z_0 = \frac{(\epsilon - \epsilon_0) E^2}{2 \rho g}. \tag{66.13}$$

This rise of a dielectric liquid in an electric field is easy to verify, because the experiment is elementary.

According to (66.10), the electric force density is $(\epsilon - \epsilon_0)\, \mathbf{grad}\, \tfrac{1}{2} E^2$. It is small near the top of the liquid in the electrified arm because, well inside the capacitor, the electric field is practically uniform (Fig. 31.1). The region of greatest non-uniformity is near the lower edge of the capacitor, and that is where the electric force density, directed upward, is maximal. It is in order to balance this upward force that a higher column of liquid is required. According to (66.13), the weight per unit area (pressure) of this added column, $\rho g(z - z_0)$, is precisely $(\epsilon - \epsilon_0)\frac{1}{2}E^2$.

Problem 66.1   A conducting sphere of radius a and density ρ_s floats in a liquid of density ρ and dielectric constant K. Determine the charge on the sphere if it is just half submerged.

Solution   The electrostatic problem is solved by a uniform surface charge density σ_0 on the upper half of the sphere, and a uniform σ on the lower half, such that the tangential electric field be continuous at the liquid–air surface:

$$\frac{\sigma_0}{\epsilon_0} = \frac{\sigma}{\epsilon}.$$

The total charge is $Q = 2\pi a^2(\sigma + \sigma_0) = 2\pi\epsilon_0 a^2(K + 1)\sigma/\epsilon$, so that the radial field at the surface of the sphere is

$$E = \frac{\sigma}{\epsilon} = \frac{\sigma_0}{\epsilon_0} = \frac{Q}{2\pi\epsilon_0 a^2(K + 1)}. \tag{1}$$

The resultant force on the sphere is

$$\oint T\mathbf{n}\,dS + \int \rho_s \mathbf{g}\,dV,$$

where

$$T\mathbf{n} = -\left(p + \tfrac{1}{2}\epsilon_0 E^2\right)\mathbf{n} + D_n\mathbf{E},$$

with the appropriate quantities on each hemisphere. The pressure in the liquid is given by (cf. (66.11))

$$p = \tfrac{1}{2}(\epsilon - \epsilon_0)E^2 - \rho gz + p_0,$$

where p_0 is the air pressure. We neglect the air density ρ_0. Noting that $D_n = \sigma_0 = \epsilon_0 E$ on the upper hemisphere and $D_n = \sigma = \epsilon E$ on the submerged one; that $\int \mathbf{n}\,dS$ over a hemisphere is a vector of magnitude πa^2 in the obvious direction; and that $\int -z\mathbf{n}\,dS$ over the hemisphere $z < 0$ has the magnitude $2\pi a^3/3$, we obtain the equilibrium condition

$$\tfrac{1}{2}(\epsilon - \epsilon_0)E^2 = \tfrac{2}{3}ag(\rho - 2\rho_s),$$

which is independent of p_0. The required formula now follows by substituting the expression (1) for E in terms of Q:

$$\frac{Q^2}{4\pi\epsilon_0} = \frac{4\pi a^5(K + 1)^2 g(\rho - 2\rho_s)}{3(K - 1)}.$$

Exercise

66.2. Two coaxial cylindrical surfaces of radii a and b are lowered vertically into a liquid dielectric, and a voltage V is established between them. Find the shape of the meniscus, that is, the height $z(r)$ to which the liquid between the surfaces rises at a distance r from the common axis. If h is the volume-averaged height, defined by

$$ h \int_a^b 2\pi r \, dr = \int_a^b z(r) 2\pi r \, dr, $$

show that

$$ \epsilon - \epsilon_0 = \frac{\rho g h (b^2 - a^2)}{V^2} \ln \frac{b}{a}. $$

67. Pyroelectricity

We have seen that the electrostatics of linear, isotropic dielectrics with constant permittivity is governed by Laplace's equation. The ensuing potential problems are similar to those that we have encountered with conductors in vacuum and can be tackled by the same methods. As in the case of conductors, the electric part of the stress is quadratic in $\mathcal{E}$. If the susceptibility χ is variable the equations and the boundary conditions are still linear and therefore relatively simple to handle, and the electric part of the stress is still quadratic in $\mathcal{E}$.

Not all dielectrics are so obliging, however. In some of them the polarization $\mathbf{P}$ is not only non-linear in the electromotive intensity $\mathcal{E}$, it even depends on the history of the material. Such are the *ferroelectrets*, which we shall discuss in the next section. The history dependence, called *hysteresis*, means that they can *not* be described by the material classes of Chapter 15. Methods have been developed for dealing with the thermodynamics of materials with memory, but they lie outside the scope of this book. Of course they are still based on the principles we have laid down, in particular on the entropy inequality (55.5), but they deal with classes of materials more general than those defined by the constitutive equations (56.1) or (58.1).

Whether they are ferroelectric or not, dielectrics exist that may be polarized even in the absence of any electromotive intensity. Such materials are the electric analogues of permanent magnets. They are called *electrets*, and when they are in a state, or phase, with $\mathcal{E} = 0$ and $\mathbf{P} \neq 0$, the state (or phase) is said to be *pyroelectric*.

Consider, for example, a pyroelectric sphere of radius a which is uniformly polarized. We wish to determine the electric field outside the sphere (where we assume a vacuum, or air, with $\epsilon = \epsilon_0$.) Obviously, we must solve a potential problem with the jump conditions $\mathbf{n} \times [\![\mathbf{E}]\!] = 0$ and $\mathbf{n} \cdot [\![\mathbf{D}]\!] = 0$ on $r = a$. This can be done by expanding the potential as a series in $r^n P_n(\cos\theta)$ inside and $r^{-(n+1)} P_n(\cos\theta)$ outside (with θ measured from the direction of $\mathbf{P}$). We prefer a short-cut, assuming that only the lowest terms do not vanish. For the field outside the sphere we try the form

$$\epsilon_0 \mathbf{E}_e = \alpha \frac{a^3}{r^3} [3(\mathbf{P} \cdot \mathbf{n})\mathbf{n} - \mathbf{P}], \qquad (67.1)$$

where α is a constant and $\mathbf{n}$ is a unit vector in the direction of the radius-vector from the centre of the sphere. Inside, we assume a uniform field, say

$$\epsilon_0 \mathbf{E} = \beta \mathbf{P}, \qquad (67.2)$$

where β is another constant. The condition $\mathbf{n} \times [[\mathbf{E}]] = 0$ gives $(\alpha + \beta)\mathbf{n} \times \mathbf{P} = 0$, or $\alpha + \beta = 0$. The condition $\mathbf{n} \cdot [[\mathbf{D}]] = 0$ gives $(2\alpha - 1 - \beta)(\mathbf{P} \cdot \mathbf{n}) = 0$ or $2\alpha - 1 - \beta = 0$. Thus $\alpha = 1/3$ and $\beta = -1/3$. We have found a solution. It is easy to verify that there is no other. Hence it is *the* solution. The external field is therefore

$$\mathbf{E}_e = \frac{1}{4\pi\epsilon_0} \frac{3(\mathcal{P} \cdot \mathbf{n})\mathbf{n} - \mathcal{P}}{r^3}, \qquad (67.3)$$

where $\mathcal{P} = \int \mathbf{P} \, dV$ is the dipole moment of the pyroelectric sphere. Inside, the electric field is $\mathbf{E} = -\mathbf{P}/(3\epsilon_0)$. It is uniform and independent of the size of the sphere. Since it is opposite to the polarization, and since we generally expect a connection between $\mathbf{E}$ and $\mathbf{P}$ through $\mathbf{P} = -\rho\varphi_{\mathcal{E}}$, an internal field satisfying $\epsilon_0 \mathbf{E} = -N\mathbf{P}$, with positive N, is called a *depolarizing field*, and N is the *depolarization factor*. We have just shown that for a sphere $N = \frac{1}{3}$.

Exercise

67.1. An electret with given, uniform, permanent polarization $\mathbf{P}$ has one of following shapes: (a) a slab normal to $\mathbf{P}$; (b) a slab parallel to $\mathbf{P}$; (c) a cylinder with its axis parallel to $\mathbf{P}$; (d) a cylinder with its axis normal to $\mathbf{P}$. Find the depolarization factor in each case.

---

68. Piezoelectricity

We have already noted that the material classes we have introduced for fluids and elastic materials in Chapter 15 cannot describe ferroelectric behaviour, since they assume that all properties depend on the present values of the arguments. They can, however, be used to illustrate pyroelectricity, as well as another phenomenon— *piezoelectricity*. Piezoelectricity is a property of many crystals—quartz being the notorious example—in which polarization arises as a result of deformation, even in the absence of $\mathcal{E}$, and stresses appear that are linear, rather than quadratic, in $\mathcal{E}$.

In a dielectric which is an elastic material, the free energy must be of the form

$$\varphi = \Phi(F^T F, \vartheta, F^T \mathcal{E}). \qquad (68.1)$$

This is the reduced form (58.10) that satisfies both the entropy inequality and Euler's second law. We have left out the dependence on $\mathbf{B}$, since $\mathcal{M} = -\rho\varphi_{\mathbf{B}} = 0$. The function Φ depends on the six independent components of the symmetric matrix $F^T F$, on the temperature ϑ, and on the three components of $F^T \mathcal{E}$. The polarization and stress involve the derivatives $\varphi_{\mathcal{E}}(F, \vartheta, \mathcal{E})$ and $\varphi_F(F, \vartheta, \mathcal{E})$, which are connected with the derivatives of Φ through the chain rule of differentiation. It is easy to show that

$$\varphi_{\mathcal{E}} = F\Phi_{F^T \mathcal{E}},$$
$$\varphi_F = 2F\Phi_{F^T F} + \mathcal{E} \otimes \Phi_{F^T \mathcal{E}}. \tag{68.2}$$

In terms of the derivatives of Φ, the polarization (58.4) and the stress (58.6) in an elastic dielectric become

$$\mathbf{P} = -\rho F \Phi_{F^T \mathcal{E}}, \tag{68.3}$$

$$T = 2\rho F \Phi_{F^T F} F^T - \tfrac{1}{2}\epsilon_0 E^2 I + \epsilon_0 \mathbf{E} \otimes \mathbf{E}. \tag{68.4}$$

Consider now a dielectric described by a Φ which is of second degree in the electromotive intensity:

$$\Phi = a_i (F^T \mathcal{E})_i + b_{ij,k}(F^T F)_{ij}(F^T \mathcal{E})_k$$
$$- \tfrac{1}{2}\epsilon_0 c_{ij}(F^T \mathcal{E})_i (F^T \mathcal{E})_j, \tag{68.5}$$

where the coefficients a, b and c are functions of the temperature; there is no restriction in assuming $b_{ij,k}$ and c_{ij} to be symmetric in i and j (in the case of the former these indices are separated from the third by a comma). If the electric field is weak and the deformation small (F being close to the identity I), the dependence of Φ on $\mathcal{E}$ is well approximated by (68.5). Thus our Φ is more than a prelude to a mathematical game.

Since the polarization and stress of (68.3)–(68.4) are linear in the derivatives of Φ, each of the three terms of (68.5) makes its individual contribution. The first and third do not contribute to T because they do not involve $F^T F$. They give rise to the following polarization:

$$\mathbf{P} = -\rho F \mathbf{a} + \epsilon_0 \rho F c F^T \mathcal{E}. \tag{68.6}$$

The first term on the right-hand side of this equation is independent of $\mathcal{E}$ and therefore pyroelectric. The second is linear in $\mathcal{E}$, with a symmetric susceptibility matrix $\chi = \rho F c F^T$.

The second term of (68.5) involves both $F^T \mathcal{E}$ and $F^T F$. It gives rise to a polarization which depends on the deformation F, but not on $\mathcal{E}$, and also to a stress which is linear in $\mathcal{E}$. Both effects are controlled by the same coefficients $b_{ij,k}$ ($6 \times 3 = 18$ at most). This is piezoelectricity. In a sufficiently weak $\mathcal{E}$ the linear, piezoelectric term in the stress will dominate the other, quadratic terms. Conversely, the electromotive intensity in a piezoelectric material varies as the stress, rather than as its square root. In piezoelectric cigarette lighters this effect is used in order to produce a spark by squeezing.

Other interesting phenomena in piezoelectrets are connected with oscillations and the propagation of elastic waves. They do not belong to electrostatics. Rather, they are governed by the following system of equations:

$$\text{div } \mathbf{D} = 0, \qquad \text{curl } \mathbf{H} = \mathbf{D}_t,$$
$$\mathbf{D} = \epsilon_0 \mathbf{E} + \mathbf{P},$$
$$\text{div } \mathbf{B} = 0, \qquad \text{curl } \mathbf{E} = -\mathbf{B}_t, \qquad (68.7)$$
$$\mathbf{H} = \mathbf{B}/\mu_0 + \dot{\mathbf{x}} \times \mathbf{P},$$
$$\rho\ddot{\mathbf{x}} = \text{div } T + \rho\mathbf{b}.$$

To this system we must add the corresponding jump conditions and substitute the expressions for $\mathbf{P}$ and T that follow from the middle, piezoelectric term of (68.5); for example, $P_i = -\rho F_{ij} b_{nm,j} (F^T F)_{nm}$. A term depending only on $F^T F$ must be added to the specific free energy (68.5) in order to account for elastic behaviour. Obviously, the system is far from simple. Progress can be made by linearization, i.e. by casting away all terms that are quadratic in $\mathbf{E}$, $\mathbf{B}$, $\dot{\mathbf{x}}$ or $F - I$; often the magnetic field is left out entirely.

The effects of pyroelectricity, linear polarization and piezoelectricity are thus seen to arise quite naturally in elastic dielectrics. Indeed we should expect them to occur simultaneously. In a very weak field the first two terms of (68.5) are the most important, and the material will exhibit pyroelectric or piezoelectric behaviour. In stronger fields the third term of (68.5) will dominate, and we expect the same material to behave like a linear dielectric. It is therefore not surprising that tables of material permittivities list quartz, a common piezoelectret, as a linear dielectric with $\epsilon = 3.85\epsilon_0$. Finally, since the coefficients a, b and c are functions of the temperature, we also expect the magnitudes of these effects to be temperature-dependent.

Exercise

68.1. A sphere is carved out of dielectric material with the constitutive relation $\mathbf{P} = \mathbf{P}_0 + (\epsilon - \epsilon_0)\mathcal{E}$ (permanent polarization $\mathbf{P}_0$ and linear dielectricity; cf. (68.6)). Determine the electric field inside the sphere.

---

69. Ferroelectricity

Ferroelectrets are materials in which the permanent polarization can be changed— and even reversed—by an electric field. They are the electric analogues of ferromagnets (but, unlike the latter, they have nothing to do with iron). Like ferromagnets, they exhibit non-linear behaviour and hysteresis. They may be strongly polarized even in the absence of any electromotive intensity, but only so long as the temperature lies below a characteristic temperature ϑ_c, called the *Curie temperature*.

Some of these properties can be shown to follow from a suitable choice[†] of the free energy φ, which determines the polarization through $\mathbf{P} = -\rho\varphi_{\mathcal{E}}$.

In order to keep things as simple as possible, we consider an electric material from the simple class of fluids (Section 56) for which φ depends on ρ, ϑ and $\mathcal{E}$. Obviously, we cannot hope to exhibit the phenomenon of hysteresis in this way, since the φ's we shall be dealing with depend on *present* values of their arguments.

We now regard $\rho\varphi_{\mathcal{E}}(\rho, \vartheta, \mathcal{E}) = -\mathbf{P}$ as an equation that (implicitly) determines $\mathcal{E}$ as a function of ρ, ϑ and $\mathbf{P}$, and then carry out a Legendre transformation (like the transformation from a Lagrangian to a Hamiltonian) to a new thermodynamic potential:

$$\rho\tilde{\varphi}(\rho, \vartheta, \mathbf{P}) = \rho\varphi + \mathbf{P} \cdot \mathcal{E}. \tag{69.1}$$

Since $\mathcal{E}$ on the right-hand side is now regarded as the solution of $\rho\varphi_{\mathcal{E}} = -\mathbf{P}$, the new thermodynamic potential $\tilde{\varphi}$ depends on ρ, ϑ and $\mathbf{P}$. Its $\mathbf{P}$ derivative is easily obtained by the chain rule:

$$\rho\tilde{\varphi}_{\mathbf{P}}(\rho, \vartheta, \mathbf{P}) = \mathcal{E}. \tag{69.2}$$

Instead of specifying an electric material by $\varphi(\rho, \vartheta, \mathcal{E})$, we can specify it by $\tilde{\varphi}(\rho, \vartheta, \mathbf{P})$; this procedure is, of course, analogous to specifying a dynamical system by a Hamiltonian rather than by a Lagrangian.

Consider now the thermodynamic potential

$$\rho\tilde{\varphi}(\rho, \vartheta, \mathbf{P}) = \rho\tilde{\varphi}_0(\rho, \vartheta) + \left[a(\vartheta - \vartheta_c)P^2 + bP^4\right]\Big/\epsilon_0, \tag{69.3}$$

where a and b are both positive and independent of ϑ or $\mathbf{P}$, and ϑ_c is a positive constant. We note that $\tilde{\varphi}$ is rotationally invariant, since it depends only on the magnitude of $\mathbf{P}$. The same will be true of φ, which according to (69.1) differs from $\tilde{\varphi}$ by a scalar product of vectors. This ensures that φ will obey the restriction imposed by Euler's second law of mechanics (cf. the statement following (56.13)). Substitution of (69.3) in (69.2) gives

$$[2a(\vartheta - \vartheta_c) + 4bP^2]\mathbf{P} = \epsilon_0\mathcal{E}. \tag{69.4}$$

The polarization $\mathbf{P}$ is parallel or antiparallel to $\mathcal{E}$, depending on the sign of the expression in the square brackets of (69.4). This is, of course, a consequence of the isotropy of the particular $\tilde{\varphi}$ we have chosen.

If $\vartheta > \vartheta_c$, $\mathbf{P}$ is always parallel to $\mathcal{E}$. When $\mathcal{E}$ is small, the relationship is linear, with a susceptibility given by

$$\chi = \frac{(2a)^{-1}}{\vartheta - \vartheta_c}. \tag{69.5}$$

[†]Due to Ginzburg (1945).

This behaviour (called *paraelectric*, in analogy with the behaviour that is called *paramagnetic* in the magnetic case), with a susceptibility χ inversely proportional to $\vartheta - \vartheta_c$, is the (ferroelectric) *Curie–Weiss law*. It is rather well substantiated by experiments. The critical temperature ϑ_c is the (ferroelectric) *Curie temperature* or the *Curie point*. When $\mathcal{E}$ is very large, (69.4) gives

$$\mathbf{P} = (4b\epsilon_0^2\mathcal{E}^2)^{-1/3}\epsilon_0\mathcal{E}. \tag{69.6}$$

In actual ferroelectrets $\mathbf{P}$ attains a constant *saturation* value for very large $\mathcal{E}$. Equation (69.6) does not lead to saturation, although it does predict that $\mathbf{P}$ will increase more slowly than $\mathcal{E}$. It is therefore not valid for very large $\mathbf{P}$. This failure is not wholly unexpected, since the $\tilde{\varphi}$ of (69.3) looks like the beginning of an expansion in powers of P^2. We should not expect it to hold when $\mathbf{P}$ is large.

Below the Curie temperature the $\mathbf{P}(\mathcal{E})$ relationship is more complicated. For $\mathcal{E} = 0$ we have, besides $\mathbf{P} = 0$, the *spontaneous* ferroelectric solution

$$P^2 = \frac{a(\vartheta_c - \vartheta)}{2b}. \tag{69.7}$$

Since it corresponds to the vanishing of the square bracket in (69.4), its direction is undetermined. Again, this is a consequence of the isotropy of (69.3). For $\mathcal{E} \neq 0$, every solution has a determinate direction, and we may assume that $\mathbf{P}$ and $\mathcal{E}$ both lie on the z axis:

$$[2a(\vartheta - \vartheta_c) + 4bP_z^2]P_z = \epsilon_0\mathcal{E}_z. \tag{69.8}$$

The resulting polarization curve is shown in Fig. 69.1. Clearly,

$$[2a(\vartheta - \vartheta_c) + 12bP_z^2]\frac{\partial P_z}{\partial \mathcal{E}_z} = \epsilon_0. \tag{69.9}$$

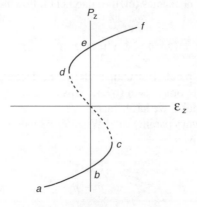

Fig. 69.1

The extrema c and d are the points at which $\partial \mathcal{E}_z / \partial P_z = 0$:

$$P_z^2 = \frac{a(\vartheta_c - \vartheta)}{6b}. \tag{69.10}$$

For each value of $\mathcal{E}_z$ between the abscissae corresponding to c and d the cubic equation (69.8) has three real roots. One of these must, however, lie on the dashed part cd of the curve, which turns out to correspond to unstable states. In order to see this, we substitute (69.3) into (69.1) and obtain

$$\rho\varphi = \rho\tilde{\varphi}(\rho, \vartheta) + [a(\vartheta - \vartheta_c)P_z^2 + bP_z^4]/\epsilon_0 - P_z\mathcal{E}_z. \tag{69.11}$$

Equation (69.8) is the same as $\partial\varphi/\partial P_z = 0$, the derivative being taken at fixed ρ, ϑ and $\mathcal{E}_z$. The polarization curve is thus the locus of extrema of φ. Equation (69.9) is the same as

$$\left(\rho\frac{\partial^2\varphi}{\partial P_z^2}\right)\frac{\partial P_z}{\partial \mathcal{E}_z} = \epsilon_0; \tag{69.12}$$

the two factors on the left-hand side must therefore have the same sign. This shows that, wherever $\partial P_z/\partial\mathcal{E}_z > 0$, φ has a minimum, and wherever $\partial P_z/\partial\mathcal{E}_z < 0$, φ has a maximum. The dashed part cd of the curve therefore corresponds to states which are unstable (cf. Section 55).

We have shown that, for $\mathcal{E}_z$ between the abscissae corresponding to c and d, there are only two stable solutions, corresponding to minima of φ. They give rise to oppositely directed polarizations. Above the Curie temperature these ferroelectric solutions disappear, and the material becomes paraelectric.

The thermodynamic potentials we have considered above were based on the class of simple fluids. Since real ferroelectric materials are crystalline or polycrystalline solids, we might (at least) try to construct free energies based on the reduced form (68.1) for elastic materials. The resulting expressions will be quite involved, and we shall not attempt to derive them. Even so, we may still expect them to lead to non-zero ferroelectric solutions (even when $\mathcal{E} = 0$) below a definite temperature. But in these solutions $\mathbf{P}$ will generally have components that are *transverse* with respect to $\mathcal{E}$, and the $\mathbf{P}(\mathcal{E})$ relation will be described by a polarization *surface* (or hypersurface) rather than by a polarization *curve*. In the simplest case, the material has a single ferroelectric *axis*, and the ferroelectric solutions can only have one of the two directions defined by this axis. Usually there are several such axes, but even if they are all equivalent—as in a crystal of cubic symmetry—the establishment of polarization along one of them will result in distortion due to electrostriction (i.e. electric stresses). This will cause some of the axes to become 'easy' and others 'hard'; these terms refer, of course, to the components of $\mathbf{P}$ that arise in response to components of $\mathcal{E}$ along these axes.

In addition to the $\mathbf{P}(\mathcal{E})$ relation, $\mathbf{P}$ and $\mathcal{E}$ must also satisfy the electromagnetic, mechanical and thermodynamic equations and jump conditions. In an electrostatic situation, these are

$$\operatorname{div}(\epsilon_0 \mathbf{E} + \mathbf{P}) = 0, \qquad \mathbf{n} \cdot [\![\epsilon_0 \mathbf{E} + \mathbf{P}]\!] = 0,$$
$$\operatorname{\mathbf{curl}} \mathbf{E} = 0, \qquad \mathbf{n} \times [\![\mathbf{E}]\!] = 0, \tag{69.13}$$

as well as the equation and jump condition of mechanical equilibrium. Taken together, these conditions turn out to be quite stringent. Imagine, for example, a sphere of uniaxial ferroelectric material which has been cooled down through its Curie temperature in the absence of an applied electric field. We choose the ferroelecric axis as the z axis and continue to apply our $\tilde{\varphi}$ of (69.3) with the understanding that $\mathbf{P} = (0, 0, P_z)$. When $\vartheta < \vartheta_c$, the material will become spontaneously polarized, with P_z equal to one of the roots of (69.7). If P_z has the same sign everywhere, we have the uniform pyroelectric sphere for which eqns (62.13) were solved in Section 67. According to (67.2), there will be a uniform depolarizing field $\mathbf{E} = -\mathbf{P}/(3\epsilon_0)$ throughout the sphere, which is independent of the sphere's size.

Now the polarization curves of all known ferroelectrets are such that this opposite field is orders of magnitude greater than the field required for a reversal of P_z. Of course, a mere flipping over of the polarization will leave the ferroelectric sphere in the same predicament. We must therefore conclude that P_z *cannot* have the same sign throughout the sphere. This is supported by observations of actual ferroelectric materials: the substance divides itself into regions or *domains* of uniform polarization in alternating directions. The shapes and sizes of these domains depend on the geometry of the sample, on the temperature and on the applied stresses and electric fields. The domain boundaries are observed to move—thus changing the domain structure—in response to changes in these parameters.

The manner in which domain formation reduces the depolarizing field can be illustrated by the study of simple domain structures. Consider a layer of thickness ℓ made of uniaxial ferroelectric material, with the axis perpendicular to the layer. Figure 69.2 shows two possible domain structures. We shall calculate the electric field for the striped structure. Since $\mathbf{P}$ is uniform (except for jumps across the domain

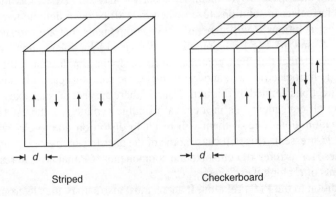

Striped   Checkerboard

Fig. 69.2

walls), eqns (69.13) reduce to a potential problem for the electric field. Clearly, P_z is a periodic function with period $2d$. We choose the upper face as the plane $z = 0$, and the x coordinate such that $P_z = -P$ for $-d < x < 0$ and $P_z = +P$ for $0 < x < d$. The Fourier expansion of P_z is then

$$P_z(x) = \sum_{n=0}^{\infty} b_n \sin \frac{(2n + 1)\pi x}{d}, \quad b_n = \frac{4P}{(2n + 1)\pi}. \qquad (69.14)$$

For the potential, we seek a solution of $V_{xx} + V_{zz} = 0$ which has the form

$$V(x, z) = \sum f_n(z) \sin \frac{(2n + 1)\pi x}{d} \qquad (69.15)$$

and remains finite as we move away from $z = 0$. This gives

$$f_n(z) = c_n e^{\mp(2n+1)\pi z/d}, \qquad (69.16)$$

the signs in the exponent referring to $z > 0$ and $z < 0$. The c_n are related to the b_n of (69.14) by the jump condition (69.13)$_2$:

$$c_n = \frac{d}{2\pi \epsilon_0 (2n + 1)} b_n. \qquad (69.17)$$

The electric field $\mathbf{E} = -\operatorname{\mathbf{grad}} V$ therefore decays exponentially away from the layer face. Inside the material it is directed so that $\mathbf{P} \cdot \mathbf{E} < 0$.

Having understood the necessity for domain formation, we are now led to ask: why does the subdivision into domains ever cease, rather than go on indefinitely, perhaps down to atomic sizes?

In order to account for this, theories of ferroelectricity postulate a surface tension along the domain walls. This means the addition, to the free energy, of a surface term $\int \alpha \, dS$, where the *surface tension coefficient* α is positive; presumably it depends on the temperature, the deformation, the polarization, etc.

We may illustrate the effect of this term with the foregoing example of striped domain structure in a layer. Since the area of each domain wall is ℓ times the depth of the layer, the term $\int \alpha \, dS$ makes a contribution of $\alpha \ell / d$ per unit area of the layer. Of the remaining terms in the free energy $\int \varphi \, dm = \int \rho \varphi \, dV$, with $\rho \varphi$ given by (69.11), only the last makes a contribution that, per unit area of the layer, depends on the domain thickness d. It is easily calculated from (69.14)–(69.17). Assuming that $d \ll \ell$ and remembering that the layer has an upper and a lower face, the contribution of $\int -P_z E_z \, dV$, per unit area, is $4P^2 d/(\pi^3 \epsilon_0) \sum (2n + 1)^{-3}$, which is proportional to d. If we add to it the surface contribution $\alpha \ell / d$, the sum is seen to have a minimum with respect to d. This is the stable value of the domain width. It is proportional to $\sqrt{\ell}$. A finite block will therefore contain fewer domains as it gets smaller.

Indeed, since the ratio of surface to volume increases as the dimensions of a body decrease, it follows that very small ferroelectric bodies must cease to be polarized: domain formation requires too much surface energy; a single-domain state is impossible because of the forbiddingly high depolarizing field; hence $\mathbf{P} = 0$ becomes a stable state. It should, however, be possible for very small bodies to form single domains if the surrounding medium is conducting (for example, a salt solution), because the depolarizing field would then be cancelled by the deposition of free charges on the surface of the body. These predictions are all borne out by experiments.

18

Magnetic materials

70. Forces on linear magnets; isentropic cooling

We have already noted in Section 53 that, in continuum mechanics, the free energy is used in a way which is similar to the use of the Lagrangian function in analytical mechanics. Consider a material for which

$$\rho\varphi = \rho\psi(\rho, \vartheta) - \tfrac{1}{2}(\mu_0^{-1} - \mu^{-1})B^2, \tag{70.1}$$

where μ may depend on ρ and ϑ. According to (56.6)–(56.7), such a material has no polarization, but it is magnetizable:

$$\mathcal{M}(\rho, \vartheta, \mathbf{B}) = (\mu_0^{-1} - \mu^{-1})\mathbf{B}. \tag{70.2}$$

If we assume that $(\mu_0^{-1} - \mu^{-1})$ is proportional to the density, the pressure becomes $\rho^2\varphi_\rho = \rho^2\psi_\rho(\rho, \vartheta)$, which is independent of $\mathbf{B}$. For the two terms containing $\mathcal{M}$ in (56.17), an easy calculation gives

$$(\mathcal{M} \cdot \mathbf{grad})\mathbf{B} + \mathcal{M} \times \mathbf{curl}\,\mathbf{B} = \tfrac{1}{2}(\mu_0^{-1} - \mu^{-1})\,\mathbf{grad}\,B^2. \tag{70.3}$$

Faraday knew that magnetic materials, like soft iron, tended to move in the direction of increasing magnetic field strength (irrespective of the *direction* of **B**). He was therefore surprised to discover that some materials, like heavy glass or bismuth, which were not then considered to be 'magnetic', tended to move in the direction of *decreasing* magnetic field strength. He inferred that those materials, which were called *diamagnetic*, must have an induced magnetization in a direction opposite to **B**. A few years later Thomson (Kelvin) suggested the relation (70.2) and characterized *paramagnetic* materials, of which soft iron was the prototype, by $\mu > \mu_0$, and diamagnetic materials by $\mu < \mu_0$.

As in the theory of linear dielectrics (Section 66), we infer that the equilibrium pressure in a linear magnetic liquid under gravity is given, to within an additive

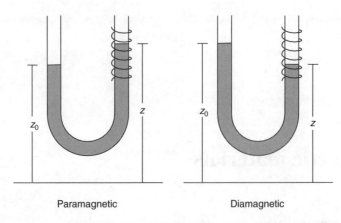

Paramagnetic   Diamagnetic

Fig. 70.1

constant, by

$$p = \tfrac{1}{2}(\mu_0^{-1} - \mu^{-1})B^2 - \rho g z. \tag{70.4}$$

This equation predicts the rise of a paramagnetic liquid, or the depression of a diamagnetic one, in a magnetic field. It provides a method for measuring the magnetic susceptibility χ_B of a liquid, so long as it is sufficiently different from the susceptibility χ_B^0 of air. If the liquid is in a U-shaped tube, and one of the arms is placed in a solenoid (magnetic field), as in Fig. 70.1, we have, according to (43.4),

$$\chi_B - \chi_B^0 = \frac{2\mu_0 \rho g (z - z_0)}{B^2}. \tag{70.5}$$

The susceptibility of solids is measured by a balance, using the force density (70.3).

Starting from the formulae (56.8)–(56.9) for the stress tensor T in an electromagnetic liquid, and proceeding as in Section 66, we obtain, in the case of a permeable fluid (with constant μ) at rest,

$$
\begin{aligned}
p &= \pi(\rho, \vartheta) + \tfrac{1}{2}\boldsymbol{\mathcal{M}} \cdot \mathbf{B}, \\
T &= -(p + \tfrac{1}{2}\mu_0^{-1}B^2 - \boldsymbol{\mathcal{M}} \cdot \mathbf{B})I + \mu_0^{-1}\mathbf{B} \otimes \mathbf{B} - \boldsymbol{\mathcal{M}} \otimes \mathbf{B} \\
&= -\left[\pi(\rho, \vartheta) + \tfrac{1}{2}\mu^{-1}B^2\right]I + \mathbf{H} \otimes \mathbf{B},
\end{aligned}
\tag{70.6}
$$

where $\pi(\rho, \vartheta)$ is the pressure in the absence of a magnetic field.

At a boundary between a permeable material and air, the pressure difference is (cf. (66.7))

$$\pi_0 - \pi(\rho, \vartheta) = \tfrac{1}{2}(\mu_0^{-1} - \mu^{-1})B_n^2 + \tfrac{1}{2}(\mu - \mu_0)H_s^2. \tag{70.7}$$

The right-hand side has been expressed in terms of the normal component B_n and the tangential component H_s, which are both continuous across the boundary. Thus, at the boundary between a paramagnetic ($\mu > \mu_0$) material and air, $\pi_0 > \pi$: the air pressure π_0 must exceed the pressure π of the material. This may be regarded as a magnetic tension pulling the matter outward. But if the material is diamagnetic $\mu < \mu_0$ and $\pi_0 < \pi$: in this case there is a magnetic *pressure*, and the material shies away from the surrounding magnetic field.

From the φ of (70.1) we obtain

$$\rho s = -\rho \varphi_\vartheta = \rho s_0(\rho, \vartheta) + \frac{\mu_\vartheta}{2\mu^2} B^2, \tag{70.8}$$

where $s_0(\rho, \vartheta) = -\psi_\vartheta(\rho, \vartheta)$ is the specific entropy in the absence of a magnetic field. Consider now an isentropic process, beginning with $\vartheta = \vartheta_1$ and $\mathbf{B} \neq 0$, during which $\mathbf{B}$ is switched off. If the change in density may be neglected, we have

$$\rho s_0(\vartheta_1) + \frac{\mu_\vartheta}{2\mu^2} B^2 = \rho s_0(\vartheta_2), \tag{70.9}$$

where ϑ_2 is the temperature at the end of the process. For paramagnetic materials the permeability μ is a decreasing function of ϑ. Hence the change $s_0(\vartheta_2) - s_0(\vartheta_1)$ must be negative. But $s_0(\vartheta)$ is an increasing function of ϑ (corresponding to a positive specific heat). Therefore ϑ_2 must be less than ϑ_1. This is the basis of the method of *paramagnetic cooling* or *cooling by isentropic demagnetization*, by which final temperatures in the range of micro Kelvin degrees can be achieved.

Magnetic materials, or magnets, are analogous to dielectric materials and present similar phenomena. The *ferromagnets* show hysteresis, sometimes with dramatic temperature dependence. We shall deal with some aspects of ferromagnetism in the next section. If we can ignore *hysteresis*, that is, the dependence of magnetization on history, we may apply the theory of Chapter 15. There is nothing, in principle, to prevent a material from being both dielectric and magnetic. Common magnets, however, show no polarization. Then $\mathbf{P} = -\rho \varphi_{\mathcal{E}} = 0$, so that the free energy is independent of the electromotive intensity $\mathcal{E}$. The same is then true of the magnetization

$$\mathcal{M} = -\rho \varphi_{\mathbf{B}}. \tag{70.10}$$

A theory of elastic magnets can be developed by assuming various forms for the reduced free energy (cf. (58.10) with $\mathcal{E} = 0$)

$$\varphi = \Phi(F^T F, \vartheta, F^T \mathbf{B}). \tag{70.11}$$

In particular, a simple expansion for weak magnetic fields (cf. Section 68) leads to a description of permanent magnets, piezomagnets and linear magnets (including the isotropic dia- and paramagnetic materials). These are, of course, analogues of pyroelectrets, piezoelectrets and linear dielectrics.

A magnet through which no currents are flowing is governed by the equations

$$\operatorname{curl} \mathbf{H} = 0, \qquad \mathbf{n} \times [\![\mathbf{H}]\!] = 0,$$
$$\operatorname{div} \mathbf{B} = 0, \qquad \mathbf{n} \cdot [\![\mathbf{B}]\!] = 0, \tag{70.12}$$
$$\mathbf{H} = \mathbf{B}/\mu_0 - \boldsymbol{\mathcal{M}}.$$

For permanent magnets these have non-trivial regular solutions, analogous to the electric fields of pyroelectrets.

Exercises

70.1. A magnet with given permanent magnetization $\boldsymbol{\mathcal{M}}$ has one of the following shapes: (a) a slab normal to $\boldsymbol{\mathcal{M}}$; (b) a slab parallel to $\boldsymbol{\mathcal{M}}$; (c) a cylinder with its axis parallel to $\boldsymbol{\mathcal{M}}$; (d) a cylinder with its axis normal to $\boldsymbol{\mathcal{M}}$; (e) a sphere. Find the *demagnetizing field* $\mathbf{H} = -N\boldsymbol{\mathcal{M}}$ inside the magnet and determine the *demagnetization factor* N in each case.

70.2. If the earth was a uniformly magnetized sphere, show that the tangent of the dip of the magnetic field at any point would be equal to twice the tangent of the magnetic latitude.

70.3. Out of material with the constitutive relation $\boldsymbol{\mathcal{M}} = \boldsymbol{\mathcal{M}}_0 + (\mu_0^{-1} - \mu^{-1})\mathbf{B}$ (permanent magnetization $\boldsymbol{\mathcal{M}}_0$ and linear magnetization) a cylinder is carved out with its axis parallel to $\boldsymbol{\mathcal{M}}_0$. The cylinder is then bent into a torus, around which n turns of wire are closely and evenly wound (as in Fig. 43.1, except that now there is an added circumferential permanent magnetization $\boldsymbol{\mathcal{M}}_0$ along the torus). A small air gap is formed in the material by cutting away a thin sector bounded by two planes through the torus axis which make a small angle α with each other. Find the magnetic field in the gap when a current i flows in the wire.

70.4. [D. J. Griffiths, *Introduction to Electrodynamics*, 2nd ed., Prentice Hall 1989.] A small torus of mass M is permanently magnetized in the direction of the axis. A number of such equal magnets can slide without friction on a vertical rod, like beads on a string (Fig. 70.2). They may be regarded as particles with magnetic moment $\mathbf{m}$ (either 'up' or 'down'). If one of these rests at the bottom of the rod, a second one, which has its magnetization oppositely directed, will be repelled and come to rest at a height h above the first one. Determine h. If a third magnet, with magnetization parallel to that of the bottom magnet, is added on top, it will be suspended at a height k above the middle magnet. Determine h and k in the latter situation.

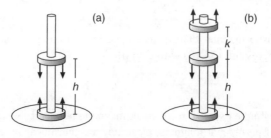

Fig. 70.2

71. Ferromagnetism

Ferromagnetic materials like iron, nickel and their alloys are the magnetic analogues of ferroelectric materials. Like the latter, they exhibit non-linear behaviour and hysteresis. They may be spontaneously magnetic, in a variety of domain structures, but only for temperatures below a characteristic temperature ϑ_c, called the (magnetic) *Curie temperature*. Above the Curie temperature, or *Curie point*, they are paramagnetic with a susceptibility χ_H that is inversely proportional to $\vartheta - \vartheta_c$; this is the (magnetic) *Curie–Weiss law*. The analogy between ferromagnetism and ferroelectricity is indeed so complete that their macroscopic treatment is governed by the same theory. There are, however, some *quantitative* differences, which we shall point out later.

If we compare the magnetostatic system of equations

$$\operatorname{div} \mu_0 (\mathbf{H} + \mathcal{M}) = 0, \qquad \mathbf{n} \cdot \mu_0 [\![\mathbf{H} + \mathcal{M}]\!] = 0,$$
$$\operatorname{curl} \mathbf{H} = 0, \qquad \mathbf{n} \times [\![\mathbf{H}]\!] = 0, \tag{71.1}$$

with the electrostatic system (69.13), it is clear that the former is obtained from the latter by the replacements $\mathbf{E} \to \mathbf{H}$, $\epsilon_0 \to \mu_0$ and $\mathbf{P} \to \mu_0 \mathcal{M}$. But this correspondence does not extend to the thermodynamic relations $\mathcal{M} = -\rho \varphi_{\mathbf{B}}$ and $\mathbf{P} = -\rho \varphi_{\mathcal{E}}$, because in $\varphi_{\mathbf{B}}$ the differentiation is with respect to $\mathbf{B}$, not $\mathbf{H}$. We shall therefore begin by replacing $\varphi(\rho, \vartheta, \mathbf{B})$ by another thermodynamic potential $\psi(\rho, \vartheta, \mathcal{G})$, where

$$\mathcal{G} = \mathbf{B}/\mu_0 - \mathcal{M} \tag{71.2}$$

is the difference between the Galilean invariants $\mathbf{B}/\mu_0$ and $\mathcal{M}$. For a stationary body in an aether frame $\mathcal{G} = \mathcal{H}$ (cf. (21.10)), and either of these vectors equals the current potential $\mathbf{H}$ (cf. (20.5)$_1$ and (21.6)$_6$). In terms of $\mathcal{G}$, the relation $\rho \varphi_{\mathbf{B}} = -\mathcal{M}$ becomes

$$\rho \varphi_{\mathbf{B}}(\rho, \vartheta, \mathbf{B}) = \mathcal{G} - \mathbf{B}/\mu_0. \tag{71.3}$$

We now regard (71.3) as an equation that (implicitly) determines $\mathbf{B}$ as a function of ρ, ϑ and $\mathcal{G}$, and then carry out a Legendre transformation to a new thermodynamic potential:

$$\rho \psi(\rho, \vartheta, \mathcal{G}) = \rho \varphi + \tfrac{1}{2} (\mathbf{B} - \mu_0 \mathcal{G})^2 / \mu_0. \tag{71.4}$$

Since $\mathbf{B}$ on the right-hand side is now regarded as the solution of (71.3), the new thermodynamic potential ψ depends on ρ, ϑ and $\mathcal{G}$. Its $\mathcal{G}$-derivative is easily obtained by the chain rule:

$$\rho \psi_{\mathcal{G}}(\rho, \vartheta, \mathcal{G}) = -\mu_0 \mathcal{M}. \tag{71.5}$$

The last result is analogous to $\rho \varphi_{\mathcal{E}} = -\mathbf{P}$. Clearly, then, we have replaced $\mathbf{B}$ by $\mathcal{G}$ as an independent variable. This replacement extends to the entropy inequality: taking the dot derivative of (71.4), we obtain

$$\rho \dot{\psi} = \dot{\rho}(\varphi - \psi) + \rho \dot{\varphi} + \mathcal{M} \cdot (\dot{\mathbf{B}} - \mu_0 \dot{\mathcal{G}})$$

or

$$-\rho \dot{\varphi} - \mathcal{M} \cdot \dot{\mathbf{B}} = -\rho \dot{\psi} - \mu_0 \mathcal{M} \cdot \dot{\mathcal{G}} + \tfrac{1}{2} \mu_0 \mathcal{M}^2 I \cdot \mathbf{grad}\, \dot{\mathbf{x}}.$$

With the help of this relation, the entropy inequality (55.5) (with $\mathbf{P} = 0$) becomes

$$-\rho \dot{\psi} - \rho s \dot{\vartheta} - (\rho \mathbf{g} - \rho \dot{\mathbf{x}} - \epsilon_0 \mathbf{E} \times \mathbf{B}) \cdot \ddot{\mathbf{x}} - \mu_0 \mathcal{M} \cdot \dot{\mathcal{G}}$$
$$+ (\tau + \tfrac{1}{2} \mu_0 \mathcal{M}^2 I) \cdot \mathbf{grad}\, \dot{\mathbf{x}} + \mathcal{J} \cdot \mathcal{E} - (\mathbf{q} \cdot \mathbf{grad}\, \vartheta)/\vartheta \geq 0. \qquad (71.6)$$

As in Section 55, we can now conclude that, if ϑ and $\dot{\mathbf{x}}$ are constant and uniform, and if, furthermore, $\mathcal{G}$ is constant and $\mathcal{E} = 0$, then ψ cannot increase. A state of a body for which ψ (assuming that it depends on further variables) is a minimum under these conditions will therefore be stable.

Equation (71.5) can be used to determine $\mathcal{G}$ as a function of ρ, ϑ and $\mathcal{M}$. Another Legendre transformation then leads to the thermodynamic potential (cf. (69.1))

$$\rho \tilde{\psi}(\rho, \vartheta, \mathcal{M}) = \rho \psi + \mu_0 \mathcal{M} \cdot \mathcal{G}. \qquad (71.7)$$

This potential has the $\mathcal{M}$-derivative (cf. (69.2))

$$\rho \tilde{\psi}_{\mathcal{M}}(\rho, \vartheta, \mathcal{M}) = \mu_0 \mathcal{G}. \qquad (71.8)$$

Instead of specifying a magnetic material by $\varphi(\rho, \vartheta, \mathbf{B})$ or by $\psi(\rho, \vartheta, \mathcal{G})$, we can specify it by $\tilde{\psi}(\rho, \vartheta, \mathcal{M})$. Consider now the thermodynamic potential (cf. (69.3))

$$\rho \tilde{\psi}(\rho, \vartheta, \mathcal{M}) = \rho \tilde{\psi}_0(\rho, \vartheta) + \mu_0[a(\vartheta - \vartheta_c)\mathcal{M}^2 + b\mathcal{M}^4], \qquad (71.9)$$

where a and b are both positive and independent of ϑ or $\mathcal{M}$, and ϑ_c is a positive constant. Substitution of (71.9) in (71.8) gives (cf. (69.4))

$$[2a(\vartheta - \vartheta_c) + 4b\mathcal{M}^2]\mathcal{M} = \mathcal{G}. \qquad (71.10)$$

We have already noted that, for a body at rest (in an aether frame), $\mathcal{G}$ is equal to the current potential $\mathbf{H}$. Thus

$$[2a(\vartheta - \vartheta_c) + 4b\mathcal{M}^2]\mathcal{M} = \mathbf{H} \qquad (71.11)$$

is the equation that determines the relationship between $\mathcal{M}$ and $\mathbf{H}$—the *magnetization curve*. The magnetization $\mathcal{M}$ is parallel or anti-parallel to $\mathbf{H}$, depending on the sign of the expression in the square brackets of (71.11). This is, of course, a consequence of the isotropy of the particular $\tilde{\psi}$ we have chosen.

If $\vartheta > \vartheta_c$, $\mathcal{M}$ is always parallel to $\mathbf{H}$. If $\mathbf{H}$ is small, the relationship is linear, with a susceptibility given by (cf. (69.5))

$$\chi_H = \frac{(2a)^{-1}}{\vartheta - \vartheta_c}. \qquad (71.12)$$

This is the (ferromagnetic) *Curie–Weiss law*. The critical temperature ϑ_c is the *Curie temperature* or the *Curie point*. For iron it is 1043 K. Below the Curie temperature the $\mathcal{M}(\mathbf{H})$ relationship is more complicated. For $\mathbf{H} = 0$ we have, besides $\mathcal{M} = 0$, the *spontaneous* ferromagnetic solution (cf. (69.7))

$$\mathcal{M}^2 = \frac{a(\vartheta_c - \vartheta)}{2b}. \qquad (71.13)$$

Since it corresponds to the vanishing of the square bracket in (71.11), its direction is undetermined. Again, this is a consequence of the isotropy of (71.9). For $\mathbf{H} \neq 0$, every solution has a determinate direction, and we may assume that $\mathcal{M}$ and $\mathbf{H}$ both lie on the z axis:

$$[2a(\vartheta - \vartheta_c) + 4b\mathcal{M}_z^2]\mathcal{M}_z = H_z \qquad (71.14)$$

(cf. (69.8)). The resulting magnetization curve is completely analogous to the ferroelectric polarization curve in Fig. 69.1. Its stable and unstable branches are similarly determined by the minima and maxima of the thermodynamic potential (cf. (69.11))

$$\rho\psi = \rho\tilde{\psi}(\rho, \vartheta) + \mu_0[a(\vartheta - \vartheta_c)\mathcal{M}_z^2 + b\mathcal{M}_z^4 - \mathcal{M}_z H_z], \qquad (71.15)$$

obtained by substituting (71.9) into (71.7). The investigation leads to the conclusion that, so long as $\mathbf{H}$ is not too large, there are two stable, oppositely directed, spontaneous magnetizations. Above the Curie temperature these ferromagnetic solutions disappear, and the material becomes paramagnetic.

Having now sufficiently pursued the analogy between ferroelectricity and ferromagnetism, we shall be content with a few general comments, especially as regards the (quantitative) *differences* between the two phenomena. Actual ferromagnets, like actual ferroelectrets, are crystalline or polycrystalline materials, and therefore have *magnetic axes*. These are, again, characterized by the relative adjectives 'easy' and 'hard'. Although *magnetostriction* certainly exists (it is responsible for the humming noise of transformers), it is generally a weaker effect, as compared with the non-magnetic stresses, than electrostriction is in ferroelectrets.

We have noted, in Section 69, that single-domain ferroelectrets cannot exist (except in a conducting medium), because the depolarizing electric fields are much too strong. In ferromagnets the $\mathcal{M}(\mathbf{H})$ relation is usually such that the demagnetizing (opposing) $\mathbf{H}$ that would result from (71.1) in a spontaneously, uniformly magnetized specimen is not strong enough to cause a reversal of $\mathcal{M}$. Single-domain ferromagnets *are* therefore possible, but are actually observed to occur only in very small specimens. Larger specimens have a magnetic domain structure. This is, again, explained as an effect of a surface free energy $\int \alpha \, dS$: although a single domain is not precluded by the magnetization curves of ferromagnets, subdivision into domains of finite size—which again turns out to vary as the square root $\sqrt{\ell}$ of the dimension of the body—corresponds to a minimum of the total free energy $\int \rho \psi \, dV + \int \alpha \, dS$. Only in ferromagnets that are smaller than a critical size, which depends on the surface tension coefficient α, does a single-domain configuration provide a minimum. In iron this critical size is about 20 nm.

If **H** and $\mathcal{M}$ are both uniform (the latter in each domain), the system (71.1) reduces to $\mathbf{n} \cdot [\![\mathcal{M}]\!] = 0$: the normal component of the magnetization must be continuous across the domain walls. Fig. 71.1 shows some actual domain structures that were observed in a 50 μm iron whisker.

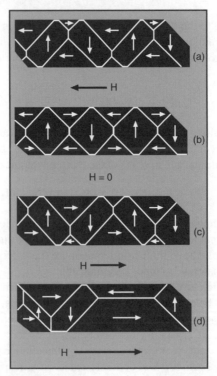

Fig. 71.1   (Adapted from R. W. DeBlois and C. D. Graham (1959) *Journal of Applied Physics*, **64**, 931.)

In a sufficiently strong applied field **H** the magnetization reaches saturation and becomes everywhere parallel to the field. This is, of course, a single-domain configuration that occurs in all ferromagnets, of any size. It does not contradict our former conclusions regarding single domains, because $-\mathcal{M} \cdot \mathbf{H}$ (the last term of (71.15)) is in this case *negative*; the minimum of the total free energy is therefore obtained with $\int \alpha \, dS = 0$.

Problem 71.1   An iron sphere of radius a is uniformly magnetized and carries a charge Q. The magnetization, which is initially $\mathcal{M}$, is gradually reduced to zero; for example, by heating the sphere up. Show that the sphere acquires angular momentum $\mathbf{L} = \frac{2}{9}\mu_0\mathcal{M}a^2Q$.[†]

Solution   Just outside the sphere, where $\mathcal{M} = 0$, the electromagnetic part of the stress tensor is

$$T = -\tfrac{1}{2}(\epsilon_0 E^2 + \mu_0 H^2)I + \mathbf{E} \otimes \mathbf{D} + \mathbf{H} \otimes \mathbf{B} + \epsilon_0 \mathbf{E} \times \mathbf{B} \otimes \dot{\mathbf{x}}.$$

The stress is

$$T\mathbf{n} = -\tfrac{1}{2}(\epsilon_0 E^2 + \mu_0 H^2)\mathbf{n} + \mathbf{E}D_n + \mathbf{H}B_n + \epsilon_0 \mathbf{E} \times \mathbf{B}\dot{x}_n.$$

The last term vanishes because there is no radial velocity. The first two terms are radial, and $\mathbf{H}B_n$ is parallel to the axis (the diameter parallel to $\mathcal{M}$). Thus the only term capable of producing a torque about the axis is the toroidal component of $\mathbf{E}D_n = \sigma \mathbf{E}$. According to Exercise 70.1, the demagnetizing factor for a sphere is $\frac{1}{3}$. Thus $\mathbf{B} = \mu_0(\mathbf{H} + \mathcal{M}) = \frac{2}{3}\mu_0\mathcal{M}$. The decrease of $\mathbf{B}$ induces a toroidal electric field, given by

$$2\pi s E_\phi = -\pi s^2 \dot{B},$$

where $s = r \sin\theta$ is the distance from the axis. (The flux $\int B \, dS$, which is an integral, is unaffected by the discontinuity of $\mathbf{B}$ at the boundary $s = a \sin\theta$.) This electric field exerts on an element dS of the surface a force $\sigma E_\phi \, dS$ and a torque $s\sigma \, dS E_\phi$, which is parallel to $\mathbf{B}$ (or $\mathcal{M}$). For the angular momentum we obtain

$$\dot{L}_z = \int s E_\phi \sigma \, dS = \int_0^\pi s E_\phi \sigma 2\pi a^2 \sin\theta \, d\theta$$

$$= -\pi \sigma a^4 \dot{B} \int_{-1}^1 \sin^2\theta \, d\cos\theta = -\frac{4\pi}{3}\sigma a^4 \dot{B}.$$

Substituting $\sigma = Q/(4\pi a^2)$ and $\int \dot{B} \, dt = -\frac{2}{3}\mu_0\mathcal{M}$, we obtain

$$\mathbf{L} = \frac{2}{9}\mu_0\mathcal{M}a^2Q.$$

[†]Sharma, N. L. (1988) *American Journal of Physics* **56**, 420, Problems 71.1 and 71.2 are examples of what has become known as Feynman's disc paradox (Feynman, R. P., Leighton, R. B. and Sands, M., *The Feynman Lectures*, Volume 2, p. 17–5, Addison-Wesley, Reading, Mass., 1964). The question, 'Where does the angular momentum come from?' constitutes the 'paradox'. Of course there is none.

Problem 71.2   The iron sphere of the last problem maintains its magnetization, but is gradually discharged by connecting the north pole through a resistance to earth. Assuming that the charge is uniformly distributed over the surface at every stage of the discharge, show that the sphere acquires the same angular momentum as in the previous Problem.

Solution   The discharge through the north pole will cause a northward surface current to flow. Let K be the surface current density. In order to determine the current $2\pi s K$ through a latitude circle of radius $s = a \sin \theta$ we use the law of (surface) charge conservation: the difference $d(2\pi s K)$ between the currents through two closely lying latitude circles must equal the rate of change of the charge $\sigma \, dS = \sigma 2\pi s a \, d\theta = -2\pi a^2 \dot\sigma \, d\cos\theta$ in the strip between them. Thus

$$2\pi s K = 2\pi a^2 \dot\sigma (c - \cos\theta),$$

where c is an integration constant. It is determined by the requirement that the limit of $2\pi s K$ at the north pole ($\cos\theta = 1$) equal the rate of decrease $-\dot Q = -4\pi a^2 \dot\sigma$ of the total charge. This yields $c = -1$. Note that the resulting $2\pi s K$ vanishes at the south pole ($\cos\theta = -1$), from which nothing is issuing. The surface current $\mathbf{K} = \mathbf{n} \times [\![\mathbf{H}]\!]$ creates a toroidal component $H_\phi = K$ just outside the sphere. Since $\mathbf{E}$ is radial outside the sphere, the toroidal stress (cf. the formula for $T\mathbf{n}$ in the last Problem) is $H_\phi B_n = K B \cos\theta$. The torque on a surface element dS is $s K B \cos\theta \, dS$. For the angular momentum we obtain

$$\dot L_z = \int s K B \cos\theta \, dS = - \int_{-1}^{1} 2\pi a^4 \dot\sigma B (1 + \cos\theta) \cos\theta \, d\cos\theta$$

$$= -2\pi a^4 \dot\sigma B \int_{-1}^{1} \cos^2\theta \, d\cos\theta = -\frac{4\pi}{3} \dot\sigma a^4 B,$$

$$\mathbf{L} = \frac{4\pi}{3} \sigma a^4 \mathbf{B} = \frac{2}{9} \mu_0 \mathcal{M} a^2 Q,$$

since $\int \dot\sigma \, dt = -Q/(4\pi a^2)$.

---

72. Superconductivity

In ordinary conductors the total current i, which enters at one end of the conductor and leaves at another, is linearly related to the electric voltage drop V between the two ends. Conduction of electricity through such conductors also leads to dissipation of energy, the rate of which is given by the volume integral of $\mathcal{J} \cdot \mathcal{E}$, which is proportional to the resistance V/i.

In 1908 Kammerlingh Onnes succeeded in liquefying helium, which provided an ideal cold bath for experiments at temperatures within a few degrees of absolute zero. When he measured the resistances of various metals at these low temperatures, he made (in 1911) an intriguing discovery: some substances, when sufficiently cooled, conduct electricity without any resistance; they become *superconductors*. The currents that these superconductors transmit are of course accompanied by magnetic fields. But these fields have a remarkable property: they do not enter the superconductor. In fact, when a material becomes superconducting, it expels any magnetic field that has previously existed inside it[†]. This magnetic field expulsion—called the *Meissner effect*—and the vanishing of resistance are not unconnected. Indeed the former implies the latter, because a magnetic field **B** that envelopes a body without penetrating it is necessarily associated with surface currents $\mathbf{K} = \mathbf{n} \times \mathbf{B}/\mu_0$ (we assume the surroundings to be non-magnetic, so that $\mathbf{H} = \mathbf{B}/\mu_0$). Unlike the volume conduction currents in ordinary conductors, these surface currents do not require any voltage drop to drive them. Hence there is no resistance. Nor is there any energy dissipation, since $\int \boldsymbol{\mathcal{J}} \cdot \boldsymbol{\mathcal{E}} \, dV$ vanishes.

We can easily calculate the magnetic part of the stress on a superconductor. Since $B_n = 0$, it is given by

$$\begin{aligned} T\mathbf{n} &= \left[-\tfrac{1}{2} B^2/\mu_0 I + \mathbf{B} \otimes \mathbf{B}/\mu_0 \right] \mathbf{n} \\ &= -\tfrac{1}{2} B^2/\mu_0 \mathbf{n} + \mathbf{H} B_n \\ &= -\tfrac{1}{2} B^2/\mu_0 \mathbf{n}. \end{aligned} \tag{72.1}$$

This is a pressure force of $\tfrac{1}{2} B^2/\mu_0 = \tfrac{1}{2}\mu_0 H^2$ per unit area.

Exercises

72.1. A non-magnetic sphere of radius a is placed in a uniform field $\mathbf{B}_0$. Prove that when the sphere becomes superconducting, the magnetic field changes to

$$\mathbf{B} = \left(1 + \frac{a^3}{2r^3} \right) \mathbf{B}_0 - \frac{3a^3}{2r^3} (\mathbf{n} \cdot \mathbf{B}_0)\mathbf{n}, \tag{72.2}$$

where $\mathbf{n} = \mathbf{r}/r$. Show that the magnetic pressure is $9B_0^2 \sin^2 \theta/(8\mu_0)$, where θ is the spherical polar angle, measured from the direction of $\mathbf{B}_0$.

72.2. Prove that the magnetic stress on a superconductor is $\tfrac{1}{2}\mathbf{K} \times \mathbf{B}$, where $\mathbf{K} = \mathbf{n} \times \mathbf{B}/\mu_0$ is the surface current. This is analogous to the electrostatic stress $\tfrac{1}{2}\sigma \mathbf{E}$ on a conductor.

The Meissner effect, of which superconductivity is a direct result, is one reason for regarding superconductors as magnetic materials. There is another. When superconductors were discovered it was hoped that they would carry arbitrarily large (surface) currents without any resistance. That would have made them ideal electromagnets, capable of creating intense magnetic fields without dissipation of energy. These hopes were shattered with the discovery that a superconductor became normal when the (tangential) magnetic field exceeded a critical value $B_c = \mu_0 H_c$, which depended on the temperature. The typical dependence of H_c on ϑ, as established by experiments, is shown in Fig. 72.1.

[†] We are confining ourselves to type I superconductors. In type II superconductors the transition and the expulsion of the magnetic field occur gradually.

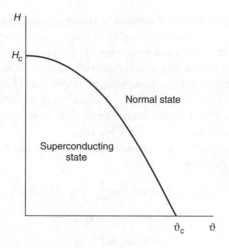

Fig. 72.1

The superconducting state is only possible for temperatures less than ϑ_c, and then only so long as the magnetic field is below the value given by the curve. For example, for tin the critical temperature is $\vartheta_c = 3.72$ K, and the largest possible magnetic field (which corresponds to zero absolute temperature) is $H_c = 25\,000$ amp m^{-1}.

The transition from the normal to the superconducting state has some further properties, which can be shown to follow from the principles laid down in Chapter 15. We assume that the superconducting and normal phases can both be described by the simple constitutive class (56.1) of fluids. Each of the two phases is then characterized by a free energy. For the normal phase, which we assume to be devoid of polarization or magnetization, we have (cf. (56.4)–(56.5) and (56.8)–(56.9)) the relations

$$s = -\varphi_\vartheta(\rho, \vartheta),$$
$$p = \rho^2 \varphi_\rho(\rho, \vartheta),$$
$$\mathbf{g} = \dot{\mathbf{x}} + \epsilon_0 \mathbf{E} \times \mathbf{B}/\rho, \tag{72.3}$$
$$T\mathbf{n} = -[p + \tfrac{1}{2}\epsilon_0 E^2 + \tfrac{1}{2}B^2/\mu_0]\mathbf{n}$$
$$+ \epsilon_0 E_n \mathbf{E} + B_n \mathbf{B}/\mu_0 + \epsilon_0 \mathbf{E} \times \mathbf{B}\dot{x}_n.$$

The relations for the superconducting phase are the same, except that the electromagnetic field vanishes. Of course the free energies of the two phases are not the same.

Consider now a surface separating a superconductor from its normal phase. We shall assume that the two phases have the same temperature ϑ. The other variables, namely the densities of the two phases and the (tangential) magnetic field, cannot then be arbitrarily specified, because they are subject to the jump conditions of Sections 51, 52 and 54. We must therefore determine the constraints that these conditions impose in the present case. The jump conditions (cf. (51.13), (54.3), (54.6) and (52.12)) are

$$[\![\rho(\dot{\mathbf{x}} - \mathbf{v})]\!] \cdot \mathbf{n} = 0, \qquad [\![\rho\mathbf{g} \otimes (\dot{\mathbf{x}} - \mathbf{v}) - T]\!]\mathbf{n} = 0,$$

$$[\![\rho\varepsilon(\dot{\mathbf{x}} - \mathbf{v}) - T^T\dot{\mathbf{x}} + \mathbf{q} + \mathcal{E} \times \mathcal{H}]\!] \cdot \mathbf{n} = 0, \tag{72.4}$$

$$[\![\rho s(\dot{\mathbf{x}} - \mathbf{v}) + \mathbf{q}/\vartheta]\!] \cdot \mathbf{n} \geq 0.$$

It should, perhaps, be noted that, with material crossing the magnetic field in the normal phase, the existence of an electric field on that side cannot be ignored. We shall find it convenient to choose a frame in which, at the instant considered, the surface is stationary, so that $v_n = 0$. The first condition then states that the mass flux $\rho\dot{x}_n$ is continuous. Turning next to $(72.4)_3$, we note that

$$\mathcal{E} \times \mathcal{H} \cdot \mathbf{n} - T^T\dot{\mathbf{x}} \cdot \mathbf{n}$$
$$= \dot{x}_n[p - \tfrac{1}{2}\epsilon_0 E^2 - \tfrac{1}{2}B^2/\mu_0] + \mathbf{n} \times \mathbf{E} \cdot \mathbf{B}/\mu_0. \tag{72.5}$$

This relation results from $\mathcal{E} = \mathbf{E} + \dot{\mathbf{x}} \times \mathbf{B}$, $\mathcal{H} = \mathbf{H} - \dot{\mathbf{x}} \times \mathbf{D} = \mathbf{B}/\mu_0 - \epsilon_0\dot{\mathbf{x}} \times \mathbf{E}$ and $(72.3)_4$. Since the surface is stationary, $\mathbf{n} \times \mathbf{E}$ is continuous, so that it vanishes on either side. Thus $(72.4)_3$ becomes

$$[\![\rho\dot{x}_n[\varepsilon - \tfrac{1}{2}\epsilon_0 E^2/\rho - \tfrac{1}{2}B^2/(\mu_0\rho) + p/\rho] + q_n]\!] = 0. \tag{72.6}$$

Using the continuity of ϑ, we now eliminate q_n between $(72.4)_4$ and (72.6) and obtain

$$\rho\dot{x}_n[\![\varepsilon - \vartheta s - \tfrac{1}{2}\epsilon_0 E^2/\rho - \tfrac{1}{2}B^2/(\mu_0\rho) + p/\rho]\!] \leq 0. \tag{72.7}$$

This inequality states that mass can only flow in the direction in which the expression in the double square brackets decreases. If we further assume that the mass flux depends continuously on ρ, ϑ, $\mathbf{E}$ and $\mathbf{B}$, it follows that the mass flux and the jump in (72.7) must vanish together (cf. the derivation of (57.3)). In equilibrium, when $\dot{x}_n = 0$, the expression in the double square brackets of (72.7) will therefore have the same value for both phases. Recalling the definition (55.7) of the free energy, we can express this condition for equilibrium in the form

$$\varphi_s(\rho_s, \vartheta) + \frac{p_s(\rho_s, \vartheta)}{\rho_s} = \varphi_n(\rho_n, \vartheta) + \frac{p_n(\rho_n, \vartheta)}{\rho_n}. \tag{72.8}$$

This is a relation between the densities of the phases and the temperature. The function

$$\gamma = \varphi(\rho, \vartheta) + \frac{\rho^2\varphi_\rho(\rho, \vartheta)}{\rho} \tag{72.9}$$

is called the (specific) *Gibbs thermodynamic potential*. If we regard the equation $p = \rho^2 \varphi_\rho(\rho, \vartheta)$ as defining the density ρ in terms of p and ϑ, the Gibbs potential becomes a function of p and ϑ. The equilibrium condition (72.8) is then

$$\gamma_s(p_s, \vartheta) = \gamma_n(p_n, \vartheta). \tag{72.10}$$

Exercise

72.3. Prove that $\gamma(p, \vartheta)$ has the partial derivatives

$$\frac{\partial \gamma(p, \vartheta)}{\partial \vartheta} = -s, \qquad \frac{\partial \gamma(p, \vartheta)}{\partial p} = \frac{1}{\rho}. \tag{72.11}$$

We have not yet made any use of the jump condition (72.4)$_2$. In equilibrium it requires $T\mathbf{n}$ to be continuous. According to (72.3)$_4$ this means

$$p_s(\rho_s, \vartheta) = p_n(\rho_n, \vartheta) + B^2/(2\mu_0), \tag{72.12}$$

which is a second relation between the densities of the phases. It states that the density of the superconductor will have a value corresponding to pressure equilibrium, not with p_n but with $p_n + B^2/(2\mu_0)$. Of course the extra pressure is just the one we have found in (72.1).

In actual experimental situations the normal state does not extend to infinity. It is usually bounded, and is therefore itself in pressure equilibrium with another material (such as the atmosphere). It is thus more practical to regard p_n, rather than ρ_n, as given (along with the temperature ϑ). We denote this pressure simply by p (e.g., one atmosphere) and combine the equilibrium conditions (72.10) and (72.12) in the form

$$\gamma_s(p + \tfrac{1}{2}\mu_0 H^2, \vartheta) = \gamma_n(p, \vartheta); \tag{72.13}$$

it is customary to use H rather than B/μ_0 in treatments of superconductivity. According to this equation, equilibrium of a superconductor with its normal state at a pressure p and temperature ϑ requires that the magnetic field have a definite, or *critical*, value critical magnetic field $H_c(p, \vartheta)$. Since, by (72.11)$_2$, γ_s has a positive derivative with respect to its first argument, a higher field will lead to $\gamma_s > \gamma_n$. That will result in a transition to the normal state. Similarly, $H < H_c(p, \vartheta)$ will result in a transition from the normal to the superconducting state. When $H = 0$ the transition, at given p, will take place at a definite critical transition temperature ϑ_c. These considerations are in accord with Fig. 72.1, which shows the critical field H_c as a function of temperature for a given 'external' pressure p.

Let us now differentiate (72.13) with respect to the temperature, keeping p fixed. Using (72.11), we obtain

$$s_n - s_s = -\frac{1}{\rho_s} \frac{\partial}{\partial \vartheta} \left[\tfrac{1}{2} \mu_0 H_c^2 (p, \vartheta) \right]. \tag{72.14}$$

In order to interpret this result, we turn back to the jump condition (72.6), which up to this point has only been used in conjunction with (72.4)$_4$. According to (72.6), mass flow with a discontinuity (of magnitude $\mathcal{L}$, say) in the expression multiplying $\rho \dot{x}_n$ is associated with a discontinuity of the heat flux q_n at the surface of separation. The discontinuity $\mathcal{L}$ is therefore called the (specific) *latent heat of transition*. Now the jumps by which the continuous $\rho \dot{x}_n$ is multiplied in (72.6) and (72.7) differ exactly by $[\![\vartheta s]\!] = \vartheta [\![s]\!]$. In the limit of equilibrium the jump in (72.7) vanishes, so that the latent heat $\mathcal{L}$ (per unit mass) becomes ϑ times the equilibrium jump of s. From (72.14) we therefore obtain Keesom's formula

$$\mathcal{L} = -\frac{\vartheta}{\rho_s} \frac{\partial}{\partial \vartheta} \left[\tfrac{1}{2} \mu_0 H_c^2 (p, \vartheta) \right] \tag{72.15}$$

for the equilibrium limit of the latent heat (per unit mass) of transition from the superconducting to the normal state. According to the typical Fig. 72.1, $\mathcal{L}$ is positive, since the slope of H_c with ϑ is negative. The latent heat vanishes at $\vartheta = 0$ and at $\vartheta = \vartheta_c$ (where $H_c = 0$). These results, too, are confirmed by experiments on superconductors.

The specific heat at constant pressure is defined by $c = \vartheta \partial s(p, \vartheta)/\partial \vartheta$. We can calculate the difference between the specific heats of the two phases at equilibrium by differentiating (72.14) at constant p. It must, of course, be remembered that s_s and ρ_s are the partial derivatives of the γ_s of (72.13) and therefore have the same arguments as the latter; and that, furthermore, $\partial s(p, \vartheta)/\partial p = \rho^{-2} \partial \rho(p, \vartheta)/\partial \vartheta$ (which follows from (72.11)). The result of this calculation is

$$c_n - c_s = -\frac{\vartheta}{\rho_s} \frac{\partial^2}{\partial \vartheta^2} \tfrac{1}{2} \mu_0 H_c^2 - \frac{\mathcal{L}}{\rho_s} \frac{\partial \rho_s}{\partial \vartheta}. \tag{72.16}$$

At $\vartheta = \vartheta_c$ both H_c and $\mathcal{L}$ vanish. Then

$$c_n - c_s = -\frac{\mu_0 \vartheta_c}{\rho_s} \left[\frac{\partial H_c(p, \vartheta_c)}{\partial \vartheta} \right]^2. \tag{72.17}$$

A pronounced jump in the specific heat at the transition temperature had, in fact, been observed before the thermodynamic theory of superconductors was developed. Equation (72.17), which is known as Rutgers's formula, relates this jump to the transition temperature, the density and the slope of H_c at $\vartheta = \vartheta_c$, all of which are measurable. Rutgers's formula has been verified by numerous experiments.

Exercises

72.4. Show, by differentiating (72.13) with respect to p, that

$$\frac{1}{\rho_n} - \frac{1}{\rho_s} = \frac{1}{\rho_s}\frac{\partial}{\partial p}\left[\tfrac{1}{2}\mu_0 H_c^2(p, \vartheta)\right]. \tag{72.18}$$

This difference between the specific volumes of the phases at equilibrium vanishes (with H_c) at $\vartheta = \vartheta_c$.

72.5. Show, by differentiating (72.13) with respect to ϑ at fixed H_c, that

$$\frac{\partial p(\vartheta, H_c)}{\partial \vartheta} = \frac{\mathcal{L}}{\vartheta(\rho_n^{-1} - \rho_s^{-1})}. \tag{72.19}$$

This is the rate of change, with respect to temperature, of the pressure needed to keep the magnetic field critical.

Consider now a spherical superconductor in a magnetic field which is uniform at infinity (Exercise 72.1). Near the sphere the enveloping magnetic field is non-uniform and has its greatest intensity at the 'equator'. If the field is gradually increased, it will at first exceed the critical value near the equator. It might be supposed that this will cause the material in an equatorial belt to become normal, the rest of the sphere remaining in the superconducting state; and that the boundary between the two phases will gradually move inward as the external magnetic field is increased, until the whole sphere becomes normal (and the field uniform). Throughout this process the field must be critical at every point on the phase boundary; it must also increase as one moves from the boundary into the normal belt, for otherwise a part of the normal region will again become superconducting.

It turns out that these requirements cannot be met with any configuration in which the superconducting region is simply connected. What happens then is a breakup of the whole sphere into a complicated structure of normal and superconducting domains, called a *mixed state*. Such structures have indeed been observed. As with ferroelectrets and ferromagnets, a surface free energy is invoked in order to account for the eventual halting of the breakup, and for its variation between different materials.

Exercise

72.6. A small permanent magnet of moment **m** is placed near a superconductor. Show that the magnetic field is obtained by adding the field of an image magnet **m**′ inside the superconductor. Deduce from the known behaviour of a pair of magnets that **m** is repelled by the superconductor.

Appendix A

Gaussian units

In the foregoing chapters we have introduced the Système International (SI) electromagnetic units and used them throughout. It will be recalled that these units were based on the definition of the ampère (cf. (15.2) and (54.7)),

$$\mu_0 = 4\pi \ 10^{-7} \ \frac{\text{newton}}{\text{amp}^2}. \tag{A.1}$$

All other SI electromagnetic units followed from this definition, together with Maxwell's equations and the various defining relations for other quantities, such as $C = Q/V$ or $R = V/i$. In terms of these units, we have also written the relation $\epsilon_0\mu_0 = 1/c^2$ in the form (cf. (15.3))

$$\frac{1}{4\pi\epsilon_0} = \frac{\mu_0 c^2}{4\pi} = 9 \times 10^9 \ \frac{\text{newton} \cdot \text{m}^2}{\text{coulomb}^2}, \tag{A.2}$$

where the factor 9 (*not* the exponent 9) stands for the square of the speed of light c in units of $10^8 \ \text{m s}^{-1}$.

Many other systems of units have been proposed during the past 150 years, but only two of them are still widely used: SI and Gaussian units. SI units are used by all engineers and technicians, and also by many physicists. The trend throughout the latter half of the twentieth century has certainly been in favour of a universal adoption of SI units. But Gaussian units are still being used by physicists, especially in atomic and particle physics. They also appear in some of the most popular textbooks and reference books, and are therefore indispensable to any physicist.

We have seen that Maxwell's equations result from the two tensor equations (6.2) and (9.11) after giving the names $\{\mathbf{H}, \mathbf{D}\}$ and $\{\mathbf{B}, \mathbf{E}\}$ to the components of the charge-current potential f and the electromagnetic field F. In the Gaussian system, these names are given differently: the new (primed) names are related to the SI names through the equality of the second and third columns in Table A.1. For example, SI charge Q and Gaussian charge Q' are related through $Q = \sqrt{4\pi\epsilon_0}\,Q'$. Note that the Gaussian quantities have dimensions that are different from those of the corresponding SI quantities. With these new names, Maxwell's equations and

Table A.1

Quantity	SI	Gaussian
Charge	$Q(q, \sigma)$	$(4\pi\epsilon_0)^{\frac{1}{2}} Q'(q', \sigma')$
Current	$i(\mathbf{j}, \mathbf{K})$	$(4\pi\epsilon_0)^{\frac{1}{2}} i'(\mathbf{j}', \mathbf{K}')$
Charge potential	$\mathbf{D}(\mathbf{P})$	$(\epsilon_0/4\pi)^{\frac{1}{2}} \mathbf{D}'(4\pi\mathbf{P}')$
Current potential	$\mathbf{H}(\mathbf{M})$	$(4\pi\mu_0)^{-\frac{1}{2}} \mathbf{H}'(4\pi\mathbf{M}')$
Electric field	$\mathbf{E}(V)$	$(4\pi\epsilon_0)^{-\frac{1}{2}} \mathbf{E}'(V')$
Magnetic field	$\mathbf{B}(\mathbf{A})$	$(\mu_0/4\pi)^{\frac{1}{2}} \mathbf{B}'(\mathbf{A}')$

the aether relations become

$$\operatorname{div} \mathbf{D}' = 4\pi q', \qquad \operatorname{curl} \mathbf{H}' = \frac{4\pi}{c}\mathbf{j}' + \frac{1}{c}\mathbf{D}'_t,$$

$$\operatorname{div} \mathbf{B}' = 0, \qquad \operatorname{curl} \mathbf{E}' = -\frac{1}{c}\mathbf{B}'_t, \qquad (\text{A.3})$$

$$\mathbf{D}' = \mathbf{E}' + 4\pi\mathbf{P}', \qquad \mathbf{H}' = \mathbf{B}' - 4\pi\mathbf{M}'.$$

Of course, in the Gaussian system these equations are written without the primes. The most striking fact about these equations is that the Gaussian fields $\mathbf{D}'$, $\mathbf{H}'$, $\mathbf{E}'$, $\mathbf{B}'$, $\mathbf{P}'$ and $\mathbf{M}'$ all have the same dimensions.

We also write down a few other obviously useful relations:

$$\mathcal{E} = \mathbf{E} + \dot{\mathbf{x}} \times \mathbf{B} = \frac{1}{\sqrt{4\pi\epsilon_0}}\left(\mathbf{E}' + \frac{1}{c}\dot{\mathbf{x}} \times \mathbf{B}'\right) = \frac{1}{\sqrt{4\pi\epsilon_0}}\mathcal{E}',$$

$$\mathcal{H} = \mathbf{H} - \dot{\mathbf{x}} \times \mathbf{D} = \frac{1}{\sqrt{4\pi\mu_0}}\left(\mathbf{H}' - \frac{1}{c}\dot{\mathbf{x}} \times \mathbf{D}'\right) = \frac{1}{\sqrt{4\pi\mu_0}}\mathcal{H}',$$

$$\mathcal{E} \times \mathcal{H} = \frac{c}{4\pi}\mathcal{E}' \times \mathcal{H}', \qquad \mathcal{J} \cdot \mathcal{E} = \mathcal{J}' \cdot \mathcal{E}', \qquad (\text{A.4})$$

$$\frac{\epsilon_0 E^2}{2} = \frac{E'^2}{8\pi}, \qquad \frac{B^2}{2\mu_0} = \frac{B'^2}{8\pi}.$$

Since the electromagnetic quantities in the Gaussian system have different dimensions, their units must be newly defined. There is a difference even in the mechanical units, for the Gaussian system uses CGS mechanical units. It is obvious from (A.2) that the Gaussian charge $Q' = Q/\sqrt{4\pi\epsilon_0}$ has the dimensions of length times the square root of force. The Gaussian unit of electric charge is called the *electrostatic unit* (esu), or the *statcoulomb*, and is defined as

$$1\,\text{esu} = 1\,\text{dyne}^{1/2} \cdot \text{cm}. \qquad (\text{A.5})$$

It follows that

$$\frac{1}{4\pi\epsilon_0} = 9 \times 10^9 \frac{\text{newton} \cdot \text{m}^2}{\text{coulomb}^2}$$

$$= 9 \times 10^{18} \frac{\text{dyne} \cdot \text{cm}^2}{\text{coulomb}^2}$$

$$= 9 \times 10^{18} \frac{\text{esu}^2}{\text{coulomb}^2}. \tag{A.6}$$

Since the esu and the coulomb have different dimensions, it is meaningless to enquire about 'the number of esu's contained in one coulomb'. What we *can* ask is, if the SI charge Q is 1 coulomb, how many esu's are there in the corresponding Gaussian Q'? In order to answer this question, we set $Q = 1$ coulomb and use the first line of Table A.1:

$$Q' = \frac{Q}{\sqrt{4\pi\epsilon_0}} = \frac{3 \times 10^9 \text{ esu}}{\text{coulomb}} \, 1 \text{ coulomb} = 3 \times 10^9 \text{ esu}. \tag{A.7}$$

This gives the first line in Table A.2.

The Gaussian unit of current is 1 statamp $= 1$ esu s^{-1}. According to (A.7), an SI current of 1 amp corresponds to a Gaussian current of 3×10^9 statamp.

The Gaussian unit of electric potential, the *statvolt*, is defined by

$$1 \text{ statvolt} = 1 \frac{\text{esu}}{\text{cm}} = 1 \frac{\text{dyne} \cdot \text{cm}}{\text{esu}}, \tag{A.8}$$

the last equality following from (A.5). In order to find the correspondence between statvolts and volts we set $V = 1$ volt $= 1$ newton m coulomb^{-1} in $V' = \sqrt{4\pi\epsilon_0} V$

Table A.2

Quantity	SI	Gaussian
Charge	1 coulomb	3×10^9 esu
Current	1 amp	3×10^9 statamp
Electric Potential	1 volt	$(300)^{-1}$ statvolt
Capacitance	1 farad	9×10^{11} statfarad
Resistance	1 ohm	$(9 \times 10^{11})^{-1}$ statohm
Conductivity	1 siemen m^{-1}	9×10^9 statohm^{-1} cm^{-1}
Electric field	1 volt m^{-1}	$(3 \times 10^4)^{-1}$ statvolt cm^{-1}
Charge potential	1 coulomb m^{-2}	$12\pi \, 10^5$ esu cm^{-2}
Polarization	1 coulomb m^{-2}	3×10^5 esu cm^{-2}
Current potential	1 amp m^{-1}	$4\pi \, 10^{-3}$ oersted
Magnetization	1 amp m^{-1}	10^{-3} oersted
Magnetic field	1 tesla	10^4 gauss
Magnetic flux	1 weber	10^8 maxwell
Inductance	1 henry	$(9 \times 10^{11})^{-1}$ stathenry

and obtain

$$V' = \frac{\text{coulomb}}{3 \times 10^9 \text{ esu}} 1 \frac{\text{newton} \cdot \text{m}}{\text{coulomb}}$$

$$= \frac{1}{300} \frac{\text{dyne} \cdot \text{cm}}{\text{esu}}$$

$$= \frac{1}{300} \text{ statvolt} \tag{A.9}$$

Thus a household voltage of 240 volts corresponds to 0.8 statvolt.

The Gaussian definition of capacitance is

$$C' = \frac{Q'}{V'} = \frac{1}{4\pi\epsilon_0} \frac{Q}{V}. \tag{A.10}$$

Its unit is the *statfarad*, defined by

$$1 \text{ statfarad} = 1 \frac{\text{esu}}{\text{statvolt}} = 1 \text{ cm}. \tag{A.11}$$

Since the statfarad is nothing other than a centimetre, the name merely serves to remind us that we are concerned with a capacitance. If we set $C = 1$ farad $= 1$ coulomb volt^{-1} in (A.10), we obtain

$$C' = 9 \times 10^{18} \frac{\text{esu}^2}{\text{coulomb}^2} \frac{\text{coulomb}^2}{\text{newton} \cdot \text{m}}$$

$$= 9 \times 10^{11} \frac{\text{esu}^2}{\text{dyne} \cdot \text{cm}}$$

$$= 9 \times 10^{11} \text{ statfarad.} \tag{A.12}$$

Gaussian resistance and conductivity are defined by $R' = V'/i'$ and $\mathcal{J}' = \sigma'\mathcal{E}'$. Their units are, respectively,

$$1 \text{ statohm} = 1 \frac{\text{statvolt}}{\text{statamp}} = 1 \frac{\text{s}}{\text{cm}},$$

$$1 \text{ (statohm} \cdot \text{cm)}^{-1} = 1 \frac{\text{statamp}}{\text{statvolt} \cdot \text{cm}} = 1 \text{ s}^{-1}. \tag{A.13}$$

The correspondences between statohms and ohms, and between (statohm cm)$^{-1}$ and siemen m^{-1}, are easily worked out and are given in the fifth and sixth lines of Table A.2.

Next, we consider the fields. The Gaussian electric field $\mathbf{E}'$ has the dimensions of Gaussian electric potential, divided by length. Its unit may be taken as 1 statvolt cm^{-1}. Since $\mathbf{E}' = \sqrt{4\pi\epsilon_0}\mathbf{E}$, we find that an SI electric field of 1 volt m^{-1} corresponds to

$$E' = \frac{\text{coulomb}}{3 \times 10^9 \text{ esu}} 1 \frac{\text{newton}}{\text{coulomb}} = \frac{1}{3 \times 10^4} \frac{\text{statvolt}}{\text{cm}}; \tag{A.14}$$

the difference between the 3×10^4 in this equation and the 300 in (A.9) results from $1\,\text{m} = 100\,\text{cm}$.

The Gaussian charge potential (or electric displacement) $\mathbf{D}'$ has the dimensions of Gaussian charge, divided by area (cf. $(A.3)_1$). Its unit may be taken as $1\,\text{esu}\,\text{cm}^{-2}$, which is the same as the electric field unit $1\,\text{statvolt}\,\text{cm}^{-1}$ (we have already noted that the Gaussian fields all have the same dimensions). Since $\mathbf{D}' = \sqrt{4\pi/\epsilon_0}\mathbf{D}$, an SI charge potential of $1\,\text{coulomb}\,\text{m}^{-2}$ corresponds to

$$D' = 4\pi\, 3 \times 10^9 \, \frac{\text{esu}}{\text{coulomb}} \, 1 \, \frac{\text{coulomb}}{10^4\,\text{cm}^2} = 12\pi\, 10^5 \, \frac{\text{esu}}{\text{cm}^2}. \tag{A.15}$$

The Gaussian polarization includes an extra factor of 4π (cf. Table A.1, or $(A.3)_5$). An SI polarization of $1\,\text{coulomb}\,\text{m}^{-2}$ therefore corresponds to a Gaussian polarization of $3 \times 10^5 \, \text{esu}\,\text{cm}^{-1}$.

In the case of the Gaussian current potentials $\mathbf{H}'$ and $\mathbf{M}'$, the field unit of $\text{esu}\,\text{cm}^{-2}$, or $\text{statvolt}\,\text{cm}^{-1}$, is called the *oersted*. According to (A.1) or (A.2) we have

$$\sqrt{4\pi\mu_0} = 4\pi\, 10^{-1} \, \frac{\text{esu}}{\text{amp}\cdot\text{cm}}. \tag{A.16}$$

Since $\mathbf{H}' = \sqrt{4\pi\mu_0}\mathbf{H}$, an SI charge potential of $1\,\text{amp}\,\text{m}^{-1}$ corresponds to

$$H' = 4\pi\, 10^{-1} \, \frac{\text{esu}}{\text{amp}\cdot\text{cm}} \, 1 \, \frac{\text{amp}}{10^2\,\text{cm}} = 4\pi\, 10^{-3} \text{ oersted}. \tag{A.17}$$

The Gaussian magnetization (like the polarization) includes an extra factor of 4π. An SI magnetization of $1\,\text{amp}\,\text{m}^{-1}$ therefore corresponds to a Gaussian magnetization of 10^{-3} oersted.

In the case of the Gaussian magnetic field $\mathbf{B}'$, the field unit of $\text{statvolt}\,\text{cm}^{-1}$ is called the *gauss*. Since $\mathbf{B}' = \sqrt{4\pi/\mu_0}\mathbf{B}$, a magnetic field of $1\,\text{tesla} = 1\,\text{volts}\,\text{s}\,\text{m}^{-2}$ corresponds to

$$B' = 10 \, \frac{\text{amp}\cdot\text{cm}}{\text{esu}} \, 1 \, \frac{\text{newton}\cdot\text{m}\cdot\text{s}}{\text{coulomb}\cdot\text{m}^2} = 10^4 \text{ gauss}. \tag{A.18}$$

The Gaussian magnetic flux unit of $\text{gauss}\,\text{cm}^2$ is called the *maxwell*. Since $1\,\text{m}^2 = 10^4\,\text{cm}^2$, an SI flux of 1 weber corresponds to a Gaussian flux of 10^8 maxwell.

Finally, we consider the inductance. In SI it is defined as magnetic flux, divided by current: $L = \Phi/i$; its unit is $1\,\text{henry} = 1\,\text{weber}\,\text{amp}^{-1}$. The Gaussian inductance is defined by $L' = \Phi'/(ci')$; note the velocity of light in the denominator. Its unit, the *stathenry*, is

$$1 \text{ stathenry} = 1 \, \frac{\text{maxwell}}{\text{cm}\cdot\text{s}^{-1}\cdot\text{statamp}} = 1 \, \frac{\text{s}^2}{\text{cm}}. \tag{A.19}$$

It is easy to show that $L' = 4\pi\epsilon_0 L$. An SI inductance of 1 henry $=$ 1 weber amp^{-1} therefore corresponds to a Gaussian inductance of

$$
\begin{aligned}
L' &= \frac{\text{coulomb}^2}{9 \times 10^{18} \text{ esu}^2} \, 1 \, \frac{\text{volt} \cdot \text{s}}{\text{amp}} \\
&= \frac{\text{coulomb}^2}{9 \times 10^{18} \text{ esu}^2} \, \frac{\text{newton} \cdot \text{m} \cdot \text{s}^2}{\text{coulomb}^2} \\
&= \frac{1}{9 \times 10^{11}} \text{ stathenry.}
\end{aligned}
\tag{A.20}
$$

Of course every 3 (including the 3 in 300, and in $12\pi = 3 \cdot 4\pi$) which appears in the third column of Table A.2 stands for the speed of light in units of 10^8 m s^{-1}, and every factor (*not* exponent) 9, for the square of this '3'.

Gaussian units are neither superior nor inferior to SI units. Atomic physicists still show a preference for the Gaussian system, but the reason for this is not obvious, for they seem to think of electric charge in terms of a multiple of the electron's charge, rather than so many esu's. Perhaps they merely prefer to have Maxwell's equations and the aether relations in the form (A.3). After all, many of the greatest works in electromagnetism have used the Gaussian system[†].

Gaussian adherents have sometimes claimed that the basic Maxwell–Lorentz aether relations $\mathbf{D}' = \mathbf{E}'$ and $\mathbf{H}' = \mathbf{B}'$ between the total (free and bound) charge and current potentials and the fields are the simplest expression of a fundamental property of space-time, certainly simpler than the SI relations $\mathbf{D} = \epsilon_0 \mathbf{E}$ and $\mathbf{H} = \mathbf{B}/\mu_0$ with the dimensional constants ϵ_0 and μ_0. Should we not then recognize these basic identities by assigning the same dimensions to the electromagnetic fields? Here, at last, we seem to have an *objective* argument in favour of the Gaussian system. Unfortunately, the claim itself is untrue, for we have seen that the aether relations characterize a special class of frames, rather than a property of space-time.

Since the Gaussian fields $\mathbf{D}', \mathbf{H}', \mathbf{E}', \mathbf{B}', \mathbf{P}'$ and $\mathbf{M}'$ all have the same dimensions, would it not be sensible to employ the same unit, for example the gauss, for all of them? It would not be tactful for an author who uses SI units to attempt an answer to this question. The choice of physical units is usually based on habit and convenience. The SI units of $\mathbf{D}$ and $\mathbf{P}$ are of course the same (coulomb m^{-2}), as are those of $\mathbf{H}$ and $\mathbf{M}$ (amp m^{-1}).

[†]Maxwell himself used the *electromagnetic* system, a variant of the Gaussian.

Appendix B

Solutions to the exercises

Chapter 2

3.1. In the sum $\epsilon^{i_1 i_2 \cdots i_n} \epsilon_{i_1 i_2 \cdots i_n}$ only those terms which correspond to permutations of $\{1, 2, \ldots, n\}$ do not vanish. There are $n!$ such terms, and each one equals 1.

3.2. $c^1 = \epsilon^{1st} a_s b_t = \epsilon^{123} a_2 b_3 + \epsilon^{132} a_3 b_2 = a_2 b_3 - a_3 b_2$. Similarly, $c^2 = a_3 b_1 - a_1 b_3$ and $c^3 = a_1 b_2 - a_2 b_1$.

3.3. Apply the last exercise to the vector product $\nabla \times \mathbf{b}$.

3.4. By the chain rule

$$\frac{\partial x^r}{\partial x^{r'}} \frac{\partial x^{r'}}{\partial x^{r''}} = \frac{\partial x^r}{\partial x^{r''}}.$$

3.5. The sum $A_{rs} + A_{sr}$ and difference $A_{rs} - A_{sr}$ are tensors. Thus $A_{(rs)} = \frac{1}{2}(A_{rs} + A_{sr})$ either vanishes in all frames, or in none, which means that A is either antisymmetric in all frames, or in none. Similarly, since $A_{[rs]} = \frac{1}{2}(A_{rs} - A_{sr})$ is a tensor, it follows that A is either symmetric in all frames, or in none.

3.6.

$$a_{r'} b^{r'} = \frac{\partial x^r}{\partial x^{r'}} a_r \frac{\partial x^{r'}}{\partial x^s} b^s = \delta_s^r a_r b^s = a_r b^r.$$

3.7. The difference

$$a_{r'} X^{r'} - a_r X^r = \left(a_{r'} \frac{\partial x^{r'}}{\partial x^r} - a_r \right) X^r$$

vanishes for arbitrary X^r. Hence

$$a_r = a_{r'} \frac{\partial x^{r'}}{\partial x^r}.$$

3.8.

$$A^{r's'} - \frac{\partial x^{r'}}{\partial x^r}\frac{\partial x^{s'}}{\partial x^s}A^{rs} = u^{r'}v^{s'} - \frac{\partial x^{r'}}{\partial x^r}\frac{\partial x^{s'}}{\partial x^s}u^r v^s = 0.$$

3.9.

$$A_{r'} = \partial_{r'}\phi = \frac{\partial x^r}{\partial x^{r'}}\partial_r\phi = \frac{\partial x^r}{\partial x^{r'}}A_r.$$

4.1. According to the definitions (4.10)–(4.11),

$$\text{dual } F \cdot \text{dual } G = \frac{1}{(n-M)!}(\text{dual } F)_{r_1 r_2 \dots r_{n-M}}(\text{dual } G)^{r_1 r_2 \dots r_{n-M}}$$

$$= \frac{F^{s_1 s_2 \dots s_M}G_{t_1 t_2 \dots t_M}}{(n-M)!M!M!}\epsilon_{s_1 s_2 \dots s_M r_1 r_2 \dots r_{n-M}}\epsilon^{r_1 r_2 \dots r_{n-M}t_1 t_2 \dots t_M}.$$

In $\epsilon_{s_1 s_2 \dots s_M r_1 r_2 \dots r_{n-M}}$, it takes M interchanges—a factor of $(-)^M$—to bring r_1 in front of s_1. Next, we transfer r_2 in front of s_1, which gives another factor of $(-)^M$. In this way we have

$$\epsilon_{s_1 s_2 \dots s_M r_1 r_2 \dots r_{n-M}} = (-)^{(n-M)M}\epsilon_{r_1 r_2 \dots r_{n-M}s_1 s_2 \dots s_M}.$$

The sum $\epsilon_{r_1 r_2 \dots r_{n-M}s_1 s_2 \dots s_M}\epsilon^{r_1 r_2 \dots r_{n-M}t_1 t_2 \dots t_M}$ has $(n-M)!$ terms, each of them equal to 1, if $(t_1 t_2 \dots t_M)$ are distinct integers (selected from the range $(1, 2, \dots n)$), and $(s_1 s_2 \dots s_M)$ is an *even* permutation of $(t_1 t_2 \dots t_M)$. It has $(n-M)!$ terms, each of them equal to -1, if $(t_1 t_2 \dots t_M)$ are distinct integers and $(s_1 s_2 \dots s_M)$ is an *odd* permutation of $(t_1 t_2 \dots t_M)$. Otherwise the sum vanishes. Thus

$$\epsilon_{r_1 r_2 \dots r_{n-M}s_1 s_2 \dots s_M}\epsilon^{r_1 r_2 \dots r_{n-M}t_1 t_2 \dots t_M} = \frac{1}{(n-M)!}\delta^{t_1 t_2 \dots t_M}_{s_1 s_2 \dots s_M},$$

where $\delta^{t_1 t_2 \dots t_M}_{s_1 s_2 \dots s_M}$, a generalization of the Kronecker δ^t_s, takes the values ± 1 or 0 in accordance with the foregoing conditions on $(t_1 t_2 \dots t_M)$ and $(s_1 s_2 \dots s_M)$. The final step is to note that, for any skew-symmetric G,

$$\delta^{t_1 t_2 \dots t_M}_{s_1 s_2 \dots s_M}G_{t_1 t_2 \dots t_M} = M!\,G_{s_1 \dots s_M}.$$

4.2. In Exercises 3.4–3.7, 'tensor' or 'scalar' applies to any kind of tensor, absolute or relative. In Exercise 3.8 one of the tensors u^r or v^r must be absolute.

4.3.

$$(\text{curl } F)^r = (\text{dual rot } F)^r = \frac{1}{2!}\epsilon^{rst}(\text{rot } F)_{st}$$

$$= \frac{\epsilon^{rst}}{2!}2\,\partial_{[s}F_{t]} = \epsilon^{rst}\partial_s F_t$$

and Exercise 3.3.

4.4. For any M-vector F, the two expressions

$$(\text{dual rot } F)^{r_1 r_2 \cdots r_{n-M-1}}$$

$$= \frac{1}{(M+1)!} \, \epsilon^{r_1 r_2 \cdots r_{n-M-1} s_1 s_2 \cdots s_{M+1}} \, (M+1) \, \partial_{[s_1} F_{s_2 \cdots s_{M+1}]}$$

$$= \frac{1}{M!} \, \epsilon^{r_1 r_2 \cdots r_{n-M-1} s_1 s_2 \cdots s_{M+1}} \partial_{s_1} F_{s_2 \cdots s_{M+1}}$$

and

$$(\text{div dual } F)^{r_1 r_2 \cdots r_{n-M-1}} = \partial_{r_{n-M}} (\text{dual } F)^{r_1 r_2 \cdots r_{n-M}}$$

$$= \partial_{r_{n-M}} \frac{1}{M!} \, \epsilon^{r_1 r_2 \cdots r_{n-M} s_1 s_2 \cdots s_M} F_{s_1 s_2 \cdots s_M}$$

differ only in the names of summation indices.

4.5. The statements in this exercise are no more that the duals of the statements (4.3)–(4.7), combined with curl = dual rot = div dual.

Chapter 3

5.1. In Exercises 3.4–3.7, 'tensor' or 'scalar' applies to any kind of tensor: absolute, relative or tensor density. In Exercise 3.8 one of the tensors u^r or v^r must be absolute.

6.1. The first three components of $s = \text{div } f$ are

$$s^r = \partial_\alpha f^{r\alpha} = \partial_s f^{rs} + \partial_4 f^{r4} = \epsilon^{rst} \partial_s H_t - \partial_4 D^r.$$

These are the components of $\mathbf{j} = \text{curl } \mathbf{H} - \partial \mathbf{D}/\partial t$. The fourth component of $s = \text{div } f$ is

$$s^4 = \partial_\alpha f^{4\alpha} = \partial_s f^{4s} = \partial_s D^s,$$

which is the same as $q = \text{div } \mathbf{D}$.

6.2. From (4.10) we find

$$\text{dual } f = \begin{pmatrix} 0 & -D^3 & D^2 & H_1 \\ D^3 & 0 & -D^1 & H_2 \\ -D^2 & D^1 & 0 & H_3 \\ -H_1 & -H_2 & -H_3 & 0 \end{pmatrix}.$$

Now $\frac{1}{2} f \cdot \text{dual } f = \frac{1}{4} f^{\alpha\beta} (\text{dual } f)_{\alpha\beta}$ is a simple double sum.

Chapter 4

9.1. From (4.11) we find

$$\text{dual } F = \begin{pmatrix} 0 & E_3 & -E_2 & B^1 \\ -E_3 & 0 & E_1 & B^2 \\ E_2 & -E_1 & 0 & B^3 \\ -B^1 & -B^2 & -B^3 & 0 \end{pmatrix}.$$

Now $\frac{1}{2}F \cdot \text{dual } F = \frac{1}{4}F_{\alpha\beta}(\text{dual } F)^{\alpha\beta}$ is a simple double sum.

9.2. $(\text{rot } F)_{\alpha\beta\gamma} = 3\partial_{[\alpha}F_{\beta\gamma]}$ is a 3-vector, but it has only four independent components, because its dual is a 1-vector. The equation rot $F = 0$ is equivalent to dual rot $F = 0$, but this is curl $F = \text{div dual } F = 0$. The components of dual F in a Euclidean frame were calculated in Exercise 9.1. The first component of div dual $F = 0$ is

$$\partial_\beta(\text{dual } F)^{1\beta} = \partial_2 E_3 - \partial_3 E_2 + \partial_4 B^1 = \left(\mathbf{curl \ E} + \frac{\partial \mathbf{B}}{\partial t}\right)_1 = 0.$$

The second and third components yield the two other components of $\mathbf{curl \ E} = -\partial\mathbf{B}/\partial t$. The fourth component of div dual $F = 0$ is

$$\partial_\beta(\text{dual } F)^{4\beta} = -\partial_r B^r = -\text{div } \mathbf{B} = 0.$$

10.1. With $F = \text{rot } A$, the first variation is

$$\delta I = \int (s \cdot \delta A - f \cdot \text{rot } \delta A)d^4x.$$

$$f \cdot \text{rot } \delta A = \frac{1}{2!}f^{\alpha\beta}\, 2\,\partial_\alpha\delta A_\beta = \text{div } c - (\text{div } f)\cdot\delta A,$$

where $c^\alpha = f^{\alpha\beta}\delta A_\beta$. After transforming $\int \text{div } c\, d^4x$ to an integral over the boundary, on which $\delta A = 0$, we are left with

$$\delta I = \int (s - \text{div } f)\cdot\delta A\, d^4x.$$

Thus $s = \text{div } f$, which is the first pair of Maxwell's equations. The functional $I[A]$ is used in constructing a Lagrangian for charged particles in an electromagnetic field.

10.2. According to $\mathbf{B} = \mathbf{curl \ A}$, the magnetic flux through any surface is

$$\int B_n\, dS = \oint_c \mathbf{A}\cdot\mathbf{ds},$$

where c is the boundary of the surface. Applying this relation to a rectangle in a plane perpendicular to the surface of discontinuity S, as in Fig. 7.2, we obtain $\mathbf{t}\times\mathbf{n}\cdot[\![\mathbf{A}]\!] = 0$, where $\mathbf{t}$ is the normal to the plane of the rectangle. Since $\mathbf{t}$ is arbitrary, we must have

$$\mathbf{n}\times[\![\mathbf{A}]\!] = 0.$$

Thus only the normal component of $\mathbf{A}$ can suffer a jump.

Integrating the relation $\mathbf{E} = -\partial\mathbf{A}/\partial t - \mathbf{grad}\, V$ along a fixed line, and then over a short time interval δt, we have

$$\delta t \int_{(t)}\mathbf{E}\cdot\mathbf{ds} = -\int_{(t+\delta t)}\mathbf{A}\cdot\mathbf{ds} + \int_{(t)}\mathbf{A}\cdot\mathbf{ds} - \delta t(V_B - V_A), \qquad (1)$$

where A and B are the ends of the line. Applying this formula to a fixed line segment that is perpendicular to the surface, with its ends on either side (Fig. B.1), and proceeding as in

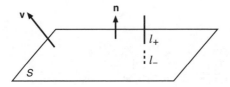

Fig. B.1

the derivation of (7.4), we have

$$\int_{(t)} \mathbf{E} \cdot \mathbf{ds} = (\mathbf{E}_+ \cdot \mathbf{n})l_+ + (\mathbf{E}_- \cdot \mathbf{n})l_-,$$

$$\int_{(t)} \mathbf{A} \cdot \mathbf{ds} = (\mathbf{A}_+ \cdot \mathbf{n})l_+ + (\mathbf{A}_- \cdot \mathbf{n})l_-,$$

$$\int_{(t+\delta t)} \mathbf{A} \cdot \mathbf{ds} = (\mathbf{A}_+ \cdot \mathbf{n})(l_+ - \delta t v_n) + (\mathbf{A}_- \cdot \mathbf{n})(l_- + \delta t v_n).$$

In the limit when the segments l_- and l_+ on either side of S vanish, eqn (1) reduces to $0 = \delta t v_n \mathbf{n} \cdot [\![\mathbf{A}]\!] - \delta t [\![V]\!]$. Thus

$$[\![V]\!] - v_n \mathbf{n} \cdot [\![\mathbf{A}]\!] = 0.$$

Chapter 5

12.1. Let the velocity of the point P in the frame $\Sigma(x)$ be $\mathbf{v}$. Then its classical 4-velocity in Σ is $(w^\alpha) = (\mathbf{v}, 1)$. Under a Galilean transformation $x^{r'} = A_r^{r'}(x^r - u^r x^4)$, $x^{4'} = x^4$ to the frame $\Sigma'(x')$, assuming that w is a contravariant vector,

$$w^{r'} = \frac{\partial x^{r'}}{\partial x^\alpha} w^\alpha = A_r^{r'} w^r - A_r^{r'} u^r = A_r^{r'}(v^r - u^r),$$

which are the components of $\mathbf{v} - \mathbf{u}$, referred to primed axes. By the Galilean law of composition of velocities these are just the components of the velocity $\mathbf{v}'$ of P in Σ'. Furthermore

$$w^{4'} = \frac{\partial x^{4'}}{\partial x^\alpha} w^\alpha = \frac{\partial x^{4'}}{\partial x^4} w^4 = w^4 = 1.$$

Thus $w' = (\mathbf{v}', 1)$, which is, by definition, the classical 4-velocity one would assign to the point P in Σ'. Hence w behaves under Galilean transformations as a contravariant vector. We can deduce, for example, that $w^\alpha A_\alpha = \mathbf{v} \cdot \mathbf{A} - V$ is a Galilean invariant (cf. the remark following (10.11)).

12.2. Use (12.11)–(12.12) and verify that $\mathbf{E}' \times \mathbf{B}'$ vanishes.

Chapter 6

14.1. We have

$$\frac{\partial}{\partial x^i}\frac{1}{r} = -\frac{x^i}{r^3} = -\frac{n_i}{r^2},$$

$$\frac{\partial^2}{\partial x^i \partial x^j}\frac{1}{r} = -\frac{\partial}{\partial x^i}\frac{x^j}{r^3} = -\frac{\delta^{ij}}{r^3} + 3\frac{x^i x^j}{r^5} = \frac{3n_i n_j - \delta^{ij}}{r^3}.$$

14.2. With the notation of Exercise 14.1,

$$D'^{ij} = \sum e[3(x^i - c^i)(x^j - c^j) - (\mathbf{r} - \mathbf{c})^2 \delta^{ij}]$$
$$= D^{ij} - 3d^i c^j - 3c^i d^j + 3\left(\sum e\right)c^i c^j + 2\mathbf{c} \cdot \mathbf{d}\delta^{ij} - \left(\sum e\right)c^2 \delta^{ij}$$
$$= D^{ij}.$$

14.3. The field is

$$\mathbf{E}^{(2)} = -\mathbf{grad}\,\frac{1}{4\pi\epsilon_0}\frac{D^{ij}n_i n_j}{2r^3} = -\frac{D^{ij}}{4\pi\epsilon_0}\mathbf{grad}\,\frac{x^i x^j}{2r^5},$$

$$\frac{\partial}{\partial x^k}\frac{x^i x^j}{2r^5} = \frac{\delta^{ik}x^j + x^i \delta^{jk}}{2r^5} - 5\frac{x^i x^j x^k}{2r^7}.$$

Thus

$$E_k^{(2)} = \frac{1}{4\pi\epsilon_0}\frac{5D^{ij}n_i n_j n_k - 2D^{kj}n_j}{2r^4}.$$

14.4. Since V is proportional to $(\cos\theta)/r^2$, $E_r/E_\theta = 2\cos\theta/\sin\theta$. The field line $r(\theta)$ therefore satisfies

$$\frac{1}{r}\frac{dr}{d\theta} = 2\frac{\cos\theta}{\sin\theta} = 2\frac{1}{\sin\theta}\frac{d\sin\theta}{d\theta},$$

or $d\ln r = 2d\ln\sin\theta = d\ln\sin^2\theta$. Thus $r \propto \sin^2\theta$.

14.5. An area dS with charge $\sigma\,dS$ at distance r from P contributes an electric field $\sigma\,dS/4\pi\epsilon_0 r^2$, but the component away from the plane requires another factor $\cos\theta = a/r$. Hence dS contributes $a\sigma\,dS/4\pi\epsilon_0 r^3$. If $dS = 2\pi b\,db$ is a ring of radius $b = \sqrt{(r^2 - a^2)}$ and width $db = r\,dr/b$, as in Fig. B.2, its contribution to the field is $a\sigma/(2\epsilon_0)(dr/r^2)$. Points of the plane that are less than r from P will therefore contribute $\sigma/(2\epsilon_0)(1 - a/r)$.

14.6. $Q(r)$ equals the flux $4\pi r^2 D(r)$ of $\mathbf{D}$ through the surface of a sphere of radius r. The electric field is $E(r) = D(r)/\epsilon_0 = -dV/dr$. Hence the potential (zero at infinity) is

$$V(r) = \int_r^\infty \frac{D(r)\,dr}{\epsilon_0} = \frac{1}{4\pi\epsilon_0}\int_r^\infty \frac{Q(r)\,dr}{r^2}.$$

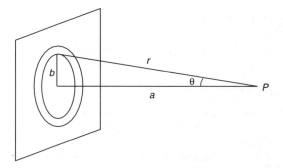

Fig. B.2

Integration by parts gives

$$V(r) = \frac{1}{4\pi\epsilon_0}\frac{Q(r)}{r} + \frac{1}{4\pi\epsilon_0}\int_r^\infty \frac{Q'(r)\,dr}{r}.$$

In the last integral $Q'(r) = 4\pi r^2 q(r)$.

14.7. Let C be the centre of the circle, A a point on the circle, and O the location of the e. The flux is $\Phi = \int E_n\,dS$, taken over any surface with the circle as its boundary. We choose the spherical cap with O as the centre of the sphere (Fig. B.3). If $r = OA$ is the radius of the cap, the flux is $e/(4\pi\epsilon_0 r^2)$ times $\int dS = r^2\Omega$, where Ω is the solid angle subtended by the circle at O. Now $\Omega = 2\pi(1 - \cos\theta)$, where $\theta = \angle AOC$. Thus $\Phi = e(1 - \cos\theta)/(2\epsilon_0)$. With $OC = 3a/4$ we find $\cos\theta = 3/5$, hence $\Phi = e/(5\epsilon_0)$. For the charge e', $O'C = 5a/12$, $\cos\theta' = 5/13$, and $\Phi' = 4e'/(13\epsilon_0)$.

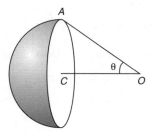

Fig. B.3

14.8. A cylinder of unit length and radius r coaxial with the straight line contains charge of amount λ. Assuming that $\mathbf{D}$ is radial, that is $\mathbf{D} = (D, 0, 0) = (\epsilon_0 E, 0, 0)$, the outward flux of $\mathbf{D}$ through the surface of this cylinder is $2\pi r D = 2\pi r\epsilon_0 E$. Thus, according to Gauss's theorem,

$$E = \frac{\lambda}{2\pi\epsilon_0 r}.$$

The potential is obtained by integrating $E = -dV/dr$:

$$V = \frac{\lambda}{2\pi\epsilon_0} \ln\frac{1}{r} + \text{const.}$$

14.9. In the dipole field

$$\mathbf{E} = \frac{1}{4\pi\epsilon_0} \frac{3(\mathbf{n}\cdot\mathbf{d})\mathbf{n} - \mathbf{d}}{r^3}$$

the distance $r = \sqrt{2}a$ is the diagonal, so that $d/r^3 = e/a^2$. The components (parallel and perpendicular to $\mathbf{d}$) of the field at the fourth corner are

$$E_\parallel = \frac{1}{4\pi\epsilon_0}\frac{e}{a^2}\left[1 + \left(\frac{3}{2} - 1\right)\right] = \frac{3}{2}\frac{1}{4\pi\epsilon_0}\frac{e}{a^2},$$

$$E_\perp = \frac{1}{4\pi\epsilon_0}\frac{e}{a^2}\left(1 + \frac{3}{2}\right) = \frac{5}{2}\frac{1}{4\pi\epsilon_0}\frac{e}{a^2}.$$

The field strength is

$$E = \sqrt{\frac{9}{4} + \frac{25}{4}}\frac{1}{4\pi\epsilon_0}\frac{e}{a^2} = \sqrt{\frac{17}{2}}\frac{1}{4\pi\epsilon_0}\frac{e}{a^2}.$$

14.10. Choosing a Cartesian frame with its axes coinciding with the principal axes of the ellipsoid

$$\frac{x^2}{a^2} + \frac{y^2}{b^2} + \frac{z^2}{c^2} \leq 1,$$

we have

$$D^{xx} = \int (3x^2 - r^2)q\, dV = q\int(2x^2 - y^2 - z^2)\, dV,$$

with similar formulae for the other components. With the substitutions

$$x = x'a, \qquad y = y'b, \qquad z = z'c,$$

the integrals extend over the unit sphere $x'^2 + y'^2 + z'^2 \leq 1$. The non-vanishing components of D are

$$D^{xx} = \frac{Q}{5}(2a^2 - b^2 - c^2), \quad D^{yy} = \frac{Q}{5}(2b^2 - c^2 - a^2), \quad D^{zz} = \frac{Q}{5}(2c^2 - a^2 - b^2),$$

where $Q = 4\pi abcq/3$ is the total charge of the ellipsoid.

15.1. In order to describe a circle of radius r around the x axis, the third charge must be at equal distances from the two positive charges. Its centripetal acceleration mv^2/r will be twice the centripetal component

$$\frac{1}{4\pi\epsilon_0}\frac{e^2}{r^2 + a^2}\frac{r}{(r^2 + a^2)^{\frac{1}{2}}}$$

of its attraction to any one of the charges at $(\pm a, 0, 0)$.

15.2. The fifth charge is in equilibrium because the resultant field of the four corner charges vanishes at the centre. In order to determine whether the equilibrium is stable, we move the charge a small distance $\mathbf{x}$ from the centre and calculate the second-order term in the potential:

$$\delta^2 V = \tfrac{1}{2} x^i x^j \frac{\partial^2}{\partial x^i \partial x^j} \sum \frac{e}{4\pi\epsilon_0 r} = \tfrac{1}{2} \frac{e}{4\pi\epsilon_0 a^3} \sum [3(\mathbf{n} \cdot \mathbf{x})^2 - x^2], \qquad (1)$$

where $\mathbf{n}$ is a unit vector directed from the centre to a corner, and the sum is over the four corners. If $\mathbf{x}$ is in the plane of the charges the last sum is of the form

$$3\left[\cos^2\theta + \cos^2\left(\theta + \frac{\pi}{2}\right) + \cos^2(\theta + \pi) + \cos^2\left(\theta + \frac{3\pi}{2}\right)\right]x^2 - 4x^2 = 2x^2.$$

The charge at the centre then acquires potential energy $e\delta^2 V = \tfrac{1}{2} m\omega^2 x^2$, where

$$\omega^2 = \frac{2e^2}{4\pi\epsilon_0 m a^3}.$$

The period of oscillations is $2\pi/\omega$.

If $\mathbf{x}$ is perpendicular to the plane of the charges all the $\mathbf{n} \cdot \mathbf{x}$ vanish: the last sum in (1) is $-4x^2$ and $\delta^2 V$ is negative. The equilibrium is therefore unstable with respect to such motions.

17.1.

$$L_{\dot{\mathbf{x}}} = m\dot{\mathbf{x}} + e\mathbf{A}(\mathbf{x}, t),$$

$$\frac{d}{dt} L_{\dot{\mathbf{x}}} = m\ddot{\mathbf{x}} + e\mathbf{A}_t + e(\dot{\mathbf{x}} \cdot \mathbf{grad})\mathbf{A},$$

$$\frac{\partial L}{\partial x^i} = e\left(\dot{x}^j \frac{\partial A_j}{\partial x^i} - \frac{\partial V}{\partial x^i}\right) = e\left[\dot{x}^j\left(\frac{\partial A_j}{\partial x^i} - \frac{\partial A_i}{\partial x^j}\right) + \dot{x}^j \frac{\partial A_i}{\partial x^j} - \frac{\partial V}{\partial x^i}\right],$$

$$L_{\mathbf{x}} = e[\dot{\mathbf{x}} \times \mathbf{curl}\, \mathbf{A} + (\dot{\mathbf{x}} \cdot \mathbf{grad})\mathbf{A} - \mathbf{grad}\, V],$$

$$\frac{d}{dt} L_{\dot{\mathbf{x}}} - L_{\mathbf{x}} = m\ddot{\mathbf{x}} - e(\mathbf{E} + \dot{\mathbf{x}} \times \mathbf{B}).$$

17.2. From (17.8) we have

$$\dot{\mathbf{x}} = \frac{\mathbf{p} - e\mathbf{A}}{m}.$$

The Hamiltonian is

$$H = \mathbf{p} \cdot \dot{\mathbf{x}} - L$$

$$= \mathbf{p} \cdot \frac{\mathbf{p} - e\mathbf{A}}{m} - \tfrac{1}{2} m \left(\frac{\mathbf{p} - e\mathbf{A}}{m}\right)^2 - e\left[\frac{\mathbf{p} - e\mathbf{A}}{m} \cdot \mathbf{A} - V\right]$$

$$= \frac{(\mathbf{p} - e\mathbf{A})^2}{2m} + eV.$$

Now the two ends of

$$\dot{\mathbf{x}} = H_{\mathbf{p}} = \frac{\mathbf{p} - e\mathbf{A}}{m},$$

are (17.8). The derivative of H with respect to x^i is

$$H_{x_i} = -\frac{e(\mathbf{p} - e\mathbf{A})_j \partial_i A_j}{m} + e\partial_i V = -e\dot{x}_j \partial_i A_j + e\partial_i V$$

$$= -e\dot{x}_j(\partial_i A_j - \partial_j A_i) - e\dot{x}_j \partial_j A_i + e\partial_i V,$$

which is the ith component of

$$H_{\mathbf{x}} = -e[\dot{\mathbf{x}} \times \mathbf{curl\ A} + (\dot{\mathbf{x}} \cdot \mathbf{grad})\mathbf{A} - \mathbf{grad\ } V]$$

$$= -e\dot{\mathbf{A}} - e(\dot{\mathbf{x}} \times \mathbf{curl\ A} - \mathbf{A}_t - \mathbf{grad\ } V)$$

$$= -e\dot{\mathbf{A}} + e(\mathbf{E} + \dot{\mathbf{x}} \times \mathbf{B}).$$

Thus $\dot{\mathbf{p}} = -H_{\mathbf{x}}$, together with (17.8), yields the equation of motion (15.4).

18.1. The first equation is

$$\mathbf{p}_k = \frac{m\dot{\mathbf{x}}}{\sqrt{1 - \dot{x}^2/c^2}} = \frac{mc^2}{\sqrt{1 - \dot{x}^2/c^2}}\frac{\dot{\mathbf{x}}}{c^2} = K\frac{\dot{\mathbf{x}}}{c^2}.$$

In order to obtain the second one, we first calculate

$$\dot{\mathbf{p}} = \frac{m\ddot{\mathbf{x}}}{(1 - \dot{x}^2/c^2)^{\frac{1}{2}}} + \frac{m\dot{\mathbf{x}}}{(1 - \dot{x}^2/c^2)^{\frac{3}{2}}}\frac{(\dot{\mathbf{x}} \cdot \ddot{\mathbf{x}})}{c^2}.$$

Now

$$\dot{\mathbf{x}} \cdot \dot{\mathbf{p}} = \frac{m(\dot{\mathbf{x}} \cdot \ddot{\mathbf{x}})}{(1 - \dot{x}^2/c^2)^{\frac{1}{2}}} + \frac{m\dot{x}^2}{(1 - \dot{x}^2/c^2)^{\frac{3}{2}}}\frac{(\dot{\mathbf{x}} \cdot \ddot{\mathbf{x}})}{c^2}$$

$$= \frac{m(\dot{\mathbf{x}} \cdot \ddot{\mathbf{x}})}{(1 - \dot{x}^2/c^2)^{\frac{3}{2}}} = \dot{K}.$$

Chapter 7

21.1.

$$\mathcal{H} = \mathbf{H} - \mathbf{v} \times \mathbf{D} = \frac{\mathbf{B}}{\mu_0} - \mathbf{M} - \mathbf{v} \times (\epsilon_0 \mathbf{E} + \mathbf{P})$$

$$= \frac{\mathbf{B}}{\mu_0} - \mathbf{v} \times \epsilon_0 \mathbf{E} - (\mathbf{M} + \mathbf{v} \times \mathbf{P}) = \frac{\mathbf{B}}{\mu_0} - \epsilon_0 \mathbf{v} \times \mathbf{E} - \mathcal{M}.$$

23.1. In the classical case we substitute

$$\mathbf{M}' = \mathbf{M} + \mathbf{v} \times \mathbf{P}, \quad \mathbf{P}' = \mathbf{P}, \quad \mathbf{B}' = \mathbf{B},$$

in the rest frame response functions

$$\mathbf{M}' = \frac{\chi_B}{\mu_0}\mathbf{B}', \quad \mathbf{P}' = 0. \tag{1}$$

The resulting classical response functions and aether relations are

$$\mathbf{M} = \frac{\chi_B}{\mu_0}\mathbf{B}, \qquad \mathbf{P} = 0,$$

$$\mathbf{D} = \epsilon_0\mathbf{E}, \qquad \mathbf{H} = \frac{1 - \chi_B}{\mu_0}\mathbf{B}.$$

In the relativistic case we substitute

$$\mathbf{P}' = \gamma(\mathbf{P} - \mathbf{v} \times \mathbf{M}/c^2) - \frac{\gamma^2}{\gamma + 1}\mathbf{v}(\mathbf{v} \cdot \mathbf{P})/c^2,$$

$$\mathbf{M}' = \gamma(\mathbf{M} + \mathbf{v} \times \mathbf{P}) - \frac{\gamma^2}{\gamma + 1}\mathbf{v}(\mathbf{v} \cdot \mathbf{M})/c^2,$$

$$\mathbf{B}' = \gamma(\mathbf{B} - \mathbf{v} \times \mathbf{E}/c^2) - \frac{\gamma^2}{\gamma + 1}\mathbf{v}(\mathbf{v} \cdot \mathbf{B})/c^2$$

in (1). In order to solve the resulting pair of equations for $\mathbf{M}$ and $\mathbf{P}$, we take the scalar product of each one by $\mathbf{v}$. The resulting relativistic response functions and aether relations are

$$\mathbf{M} = \frac{\chi_B}{\mu_0(1 - v^2/c^2)}[\mathbf{B} - \mathbf{v}(\mathbf{v} \cdot \mathbf{B})/c^2 - \mathbf{v} \times \mathbf{E}/c^2],$$

$$\mathbf{P} = \epsilon_0\frac{\chi_B}{1 - v^2/c^2}\mathbf{v} \times (\mathbf{B} - \mathbf{v} \times \mathbf{E}/c^2),$$

$$\mathbf{D} = \epsilon_0\Big[\mathbf{E} + \frac{\chi_B}{1 - v^2/c^2}\mathbf{v} \times (\mathbf{B} - \mathbf{v} \times \mathbf{E}/c^2)\Big],$$

$$\mathbf{H} = \frac{1}{\mu_0}\Big[\mathbf{B} - \frac{\chi_B}{1 - v^2/c^2}[\mathbf{B} - \mathbf{v}(\mathbf{v} \cdot \mathbf{B})/c^2 - \mathbf{v} \times \mathbf{E}/c^2]\Big].$$

23.2. In the classical case we substitute

$$\mathbf{j}' = \mathbf{j} - q\mathbf{v}, \qquad \mathbf{E}' = \mathbf{E} + \mathbf{v} \times \mathbf{B}$$

in the rest frame relation $\mathbf{j}' = C\mathbf{E}'$. The resulting classical Ohm's law is

$$\mathcal{J} = C\mathcal{E},$$

where $\mathcal{J} = \mathbf{j} - q\mathbf{v}$ is the conduction current density and $\mathcal{E} = \mathbf{E} + \mathbf{v} \times \mathbf{B}$ is the electromotive intensity. In the relativistic case we substitute

$$\mathbf{j}' = \mathbf{j} + \Big[\frac{\gamma^2}{\gamma + 1}(\mathbf{v} \cdot \mathbf{j})/c^2 - \gamma q\Big]\mathbf{v},$$

$$\mathbf{E}' = \gamma(\mathbf{E} + \mathbf{v} \times \mathbf{B}) - \frac{\gamma^2}{\gamma + 1}\mathbf{v}(\mathbf{v} \cdot \mathbf{E})/c^2$$

in the rest frame relation $\mathbf{j}' = C\mathbf{E}'$ and obtain the relativistic Ohm's law

$$\mathcal{J} = C\gamma[\mathcal{E} - \mathbf{v}(\mathbf{v} \cdot \mathcal{E})/c^2].$$

Note that in either case $\mathcal{J}$ vanishes if, and only if, $\mathcal{E}$ does.

Chapter 8

26.1. Equality between $2\alpha/r$ and $\frac{1}{2}\sigma E$ gives

$$\frac{2\alpha}{r} = \frac{1}{2}\frac{Q}{4\pi r^2}\frac{Q}{4\pi\epsilon_0 r^2},$$

from which we obtain $r^3 = Q^2/(64\pi^2\epsilon_0\alpha)$.

26.2. The field is radial everywhere, and must vanish in the conducting shell. The same holds for the charge potential $\mathbf{D} = \epsilon_0\mathbf{E}$. Since $\oint D\,dS = 4\pi r^2 D$ must equal the total charge inside r, and $\mathbf{n}\cdot[\![\mathbf{D}]\!] = \sigma$, we conclude that there must be an induced charge $-e$ on the inner surface $r = a$, and a charge $Q + e$ on the outer surface $r = b$ of the shell. The field is $e/(4\pi\epsilon_0 r^2)$ in the inner region $0 < r < a$. It vanishes inside the shell, and outside it is $(Q + e)/(4\pi\epsilon_0 r^2)$.

27.1. For any sphere with centre at P

$$\int \Delta V\,d^3x = \oint \frac{\partial V}{\partial n}\,dS.$$

If V has a minimum at P then, for a sufficiently small sphere, $\partial V/\partial n$ must be positive. Hence ΔV must be positive at P. Similarly, if V has a maximum at P, then ΔV must be negative at P.

27.2. V cannot be constant, otherwise $\sigma = -\epsilon_0\partial V/\partial n$ would vanish everywhere and the conductors would all be uncharged. By considering $\oint \partial V/\partial n\,dS = -\oint D_n\,dS/\epsilon_0$ over a large sphere, which includes all the conductors, it follows that V cannot assume a minimum (maximum) at infinity, unless the net charge on the conductors is positive (negative). Thus under the present circumstances V must assume its minimum on one of the conductors, say S_1, and its maximum on another, say S_2. Let P be any point on S_1. Then the normal derivative $\partial V/\partial n$ must be positive, and σ negative, at P. Thus S_1 is negatively charged everywhere. Similarly S_2 is positively charged everywhere.

27.3. Let $S_1, S_2, \ldots$ be the surfaces of the conductors, and let S be a large sphere which includes all the conductors. Let V be the potential corresponding to charges $Q_1, Q_2, \ldots$, and V' be the potential corresponding to charges $Q_1', Q_2', \ldots$ on the conductors. We apply Green's second identity

$$\int \left(V\Delta V' - V'\Delta V\right)d^3x = \int \left(V\frac{\partial V'}{\partial n} - V'\frac{\partial V}{\partial n}\right)dS, \tag{1}$$

(cf. (27.2)) to the region outside the conductors and inside S. The left-hand side must vanish, since V and V' both satisfy Laplace's equation. The right-hand side is a sum over $S_1, S_2, \ldots$ and S, in which the normal $\mathbf{n}$ points out of the region, hence *into* the conductors and out of S. Since $V = V_1$ on S_1, and $Q_1 = \oint \epsilon_0\partial V/\partial n\,dS_1$, and similarly for V_1' and Q_1', the contribution from S_1 to the right-hand side of (1) is $(V_1 Q_1' - V_1' Q_1)/\epsilon_0$. Thus

$$\sum Q_a V_a' - \sum Q_a' V_a = \epsilon_0 \oint \left(V\frac{\partial V'}{\partial n} - V'\frac{\partial V}{\partial n}\right)dS,$$

the last integral being over the large sphere S. But S includes all the conductors, so that rV and rV', and $r^2\partial V/\partial r$ and $r^2\partial V'/\partial r$, are bounded on S. If we let S recede to infinity, we therefore get

$$\sum Q_a V_a' = \sum Q_a' V_a. \tag{2}$$

If Q_1 (say) is a point charge, we surround it by a small sphere S_1 and apply Green's second identity (1). The left-hand side vanishes again, since the singularity has been excluded. In evaluating the contribution from S_1 to the right-hand side of (1), we write $V = W + Q_1/(4\pi\epsilon_0 r)$ and $V' = W' + Q_1'/(4\pi\epsilon_0 r)$. Noting that W and W', which are the potentials due to all charges except Q_1, are regular, the contribution, after letting the small sphere S_1 collapse onto the point charge, is $(W_1 Q_1' - W_1' Q_1)/\epsilon_0$. The proof proceeds as before, with the result that, in (2), V_a and V_a' at a point charge are the W's, that is, the potentials due to all other charges (including point charges) except for Q_a and Q_a'.

27.4. Apply eqn (2) of the solution to Exercise 27.3 to the case $Q_1 = 1$, $Q_2 = 0$, $Q_3 = Q_4 = \ldots = 0$ and $Q_1 = 0$, $Q_2' = 1$, $Q_3' = Q_4' = \ldots = 0$. The result is $V_1' = V_2$, from which the two statements follow. The statements hold whether or not there are further uncharged conductors $3, 4, \ldots$ present.

28.1. The displacement $\mathbf{D}$, which has only a radial component D in this exercise, satisfies the jump condition $\mathbf{n} \cdot [\![\mathbf{D}]\!] = 0$ on each of the charged spheres, and $\operatorname{div}\mathbf{D} = r^{-2}d(r^2 D)/dr = 0$ elsewhere. Hence $r^2 D$ is constant, except for jumps at $r = a$, b or c. The constant $r^2 D$ is zero in the central ball $r < a$, for otherwise D would be singular at the centre. At $r = a$, D must jump by an amount equal to the surface charge density σ_a. Hence $D = a^2\sigma_a/r^2$ for $a < r < b$. Thus

$$V_a - V_b = \int_a^b E\,dr = \int_a^b \frac{D}{\epsilon_0}\,dr = \int_a^b \frac{a^2\sigma_a}{\epsilon_0 r^2}\,dr = \frac{a^2\sigma_a}{\epsilon_0}\left(\frac{1}{a} - \frac{1}{b}\right). \tag{1}$$

At $r = b$, D must again jump, this time by an amount σ_b. Hence $D = (a^2\sigma_a + b^2\sigma_b)/r^2$ for $b < r < c$. Thus

$$V_b - V_c = \frac{a^2\sigma_a + b^2\sigma_b}{\epsilon_0}\left(\frac{1}{b} - \frac{1}{c}\right). \tag{2}$$

There is another jump at $r = c$, and it serves to fix σ_c, through the requirement that D vanish for $r > c$, because the capacitor carries no net charge. But this is not needed for the determination of the capacity

$$C = \frac{Q_b}{V_b - V_a} = \frac{4\pi b^2\sigma_b}{V_b - V_a}. \tag{3}$$

Now, since the inner and outer spheres are connected, $V_a = V_c$. Hence the sum of (1) and (2) must vanish:

$$\frac{a^2\sigma_a}{\epsilon_0}\left(\frac{1}{a} - \frac{1}{b}\right) + \frac{a^2\sigma_a + b^2\sigma_b}{\epsilon_0}\left(\frac{1}{b} - \frac{1}{c}\right) = 0. \tag{4}$$

The capacity is obtained by substituting the voltage (1) in (3), and using (4) to express the ratio $b^2\sigma_b/a^2\sigma_a$ in terms of a, b and c.

29.1. The potential corresponding to a charged sphere in a uniform external field $\mathbf{E}_0$ is (26.11). The surface charge density has been calculated in the solution of Problem 26.1:

$$\sigma = \frac{Q}{4\pi a^2} + 3\epsilon_0(\mathbf{E}_0 \cdot \mathbf{n}).$$

It is least when $\mathbf{E}_0 \cdot \mathbf{n} = -E_0$. Thus, for $\sigma \geq 0$, Q must be at least $12\pi\epsilon_0 a^2 E_0$.

29.2. Except for the contribution $e/(4\pi\epsilon_0 R)$ from the point charge, V must be a non-singular harmonic function. We try

$$4\pi\epsilon_0 V = \frac{e}{R} + e \sum_{n=1}^{\infty} \frac{A_n P_n(\cos\theta)}{r^{n+1}}.$$

The sum does *not* include an $n = 0$ term: such a term would correspond to a non-zero charge eA_1 on the sphere. In this exercise we shall need all the P_n's. The only condition is that V be constant on $r = a$. For $r < x$ we have

$$\frac{1}{R} = \frac{1}{\sqrt{x^2 + r^2 - 2xr\cos\theta}} = \sum_{n=0}^{\infty} \frac{r^n}{x^{n+1}} P_n(\cos\theta).$$

Since the P_n are linearly independent, the constancy of V on $r = a$ requires

$$\frac{a^n}{x^{n+1}} + \frac{A_n}{a^{n+1}} = 0,$$

that is, $A_n = -a^{2n+1}/x^{n+1}$. Thus

$$4\pi\epsilon_0 V = \frac{e}{R} - e \sum_{n=1}^{\infty} \frac{a^{2n+1} P_n(\cos\theta)}{x^{n+1} r^{n+1}}.$$

In order to show that this may be written as

$$4\pi\epsilon_0 V = \frac{e}{R} - \frac{ea/x}{R_1} + \frac{ea/x}{r},$$

where R_1 is the distance from the point along OP that is at distance $x' = a^2/x$ from O, we need only note that, for $r > x'$,

$$\frac{1}{R_1} = \frac{1}{\sqrt{x'^2 + r^2 - 2x'r\cos\theta}} = \sum_{n=0}^{\infty} \frac{x'^n}{r^{n+1}} P_n(\cos\theta)$$

$$= \sum_{n=0}^{\infty} \frac{a^{2n}}{x^n r^{n+1}} P_n(\cos\theta).$$

If we multiply this by $-ea/x$ we get the correct sum, except that the $n = 0$ term must be cancelled, that is, a term $ea/(xr)$ must be added.

29.3. The dipole d at P may be regarded as the limit $el \to d$ of a pair of charges, e at $x + l$ and $(-e)$ at x. Denoting the solution of the last exercise by $V = ef(r; x)$, we take the limit of $ef(r; x + l) + (-e)f(r; x)$, that is, $d \cdot \partial f/\partial x$. The result is

$$4\pi\epsilon_0 V = \frac{\mathbf{d} \cdot \mathbf{R}}{R^3} + \mathbf{d} \cdot \sum_{n=1}^{\infty} \frac{(n+1)a^{2n+1} P_n(\cos\theta)}{x^{n+2} r^{n+1}}.$$

29.4. We fix the coefficients A_n and B_n by requiring that $V = 0$ on each sphere. At the inner sphere $r < c$ and we expand $1/R$ as in Exercise 29.3:

$$4\pi\epsilon_0 V = e \sum \frac{r^n}{c^{n+1}} P_n(\cos\theta) + \sum \left(A_n r^n + \frac{B_n}{r^{n+1}}\right) P_n(\cos\theta).$$

The requirement $V(a) = 0$ leads to

$$e \frac{a^n}{c^{n+1}} + A_n a^n + \frac{B_n}{a^{n+1}} = 0. \tag{1}$$

At the outer sphere $r < c$ and we expand $1/R$ in powers of c/r:

$$4\pi\epsilon_0 V = e \sum \frac{c^n}{r^{n+1}} P_n(\cos\theta) + \sum \left(A_n r^n + \frac{B_n}{r^{n+1}}\right) P_n(\cos\theta).$$

The requirement $V(b) = 0$ leads to

$$e \frac{c^n}{b^{n+1}} + A_n b^n + \frac{B_n}{b^{n+1}} = 0. \tag{2}$$

Equations (1) and (2) now determine A_n and B_n.

29.5. Let O be the centre of the circular wire. At a point P on the axis (the perpendicular through O to the plane of the wire) which is at a distance z from O the potential is

$$V = \frac{1}{4\pi\epsilon_0} \int \frac{q\, d^3x}{r} = \frac{1}{4\pi\epsilon_0} Q \sqrt{(a^2 + z^2)}, \tag{1}$$

since all points of the wire are at the same distance $r = \sqrt{(a^2 + z^2)}$ from P. Hence, by carrying out the appropriate expansions in powers of z/a, or of a/z, we obtain

$$\text{for } |z| < a: \quad V = \frac{Q}{pi\epsilon_0}\left(\frac{1}{a} - \frac{1}{2}\frac{z^2}{a^3} + \frac{1\cdot 3}{2\cdot 4}\frac{z^4}{a^5} - \frac{1\cdot 3\cdot 5}{2\cdot 4\cdot 6}\frac{z^6}{a^7} + \cdots\right);$$

$$\text{for } |z| > a: \quad V = \frac{Q}{4\pi\epsilon_0}\left(\frac{1}{z} - \frac{1}{2}\frac{a^2}{z^3} + \frac{1\cdot 3}{2\cdot 4}\frac{a^4}{z^5} - \frac{1\cdot 3\cdot 5}{2\cdot 4\cdot 6}\frac{a^6}{z^7} + \cdots\right). \tag{2}$$

In order to determine V at points off the axis, we note that it must be of the general form

$$V = \frac{1}{4\pi\epsilon_0} \sum \left(A_n r^n + \frac{B_n}{r^{n+1}}\right) P_n(\cos\theta), \tag{3}$$

where r is the distance from O and θ is measured from the axis. On the axis $P_n(\cos\theta) = P_n(1) = 1$ and $r = z$, and (3) must reduce to (2). Thus, on or off the axis, V is given by

$$V = \frac{Q}{4\pi\epsilon_0}\left(\frac{1}{a}P_0 - \frac{1}{2}\frac{r^2}{a^3}P_2 + \frac{1\cdot 3}{2\cdot 4}\frac{r^4}{a^5}P_4 - \frac{1\cdot 3\cdot 5}{2\cdot 4\cdot 6}\frac{r^6}{a^7}P_6 + \cdots\right) \quad \text{for } r < a,$$

$$V = \frac{Q}{4\pi\epsilon_0}\left(\frac{1}{r}P_0 - \frac{1}{2}\frac{a^2}{r^3}P_2 + \frac{1\cdot 3}{2\cdot 4}\frac{a^4}{r^5}P_4 - \frac{1\cdot 3\cdot 5}{2\cdot 4\cdot 6}\frac{a^6}{r^7}P_6 + \cdots\right) \quad \text{for } r > a.$$

30.1. At the plane the resultant field is normal to the plane and of magnitude

$$E_n = 2 \frac{x}{r} \frac{1}{4\pi\epsilon_0} \frac{e}{r^2}.$$

We obtain the charge on the plane by summing over annuli of area $2\pi s\, ds$ and surface charge density $\sigma = \epsilon_0 E_n$:

$$\int_0^\infty 2\pi s\, ds\sigma = e \int_0^\infty \frac{xs\, ds}{r^3} = e \int_0^\infty \frac{xs\, ds}{(x^2+s^2)^{\frac{3}{2}}} = e.$$

30.2. Let O be the centre of the sphere and P the position of e. The potential

$$V = \frac{1}{4\pi\epsilon_0} \left(\frac{e}{r} - \frac{e'}{r'} \right)$$

is to vanish at any point C on the sphere. Hence

$$\frac{e}{(a^2 + x^2 - 2ax\cos\theta)^{\frac{1}{2}}} = \frac{e'}{(a^2 + x'^2 - 2ax'\cos\theta)^{\frac{1}{2}}}, \tag{1}$$

where $\theta = \angle COP$. This must hold for any C, that is, for any $\cos\theta$. We choose $\cos\theta = \pm 1$, which leads to the two equations

$$\frac{e}{x-a} = \frac{e'}{a-x'}, \qquad \frac{e}{x+a} = \frac{e'}{a+x'}.$$

These have the solution

$$e' = \frac{a}{x}e, \qquad x' = \frac{a^2}{x}.$$

With these values for e' and x' eqn (1) is easily seen to hold for any $\cos\theta$. For an alternative proof see Exercise 29.2.

30.3. The field is a superposition of the field $\lambda/(2\pi\epsilon_0 r)$ of the line charge λ and the field $\lambda'/(2\pi\epsilon_0 r')$ of an image line charge $\lambda' = -\lambda$ at the image point. Its resultant at P is normal to the plane:

$$E_n = -2\frac{a}{r} \frac{\lambda}{2\pi\epsilon_0 r}.$$

The surface charge density is $\sigma = \epsilon_0 E_n$. The charge $Q(r)$, per unit length along the line charge, induced on the part of the plane within a distance r from the line charge, is $\int_{-s}^{s} \sigma\, ds$, where $s = \sqrt{(r^2 - a^2)}$. Thus

$$Q(r) = -\int_{-s}^{s} 2\frac{a}{r} \frac{\lambda}{2\pi r}\, ds = -\frac{\lambda a}{\pi} \int_{-s}^{s} \frac{ds}{a^2 + s^2} = -2\frac{\lambda}{\pi} \arctan \frac{s}{a}.$$

This equals $\frac{1}{2}\lambda$ when $\arctan(s/a) = \frac{1}{4}\pi$, that is, when $s = \sqrt{(r^2 - a^2)} = a$, or $r = a\sqrt{2}$.

30.4. The potential problem is solved by introducing an image charge $e' = -e$ for each one of the charges. Each of the two charges is repelled by the other one with a force $f = e^2/(4\pi\epsilon_0 a^2)$, and attracted by the two image charges by forces f and $\frac{1}{2}f$. The resultant is easily found to be $\frac{3}{2}f$. The attraction by the two images of course stands for the actual attraction by the charge induced on the plane. If one of the charges is reversed its image, too, must be reversed. But the other image is unchanged. Thus the net induced charge on the plane becomes zero, and its distribution is quite different. The attractive force it exerts on each one of the charges is weaker, and not in the same direction.

30.5. The potential problem is solved by introducing an image charge $e' = (a/x')e$ at a distance $x' = a^2/x$ from the centre of the cavity. This image exerts a force

$$\frac{1}{4\pi\epsilon_0}\frac{ee'}{(x'-x)^2} = \frac{1}{4\pi\epsilon_0}\frac{e^2 a x}{(a^2-x^2)^2}.$$

30.6. We regard the dipole as the limit $el \to d$ of a pair of charges, e at $x + l$ and $(-e)$ at x. For e we introduce an image charge $e_1 = (-ae)/(x+l)$ at $x_1 = a^2/(x+l)$; for $(-e)$, an image charge $e_2 = ae/x$ at $x_2 = a^2/x$. For any pair of charges e_1 at $\mathbf{x}_1$ and e_2 at $\mathbf{x}_2$,

$$\frac{e_1}{|\mathbf{x}-\mathbf{x}_1|} + \frac{e_2}{|\mathbf{x}-\mathbf{x}_2|}$$

$$= \frac{e_1}{|\mathbf{x}-\mathbf{x}_1|} + \frac{e_2}{|\mathbf{x}-\mathbf{x}_1|} - e_2(\mathbf{x}_2-\mathbf{x}_1)\cdot\mathbf{grad}\,\frac{1}{|\mathbf{x}-\mathbf{x}_1|}$$

$$= \frac{e_1+e_2}{|\mathbf{x}-\mathbf{x}_1|} + \frac{e_2(\mathbf{x}_2-\mathbf{x}_1)\cdot(\mathbf{x}-\mathbf{x}_1)}{|\mathbf{x}-\mathbf{x}_1|^3} + O(e_2|\mathbf{x}_2-\mathbf{x}_1|^2).$$

Thus, to the first order in $|\mathbf{x}_2-\mathbf{x}_1|$—which is all we need, since we shall take the limit $l \to 0$—two charges e_1 at $\mathbf{x}_1$ and e_2 at $\mathbf{x}_2$ are equivalent to a charge $e_1 + e_2$ and a dipole of moment $e_2(\mathbf{x}_2-\mathbf{x}_1)$, both at $\mathbf{x}_1$. Setting $e = d/l$ and taking the limit $l \to 0$, the total image charge is

$$e' = e_1 + e_2 = -\frac{ae}{x+l} + \frac{ae}{x} = \frac{ad}{x(x+l)} \to \frac{ad}{x^2}.$$

The image dipole moment is

$$d' = e_2(x_2 - x_1) = \frac{ae}{x}\left(\frac{a^2}{x} - \frac{a^2}{x+l}\right) = \frac{a^3 d}{x^2(x+l)} \to \frac{a^3 d}{x^3}.$$

30.7. Regarding the dipole as the limit of a pair of charges, it is clear that the image is a dipole $\mathbf{d}'$ of equal strength $d' = d$, though not necessarily in the same direction. The force on the dipole is $\mathbf{f} = (\mathbf{d}\cdot\mathbf{grad})\mathbf{E}$, where

$$\mathbf{E} = \frac{1}{4\pi\epsilon_0}\frac{3(\mathbf{d}'\cdot\mathbf{n})\mathbf{n} - \mathbf{d}'}{r^3},$$

is the field due to the image dipole. In this formula, $\mathbf{r}$ is the vector from $\mathbf{d}'$ to $\mathbf{d}$, and $\mathbf{n}$ a unit vector in the direction of $\mathbf{r}$. By carrying out the differentiation we obtain (cf. Problem 15.3)

$$\mathbf{f} = \frac{1}{4\pi\epsilon_0}\frac{3(\mathbf{d}\cdot\mathbf{d}')\mathbf{n} + 3(\mathbf{d}'\cdot\mathbf{n})\mathbf{d} + 3(\mathbf{d}\cdot\mathbf{n})\mathbf{d}' - 15(\mathbf{d}'\cdot\mathbf{n})(\mathbf{d}\cdot\mathbf{n})\mathbf{n}}{r^4}.$$

In this exercise the force does not matter, since **d** is fixed and only allowed to rotate about its centre. The torque, relative to the dipole centre, is (cf. Problem 15.2)

$$\mathbf{t} = \mathbf{d} \times \mathbf{E} = \frac{1}{4\pi \epsilon_0} \frac{3(\mathbf{d}' \cdot \mathbf{n})\mathbf{d} \times \mathbf{n} - \mathbf{d} \times \mathbf{d}'}{r^3}.$$

Of course the orientation of the image **d**′ depends on that of **d**. It is easily seen that this formula provides a restoring torque only when **d** is perpendicular to, and pointing away from, the plane. If **d** is at an angle θ with respect to this direction, **d**′ is at an angle $(-\theta)$, and the restoring torque is

$$t = -\frac{1}{4\pi \epsilon_0} \frac{d^2 \sin 2\theta}{r^3}.$$

This equals $I\ddot\theta$. Since the distance between the dipole and its image is $r = 2a$, the circular frequency for small oscillations will be

$$\omega^2 = \frac{1}{4\pi \epsilon_0} \frac{2d^2}{8a^3 I}.$$

The period is $2\pi/\omega$.

30.8. If the sphere carries a charge Q, the potential and the field at any point P are

$$V = \frac{1}{4\pi \epsilon_0} \left(\frac{e}{R} - \frac{ea/x}{R'} + \frac{ea/x + Q}{r} \right),$$

$$\mathbf{E} = -\operatorname{grad} V = \frac{1}{4\pi \epsilon_0} \left(\frac{e}{R^3}\mathbf{R} - \frac{ea/x}{R'^3}\mathbf{R}' + \frac{ea/x + Q}{r^3}\mathbf{r} \right),$$

where **R** is the vector from e to P, **R**′ the vector from the image charge $(-ae/x)$, and **r** the vector from the centre. The surface charge density is

$$\sigma = \epsilon_0 E_n = \frac{1}{4\pi} \left(\frac{e}{R^3} R_n - \frac{ea/x}{R'^3} R'_n + \frac{ea/x + Q}{r^3} \right),$$

If it is to be positive everywhere we must have

$$\frac{Q}{a^2} > \frac{ea/x}{R'^3} R'_n - \frac{e}{R^3} R_n - \frac{ea/x + Q}{r^3}.$$

The right-hand side is greatest at the point on the sphere that is closest to e and e'. For this point $R = x - a$, $R_n = -R$, $R' = a - x' = a - a^2/x$ and $R'_n = R'$. Substituting these values we get $Q > ea[(x+a)/(x-a)^2 - 1/x]$.

31.1. From the Cauchy–Riemann relations

$$\operatorname{grad} U \cdot \operatorname{grad} V = U_x V_x + U_y V_y = -V_y V_x + V_x V_y = 0.$$

But **grad** U is perpendicular to $U(x, y) = \text{const.}$, and **grad** V to $V(x, y) = \text{const.}$

31.2. From the Cauchy–Riemann relations $w'(z) = V_x + iU_x = V_x - iV_y = -E_x + iE_y$. Thus the complex number $w'(z)$ has the absolute value $\sqrt{(E_x^2 + E_y^2)} = E$ and the argument $\arctan(-E_y/E_x) = -\arctan(E_y/E_x)$.

31.3. If (dx, dy) is an element of c, the outward pointing normal to c is $(dy, -dx)$. The element of electric field flux, per unit length in the z direction, is therefore

$$
\begin{aligned}
\mathbf{E} \cdot (dy, -dx) &= (-V_x, -V_y) \cdot (dy, -dx) \\
&= (-U_y, U_x) \cdot (dy, -dx) \\
&= -U_x dx - U_y dy.
\end{aligned}
$$

For a closed c the flux is

$$
\oint (-U_x dx - U_y dy) = -\delta(U).
$$

The charge inside c is ϵ_0 times the electric field flux out of c.

31.4. The complex potential $w = -E_0(z - a^2/z)$ has no singularities outside the circle $|z| = a$, except at infinity, where it tends to the potential $-E_0 z$ of a uniform field. On the circle $a^2/z = \bar{z}$, and $w = V + iU = -E_0(z - \bar{z})$, which is purely imaginary. Thus $V = 0$ on the cylinder.

31.5. By the circle theorem

$$
w = -\frac{\lambda}{2\pi \epsilon_0} \ln(z - z_0) + \frac{\lambda}{2\pi \epsilon_0} \ln\left(\frac{a^2}{z} - \bar{z}_0\right)
$$

is the complex potential of a line of charge λ per unit length at the position z_0 outside a conducting cylinder of radius a at zero potential. The second term inserts a line of charge $-\lambda$ per unit length at the inverse point $a^2/\bar{z}_0$. Thus the conducting cylinder acquires a surface charge σ. By Gauss's theorem, the total charge $\int \sigma \, dS$ on the cylinder is $-\lambda$ per unit length. In order to describe an uncharged conducting cylinder, we add a line charge of strength λ at the origin (that is, along the axis of the cylinder), and obtain

$$
w = -\frac{\lambda}{2\pi \epsilon_0} \left[\ln(z - z_0) - \ln\left(\frac{a^2}{z} - \bar{z}_0\right) + \ln z \right].
$$

The added term changes the potential V but does not disturb its constancy on the cylinder.

31.6. According to the last exercise, the potential problem for a conducting cylinder of radius R and charge $-\lambda$ per unit length, in the field of a parallel line of charge λ per unit length at distance d from the axis, is solved by introducing an image line of charge $-\lambda$ per unit length inside the cylinder, at a distance R^2/d from the axis towards the line charge λ. Obviously, by interchanging the line charge and its image, we solve the potential problem for a second cylinder, of charge λ per unit length, in the field of a parallel line of charge $-\lambda$ per unit length.

The problem of two parallel conducting cylinders, of radius R a distance $2D$ apart, with charges $\pm\lambda$ per unit length, is thus solved by introducing two image lines of charge $\pm\lambda$, each one at a distance $b = R^2/(2D - b)$ from the corresponding cylinder axis (in the direction of the other cylinder). This quadratic equation has the solution $b = D - \sqrt{(D^2 - R^2)}$. In order to find the voltage (the potential difference) between the two cylinders we may choose any two points, one on each cylinder. The two closest points, a distance $2(D - R)$ apart, are convenient, and yield a voltage

$$
V = 2\frac{\lambda}{2\pi \epsilon_0} \ln \frac{2D - R - b}{R - b} = \frac{\lambda}{\pi \epsilon_0} \ln\left[D/R + \sqrt{(D/R)^2 - 1} \right].
$$

Noting the identity $\ln[x + \sqrt{(x^2 - 1)}] = \cosh^{-1} x$, we obtain the capacity per unit length

$$C = \frac{\lambda}{V} = \frac{\pi \epsilon_0}{\cosh^{-1}(D/R)}.$$

Chapter 9

32.1. The statements in Exercises 27.1, 27.2 and 27.3 followed from the constancy of the aether constant ϵ_0, and did not depend on its value. They all continue to hold when ϵ_0 is replaced by ϵ, so long as the latter is uniform.

32.2. The jump conditions $\mathbf{n} \times [\![\mathbf{E}]\!] = 0$ and $\mathbf{n} \cdot [\![\mathbf{D}]\!] = 0$, with $\mathbf{D} = \epsilon \mathbf{E}$, yield

$$E_1 \sin \theta_1 = E_2 \sin \theta_2 \qquad \epsilon_1 E_1 \cos \theta_1 = \epsilon_2 E_2 \cos \theta_2.$$

From these it follows that $\epsilon_1 \cot \theta_1 = \epsilon_2 \cot \theta_2$.

32.3. Outside $r = b$ the potential is of the form $V_e = -E_0 \cos \theta + \sum A_n r^{-(n+1)} P_n$. Assuming that only the first term of the sum (a dipole) survives, the field is

$$\mathbf{E}_e = \mathbf{E}_0 + \alpha \left(\frac{b}{r}\right)^3 [3(\mathbf{n} \cdot \mathbf{E}_0)\mathbf{n} - \mathbf{E}_0].$$

In the dielectric, for $a < r < b$, the potential must have the form $V = \sum(B_n r^n + C_n r^{-(n+1)}) P_n$. Assuming, similarly, that there is only one term ($n = 1$) of each kind, the field is

$$\mathbf{E} = \beta \left(\frac{b}{r}\right)^3 [3(\mathbf{n} \cdot \mathbf{E}_0)\mathbf{n} - \mathbf{E}_0] + \gamma \mathbf{E}_0.$$

The jump condition $\mathbf{n} \times [\![\mathbf{E}]\!] = 0$ at $r = b$ and the condition $\mathbf{n} \times \mathbf{E} = 0$ on the conductor $r = a$ yield the equations

$$1 - \alpha = \gamma - \beta, \qquad \gamma - (b/a)^3 \beta = 0. \tag{1}$$

The jump condition $\mathbf{n} \cdot [\![\mathbf{D}]\!] = 0$ at $r = b$ yields

$$\epsilon_0(1 + 2\alpha) = \epsilon(2\beta + \gamma). \tag{2}$$

We solve the three equations (1) and (2) for α, β and γ, and substitute the resulting β and γ in the formula

$$\sigma = D_n = \epsilon[2(b/a)^3 \beta + \gamma](\mathbf{n} \cdot \mathbf{E}_0)$$

for the surface charge density on $r = a$, obtaining

$$\sigma = \frac{9\epsilon \epsilon_0 b^3 (\mathbf{n} \cdot \mathbf{E}_0)}{(\epsilon + 2\epsilon_0)b^3 + 2(\epsilon - \epsilon_0)a^3}.$$

Calling the direction of $\mathbf{E}_0$ 'north', the surface charge density σ is positive over the northern hemisphere, where $(\mathbf{n} \cdot \mathbf{E}_0) > 0$. Over this hemisphere $\int (\mathbf{n} \cdot \mathbf{E}_0) \, dS = 2\pi \int_0^1 E_0 a^2$

$\times \cos \theta \, d(\cos \theta) = \pi a^2 E_0$. Thus the amount of positive charge induced on the northern hemisphere is

$$\frac{9\pi \epsilon \epsilon_0 a^2 b^3 E_0}{(\epsilon + 2\epsilon_0)b^3 + 2(\epsilon - \epsilon_0)a^3}.$$

There is an equal negative charge induced on the southern hemisphere.

32.4. Near the centre the potential must be indistinguishable from that of a dipole in an infinite medium of dielectric constant K:

$$V = \frac{1}{4\pi\epsilon} \frac{\mathbf{p} \cdot \mathbf{r}}{r^3} = \frac{p \cos \theta}{4\pi\epsilon_0 K r^2},$$

where the polar angle θ is measured from the direction of $\mathbf{p}$. Except for this term, the potential inside the sphere must be a regular harmonic function:

$$V = \frac{p}{4\pi\epsilon_0 K} \left[\frac{\cos \theta}{r^2} + \sum A_n r^n P_n(\cos \theta) \right].$$

Outside the sphere, the potential can differ from the dipole term only by a harmonic function, regular at infinity:

$$V_e = \frac{p}{4\pi\epsilon_0 K} \left[\frac{\cos \theta}{r^2} + \sum \frac{B_n}{r^{n+1}} P_n(\cos \theta) \right].$$

The A_n's and B_n's are determined by requiring the continuity of $V(r, \theta)$ and $\mathbf{n} \cdot \mathbf{D}$ at $r = a$, that is,

$$V(a, \theta) = V_e(a, \theta), \qquad \epsilon_0 \frac{\partial V(a, \theta)}{\partial r} = K\epsilon_0 \frac{\partial V_e(a, \theta)}{\partial r}.$$

Thus

$$V = \frac{p \cos \theta}{4\pi\epsilon_0 K r^2} \left[1 + \frac{2(K-1)}{K+2} \frac{r^3}{a^3} \right],$$

$$V_e = \frac{p \cos \theta}{4\pi\epsilon_0 r^2} \frac{3}{K+2}.$$

32.5. On the area $a(a - x)$ still covered by the dielectric, the surface charge density is $\sigma = D = \epsilon V/d$; over the remaining area ax, $\sigma = D = \epsilon_0 V/d$. The total charge on the positive plate is

$$Q(x) = a\frac{V}{d}[\epsilon_0 x + \epsilon(a - x)].$$

The current is

$$i = -\frac{dQ}{dt} = -Q'(x)\dot{x} = a\frac{V}{d}(\epsilon - \epsilon_0)v.$$

33.1. The equation div $\mathbf{D} = d(r^2 D)/(r^2 dr) = 0$ leads to $D = \text{const.}/r^2$. The constant is zero in the ball $r < 0$ because D is not singular at the centre. The jump condition $\mathbf{n} \cdot [\![\mathbf{D}]\!] = 0$ at the inner conductor $r = a$ gives $D = a^2\sigma_a/r^2$ for $a < r < d$. At $r = d$, D jumps back to zero, because $b^2\sigma_b = -a^2\sigma_a$. The voltage is

$$V = \int_a^d E\,dr = \int_a^b \frac{D}{\epsilon_0}\,dr + \int_b^c \frac{D}{\epsilon}\,dr + \int_c^d \frac{D}{\epsilon_0}\,dr$$

$$= a^2\sigma_a\left[\frac{1}{\epsilon_0}\left(\frac{1}{a} - \frac{1}{b}\right) + \frac{1}{\epsilon}\left(\frac{1}{b} - \frac{1}{c}\right) + \frac{1}{\epsilon_0}\left(\frac{1}{c} - \frac{1}{d}\right)\right]$$

$$= \frac{a^2\sigma_a}{\epsilon_0}\left[\left(\frac{1}{a} - \frac{1}{b}\right) + \frac{\epsilon_0 - \epsilon}{\epsilon}\left(\frac{1}{c} - \frac{1}{d}\right)\right].$$

The capacity is given by $C = Q/V = 4\pi a^2\sigma_a/V$.

33.2. As in the last exercise, $D = a^2\sigma_a/r^2$. Thus the field is $E(r) = a^2\sigma_a/(\epsilon r^2) = Q_a/(4\pi\epsilon r^2)$. Its absolute value is greatest at $r = a$. The voltage is

$$V = \left|\int E\,dr\right| = \frac{|Q_a|}{4\pi\epsilon}\left(\frac{1}{a} - \frac{1}{b}\right) = \frac{a}{b}|E(a)|(b - a).$$

Breakdown will occur when $|E(a)| = E_0$, that is, when $V = E_0 a(b - a)/b$.

33.3. If $V > E_0 a(b - a)/b$ charge will flow into the dielectric, and there will be a layer $a < r < c < b$ of constant field E_0 containing a volume charge density

$$q(r) = \epsilon\,\text{div}\,\mathbf{E} = \frac{\epsilon}{r^2}\frac{d}{dr}r^2 E_0 = \frac{2\epsilon E_0}{r}.$$

The voltage across this layer will be $E_0(c - a)$. When breakdown is complete, $c = b$ and $V = E_0(b - a)$.

33.4. According to Exercise 33.1 (with $c = a$ and $d = b$) the capacity of a spherical capacitor, of inner radius a and outer radius b, with material of uniform dielectric constant $K = \epsilon/\epsilon_0$, is

$$C = \frac{4\pi\epsilon_0 K ab}{b - a}.$$

In the present exercise the displacement and field are again $D = a^2\sigma_a/r^2$ and $E = D/\epsilon = a^2\sigma_a/(K\epsilon_0 r^2)$. Since $\mathbf{E}$, which is parallel to the plane dividing the dielectric from the air, must be continuous, σ_a must be proportional to the dielectric constant (1 or K) of the adjacent dielectric. Thus the total charge on the inner sphere is $Q = 2\pi a^2\sigma_1 + 2\pi a^2\sigma_K$, and the voltage is

$$V = \int_a^b E\,dr = \frac{a^2\sigma_1}{\epsilon_0}\left(\frac{1}{a} - \frac{1}{b}\right).$$

Since $\sigma_K = K\sigma_1$, the capacity is

$$C = \frac{Q}{V} = \frac{2\pi\epsilon_0(1 + K)ab}{b - a} = \frac{4\pi\epsilon_0\frac{1}{2}(1 + K)ab}{b - a}.$$

33.5. Since there is no surface charge, both $\mathbf{n} \cdot \mathbf{D}$ and $\mathbf{n} \times \mathbf{E}$ must be continuous. To each point charge e there corresponds a radially directed charge potential $D = e/(4\pi r^2)$. Continuity of $\mathbf{n} \cdot \mathbf{D}$ gives

$$e - e' = e''.$$

In each region $\mathbf{E}$ is $\mathbf{D}$, divided by the appropriate ϵ. Continuity of $\mathbf{n} \times \mathbf{E}$ gives

$$\frac{e + e'}{\epsilon_0} = \frac{e''}{\epsilon}.$$

These equations determine e' and e''.

33.6. Since $q = 0$ for $r > 0$, $r^2 \epsilon dV/dr$ must be a constant. For a point charge the constant is $-e/(4\pi)$. Thus

$$\frac{dV}{dr} = -\frac{e}{4\pi \epsilon r^2} = -\frac{e}{4\pi \epsilon_0 r^2 (1 + a/r)} = \frac{e}{4\pi \epsilon_0 a}\left(\frac{1}{r + a} - \frac{1}{r}\right).$$

33.7. The potential in the cavity is of the form

$$V = \frac{1}{4\pi \epsilon_0}\left[\frac{e}{R} + \sum A_n r^n P_n(\cos\theta)\right],$$

where R is the distance from e. In the dielectric we write

$$V = \frac{1}{4\pi \epsilon}\left[\frac{e}{R} + \sum \frac{B_n}{r^{n+1}} P_n(\cos\theta)\right].$$

At $r = a$, V and $\epsilon \partial V/\partial r$ must be continuous. Since $x < a$ we use the expansion

$$\frac{1}{R} = \sum \frac{x^n}{r^{n+1}} P_n,$$

and obtain

$$\frac{\epsilon}{\epsilon_0} a^{2n+1} A_n - B_n = -\frac{\epsilon - \epsilon_0}{\epsilon_0} e x^n,$$

$$n a^{2n+1} A_n + (n + 1) B_n = 0.$$

These equations determine the A's and B's. We only need

$$A_n = -\frac{(n + 1)(\epsilon - \epsilon_0)}{(n + 1)\epsilon + n\epsilon_0} \frac{e x^n}{a^{2n+1}}.$$

The electric field in the cavity is therefore

$$E = -\frac{dV}{dr} = \frac{1}{4\pi \epsilon_0}\left[\frac{e}{R^3}\mathbf{R} - \sum n A_n r^{n-1} P_n\right].$$

The force, to which the self-term does not contribute, is

$$F = -\frac{e}{4\pi \epsilon_0} \sum n A_n r^{n-1} P_n = \frac{e^2}{4\pi \epsilon_0} \sum \frac{n(n + 1)(\epsilon - \epsilon_0)}{(n + 1)\epsilon + n\epsilon_0} \frac{x^{2n-1}}{a^{2n+1}} P_n$$

$$= \frac{e^2}{4\pi \epsilon_0} \frac{2(\epsilon - \epsilon_0)}{2\epsilon + \epsilon_0} \frac{x}{a^3} + O(x^3).$$

Chapter 10

36.1. The equation

$$\operatorname{div} \mathbf{j} = \frac{1}{r^2} \frac{d}{dr} r^2 j = 0$$

has the solution $j = \mathrm{const} r^2$. The constant is determined by $4\pi a^2 j = i$. Thus $j = i/(4\pi r^2)$.

$$V = \int_a^b E \, dr = \int_a^b \frac{i}{4\pi \sigma r^2} \, dr = \frac{i}{4\pi \sigma} \left(\frac{1}{a} - \frac{1}{b} \right)$$

$$R = \frac{V}{i} = \frac{b-a}{4\pi \sigma ab}.$$

We obtain the capacity from the formula $RC = \epsilon/\sigma$. Thus

$$C = \frac{4\pi \epsilon ab}{b-a}.$$

36.2. Except for the term $Az = Ar \cos\theta$, which corresponds to a uniform field, V must be regular at infinity. Thus

$$V = Ar \cos\theta + \sum \frac{B_n}{r^{n+1}} P_n.$$

Since there is no flow through the cavity wall,

$$j_r = \sigma E_r = -\sigma \frac{\partial V}{\partial r} = \sigma \left[-A \cos\theta + \sum (n+1) B_n r^{-n-2} P_n \right]$$

must vanish at $r = a$. The only surviving B_n is $B_1 = \frac{1}{2} a^3 A$. From the potential

$$V = A \left(r + \frac{a^3}{2r^2} \right) \cos\theta$$

we obtain the field components

$$E_r = -\frac{\partial V}{\partial r} = -A \left(1 - \frac{a^3}{r^3} \right) \cos\theta,$$

$$E_\theta = -\frac{1}{r} \frac{\partial V}{\partial \theta} = A \left(1 + \frac{a^3}{2r^3} \right) \sin\theta.$$

The field lines are given by $(dr, r\,d\theta) \parallel (E_r, E_\theta)$, or

$$E_\theta \, dr - r E_r \, d\theta = 0. \tag{1}$$

Since $\mathbf{j} = \sigma \mathbf{E}$, this is also the equation for the flow lines. Consider now the function

$$f(r, \theta) = A \frac{r^3 - a^3}{r} \sin^2\theta.$$

Its derivatives are

$$\frac{\partial f}{\partial r} = 2r \sin \theta (E_\theta), \qquad \frac{\partial f}{\partial \theta} = 2r \sin \theta (-r E_r).$$

The equation $f = $ const. or

$$\frac{\partial f}{\partial r} dr + \frac{\partial f}{\partial \theta} d\theta = 0$$

is thus the same as eqn (1) for the field lines. The flow lines are therefore given by

$$\frac{r^3 - a^3}{r} \sin^2 \theta = \text{const.}$$

36.3. In view of the analogy between the system (36.1)–(36.3) and (36.5)–(36.7), this exercise is a restatement of Exercise 27.2.

36.4. This exercise is a restatement of Exercise 27.3.

36.5. If we denote by ϕ the rotation angle about the y axis, then (x, ϕ, y) constitute cylindrical coordinates. The field $\mathbf{B} = (0, B_\phi, 0)$ with $B_\phi = A/x$ (cf. (13.3)) corresponds to a linear current flowing along the y axis, and satisfies both div $\mathbf{B} = 0$ and **curl B** $= 0$. An electric field $\mathbf{E} = (0, E_\phi, 0)$ with $E_\phi = A/x$ thus satisfies div $\mathbf{E} = $ **curl E** $= 0$. Its circular lines are normal to the planes $\phi = $ const. Such an $\mathbf{E}$ therefore satisfies all the conditions of our exercise. The voltage between the electrodes is $V = \pi x E_\phi = \pi A$ and the current leaving an electrode is

$$i = \int j_\phi \, dS = \int \sigma E_\phi \, dS = \sigma A \int \frac{dS}{x}.$$

For the resistance we obtain

$$\frac{1}{R} = \frac{i}{V} = \frac{\sigma}{\pi} \int \frac{dS}{x}.$$

36.6. In spherical coordinates (r, θ, ϕ) the Laplacian of a potential $V(\theta)$ is

$$\Delta V = \frac{1}{r^2 \sin \theta} \frac{\partial}{\partial \theta} \left(\sin \theta \frac{\partial V}{\partial \theta} \right).$$

For the potential $V = A \ln \cot \frac{1}{2}\theta$, $\partial V/\partial \theta = -A/\sin \theta$ and therefore $\Delta V = 0$. An electric field, the only non-zero component of which is $E_\theta = A/\sin \theta$, therefore solves our potential problem. The constant must be chosen such that the same current i flows through any cone $\theta = $ const. This results in $j_\theta = i/(2\pi a t \sin \theta)$. The voltage is

$$V = \int E_\theta a \, d\theta = \int_{c/a}^{\pi - c/a} \frac{i}{2\pi \sigma a t \sin \theta} a \, d\theta.$$

Thus

$$R = \frac{V}{i} = \frac{1}{2\pi \sigma t} \int_{c/a}^{\pi - c/a} \frac{d\theta}{\sin \theta} = \frac{1}{\pi \sigma t} \ln \cot \frac{c}{2a}.$$

36.7. The surface charge density is

$$\mathbf{n} \cdot [\![\mathbf{D}]\!] = \mathbf{n} \cdot [\![\epsilon \mathbf{E}]\!] = \mathbf{n} \cdot \left[\!\left[\frac{\epsilon}{\sigma}\mathbf{j}\right]\!\right] = \left[\!\left[\frac{\epsilon}{\sigma}\right]\!\right] j_n,$$

since $\mathbf{n} \cdot \mathbf{j} = j_n$ is continuous.

36.8. The current density through slab 1 is $j_1 = \sigma_1 E_1 = \sigma_1 V_1/d_1$. Similarly $j_2 = \sigma_2 V_2/d_2$. The two current densities must be equal: $j_1 = j_2 = j$. In terms of the total voltage $V = V_1 + V_2 = jd_1/\sigma_1 + jd_2/\sigma_2$ we obtain

$$j = \frac{\sigma_1 \sigma_2 V}{\sigma_2 d_1 + \sigma_1 d_2}.$$

According to the last exercise the surface charge density on the boundary between the slabs is

$$\frac{\sigma_1 \epsilon_2 - \sigma_2 \epsilon_1}{\sigma_2 d_1 + \sigma_1 d_2} V,$$

provided the current flows from slab 1 to slab 2. Otherwise, the surface charge density has the opposite sign.

36.9. By symmetry, if a current i enters and leaves at opposite ends of the cube (A and C in Fig. B.4), it divides into three equal currents $i/3$ at these opposite ends. Each of these currents $i/3$ again divides into two equal currents $i/6$. By following any route, consisting of three consecutive wires, from the point of entry to the point of exit, we obtain $V = 2(i/3)R + (i/6)R = (5R/6)i$. Thus the resistance V/i is $5R/6$. If a current i enters and leaves at two ends A and B of one wire, let i' be the current flowing through this wire. By symmetry, a current $(i - i')/2$ will flow through each of the two other wires leaving A, or reaching B. Each of the latter currents will divide further, into a part i'' flowing parallel to AB and another part $(i - i')/2 - i''$. The direct route AB gives a voltage $i'R$, and this must

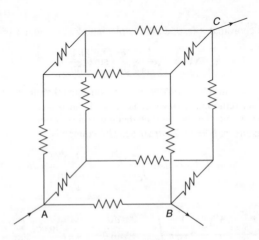

Fig. B.4

be the same as $2(i - i')R/2 + i''R$. Thus $i'' = 2i' - i$. The current through each of the wires is now known in terms of i and i'. By equating the voltages along the direct route AB and the farthest route we obtain $i' = 7i/12$. Hence $V = i'R = (7R/12)i$, and the resistance V/i is $7R/12$.

36.10. If the resistances are joined in series $V = \sum R_a i = i \sum R_a$, since the same current flows through each. Thus $R = V/i = \sum R_a$.

If the resistances are joined in parallel $i = \sum (V/R_a) = V \sum (1/R_a)$, since the voltage is V across each one. Thus

$$\frac{1}{R} = \frac{i}{V} = \sum \frac{1}{R_a}.$$

36.11. Over a length dx of the cable, with resistance $R dx$, the potential falls by an amount $dV = -(R dx)i$. Over the same length dx the leak conductance—a resistance $(G dx)^{-1}$ between cable and ground—causes a leak current $V/(G dx)^{-1} = G V dx$ to ground. The current along the cable therefore decreases by an amount $di = -V(G dx)$. We thus obtain the two differential equations

$$\frac{dV}{dx} = -Ri \qquad \frac{di}{dx} = -GV.$$

From these it follows that

$$\frac{d^2 V}{dx^2} = -R \frac{di}{dx} = GRV,$$

which has the solution $V(x) = A \cosh \sqrt{GR}x + B \sinh \sqrt{GR}x$. The constants are determined by $V(0) = V_0$ and $V(l) = 0$.

37.1.

$$\dot{V} = \frac{\dot{Q}}{C} = \frac{i}{C} = \frac{V}{RC} = \frac{V}{\tau},$$

which has the solution $V(t) = (1 - e^{-t/\tau})V(0)$. Since $V(0) = \mathcal{E}$, we obtain

$$Q(t) = CV(t) = (1 - e^{-t/\tau})C\mathcal{E}.$$

37.2. It must be understood that the sum $H = \sum i_a(R_a i_a - 2\mathcal{E}_a)$ is subject to Kirchhoff's first rule $\sum i = 0$ at each junction. Even so, it is clear that H can have no maximum, only a minimum. The minimum principle, which is already subject to Kirchhoff's first rule, should then yield the second rule, that is, $\sum (Ri - \mathcal{E}) = 0$ for each loop.

In a single loop the same i flows through all branches and

$$\delta H = \delta \sum i(Ri - 2\mathcal{E}) = 2\delta i \sum (Ri - \mathcal{E}) = 0$$

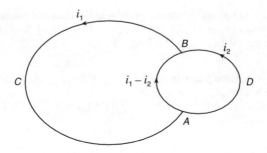

Fig. B.5

yields Kirchhoff's second rule $\sum(Ri - \mathcal{E}) = 0$. If we now add a second loop and apply Kirchhoff's first rule to the two junctions, as in Fig. B.5, we obtain

$$H = \sum_{BCA} i_1(Ri_1 - 2\mathcal{E}) + \sum_{AB}(i_1 - i_2)(R(i_1 - i_2) - 2\mathcal{E}) + \sum_{ADB} i_2(Ri_2 - 2\mathcal{E}),$$

$$\delta H = 2\delta i_1 \sum_{BCA}(Ri_1 - \mathcal{E}) + 2(\delta i_1 - \delta i_2)\sum_{AB}(R(i_1 - i_2) - \mathcal{E})$$

$$+ 2\delta i_2 \sum_{ADB}(Ri_2 - \mathcal{E})$$

$$= 2\delta i_1 \sum_{ABCA}(Ri - \mathcal{E}) + 2\delta i_2 \sum_{ADBA}(Ri - \mathcal{E}).$$

Now $\delta H = 0$ yields Kirchhoff's second rule $\sum(Ri - \mathcal{E}) = 0$ for each one of the loops. The last result can serve as a basis for a proof of the minimum principle by induction on the number of loops.

37.3. Let R be the cable resistance per unit length, and r_C the resistance connecting C to earth. If l is the total length of the cable and x the unknown length between A and C, the resistance between A and C is Rx and between C and B, $R(l - x)$. When B is put to earth and an emf applied at A, the total resistance is

$$R_A = Rx + \frac{R(l - x)r_C}{R(l - x) + r_C}.$$

When A is put to earth and the emf applied at B, the total resistance is

$$R_B = R(l - x) + \frac{Rxr_C}{Rx + r_C}.$$

The two equations can be solved for the two unknowns r_C and x.

38.1. When $\mathcal{J} = 0$, $\mathcal{J} = \sigma\mathcal{E}$ yields $\mathbf{E} + \mathbf{v} \times \mathbf{B} = 0$. Taking the divergence, we obtain

$$\text{div}\,\mathbf{E} + \mathbf{B} \cdot \mathbf{curl}\,\mathbf{v} - \mathbf{v} \cdot \mathbf{curl}\,\mathbf{B} = \text{div}\,\mathbf{E} + 2\mathbf{\Omega} \cdot \mathbf{B} = 0,$$

since $\mathbf{curl}\,\mathbf{v} = \mathbf{curl}\,\mathbf{\Omega} \times \mathbf{r} = 2\mathbf{\Omega}$ and $\mathbf{curl}\,\mathbf{B} = 0$. Thus

$$q = \text{div}\,\mathbf{D} = \text{div}\,\epsilon_0\mathbf{E} = -2\epsilon_0\mathbf{\Omega} \cdot \mathbf{B}.$$

In the disc $\mathbf{D} = \epsilon_0 \mathbf{E} = -\epsilon_0 \mathbf{v} \times \mathbf{B}$. Hence $\mathbf{D}$ has only a radial component $D = -\epsilon_0 r \Omega \cdot \mathbf{B}$. In the region $r < a$, div $\mathbf{D} = 0$ requires $\mathbf{D}$ (and $\mathbf{E} = \mathbf{D}/\epsilon_0$) to vanish. Thus the surface charge density on the inner face $r = a$ is

$$\sigma_a = D(a) - 0 = -\epsilon_0 a \Omega \cdot \mathbf{B}.$$

Since the disc was neutral to begin with, the total charge,

$$2\pi a L \sigma_a + 2\pi b L \sigma_b + \pi (b^2 - a^2) L q,$$

must vanish. This determines the surface charge density

$$\sigma_b = \epsilon_0 b \Omega \cdot \mathbf{B}$$

on the outer face $r = b$. Outside the disc, the radial solution of div $\mathbf{D}_e = 0$ is of the form $D_e = \alpha/r$. The constant α is determined by $D_e(b) - D(b) = \sigma_b$. Hence $\alpha = 0$, and $\mathbf{D}_e = 0$. Thus $\mathbf{E} = \mathbf{D}_e/\epsilon_0 = 0$ outside the disc.

38.2. According to Faraday's law of induction (25.12)

$$\int_{t_1}^{t_2} dt \oint \mathcal{E} \cdot \mathbf{ds} = -\Delta\Phi.$$

The reason for the change of magnetic flux—local time variation of $\mathbf{B}$, motion of the circuit through a non-uniform $\mathbf{B}$, a change in the shape of the closed circuit, or any combination of these—does not matter. Substituting Ohm's law $\mathcal{E} = \mathcal{J}/\sigma$ we obtain

$$-\Delta\Phi = \int_{t_1}^{t_2} dt \oint \frac{\mathcal{J} \cdot \mathbf{ds}}{\sigma} = \int_{t_1}^{t_2} i \, dt \oint \frac{ds}{\sigma A},$$

where A is the cross-section of the circuit. The last line integral is the resistance R of the circuit. The charge $\int i \, dt$ is therefore $-\Delta\Phi/R$. The sign determines the direction of flow.

38.3. The locomotive provides a conducting path between the rails. In a steady state (when the velocity has been maintained constant throughout a time comparable to the infinitesimal relaxation time) the current $\mathbf{j} = \sigma(\mathbf{E} + \mathbf{v} \times \mathbf{B})$ will vanish. Then $\mathbf{E} = -\mathbf{v} \times \mathbf{B}$, which leads to a voltage

$$-\int \mathbf{E} \cdot \mathbf{ds} = \int \mathbf{v} \times \mathbf{B} \cdot \mathbf{ds} = \mathbf{B} \cdot \int \mathbf{ds} \times \mathbf{v},$$

where the integration path is from one rail to the other. Now $\int \mathbf{ds} \times \mathbf{v}$ is a vertical vector of magnitude vl, and from this the result follows.

Chapter 11

40.1. We use the formula (40.7)$_1$,

$$\mathbf{A} = \frac{\mu_0}{4\pi} i \int \frac{\mathbf{ds}}{r}.$$

The square of the distance from (r, ϕ, z)—at which we seek **A**—to a point $(a, \phi', 0)$ on the circle is

$$(r \cos \phi - a \cos \phi')^2 + (r \sin \phi - a \sin \phi')^2 + z^2$$
$$= r^2 - 2ar(\cos \phi \cos \phi' + \sin \phi \sin \phi') + a^2 + z^2$$
$$= a^2 + r^2 + z^2 - 2ar \cos(\phi - \phi').$$

The vector **ds** has length $a\, d\phi'$, but the vectorial summation requires its components $-a \sin(\phi - \phi')\, d\phi'$ and $a \cos(\phi - \phi')\, d\phi'$. Thus

$$A_r = -\frac{\mu_0}{4\pi} i \int_0^{2\pi} \frac{a \sin \psi\, d\psi}{(a^2 + r^2 + z^2 - 2ar \cos \psi)^{\frac{1}{2}}},$$

$$A_\phi = \frac{\mu_0}{4\pi} i \int_0^{2\pi} \frac{a \cos \psi\, d\psi}{(a^2 + r^2 + z^2 - 2ar \cos \psi)^{\frac{1}{2}}},$$

$$A_z = 0.$$

A_r vanishes because its integrand is odd. For a small loop $a^2 \ll r^2 + z^2$ and

$$\frac{1}{(a^2 + r^2 + z^2 - 2ar \cos \psi)^{\frac{1}{2}}} \approx \frac{1}{(r^2 + z^2)^{\frac{1}{2}}} \left[1 + \frac{ar}{(r^2 + z^2)^{\frac{1}{2}}} \cos \psi \right].$$

Thus

$$A_\phi = \frac{\mu_0}{4\pi} \frac{\pi a^2 i r}{(r^2 + z^2)^{\frac{3}{2}}}.$$

In the formula (40.13),

$$\mathbf{A} = \frac{\mu_0}{4\pi} \frac{\mathbf{m} \times \mathbf{r}}{r^3},$$

the magnetic moment, of magnitude $\pi a^2 i$, is in the z direction. The vector **r** is of length $(r^2 + z^2)^{\frac{1}{2}}$, and the sine of the angle it makes with the z direction is $r/(r^2 + z^2)^{\frac{1}{2}}$. This yields

$$A_r = 0, \quad A_\phi = \frac{\mu_0}{4\pi} \pi a^2 i \frac{r}{(r^2 + z^2)^{\frac{3}{2}}}, \quad A_z = 0,$$

in agreement with the previous result.

40.2. According to (40.9) the magnetic field at the centre of the coil is normal to the plane of the coil, of magnitude

$$B_c = \frac{\mu_0 n i}{2a}.$$

According to the torque formula $\mathbf{t} = \mathbf{m} \times \mathbf{B}$ the compass needle will align itself along the resultant magnetic field. Thus

$$\tan \theta = \frac{B_c}{B} = \frac{\mu_0 n i}{2a B}.$$

40.3. In the formula $\mathbf{f} = i \oint \mathbf{ds} \times \mathbf{B}$, only the path inside the field region makes a contribution. The integral $\int \mathbf{ds}$ over this path is the vector $\mathbf{s}$ from P to Q. Thus $\mathbf{f} = i\mathbf{s} \times \mathbf{B}$: the loop is drawn into the magnetic field region by a force of magnitude ilB, where $l = |\mathbf{s}|$.

40.4. Since the current $i = j2\pi rL$ flowing through the disc is radial and the magnetic field is perpendicular to disc, the force $\mathbf{j} \times \mathbf{B}\, d^3x$ on a volume element d^3x of the disc is in the ϕ direction. The torque is obtained by multiplying by the distance r from the axis, and integrating over the disc:

$$\int_a^b rj B 2\pi r L\, dr = \int_a^b ri B\, dr = \tfrac{1}{2}i B(b^2 - a^2).$$

When the current i is imposed, this electromagnetic torque causes the disc to turn. The homopolar generator then becomes a dc-motor.

41.1. We assume $\mathbf{H} = (0, H_\phi(r), 0)$ and apply $\oint_c \mathbf{H} \cdot \mathbf{ds} = \int j_n\, dS$ to a circle of radius r which is perpendicular to the axis of the wire. Since the current is uniform $j_z = i/(\pi a^2)$. For $r < a$ we obtain $2\pi r H_\phi(r) = \pi r^2 j_z$, or $H_\phi = ri/(2\pi a^2)$. For $r > a$ we obtain $2\pi r H_\phi(r) = i$, or $H_\phi = i/(2\pi r)$. Thus $H_\phi(r)$ is continuous. Any field $\mathbf{B} = (0, \mu_0 H_\phi(r), 0)$ satisfies div $\mathbf{B} = 0$.

41.2. According to (41.7) and Fig. 40.1

$$\Omega_{\text{axis}} = \tfrac{1}{2}i\left[1 - \frac{z}{(a^2 + z^2)^{\frac{1}{2}}}\right]. \tag{1}$$

Since Ω is an axisymmetric solution of Laplace's equation, it has the general form

$$\Omega = \tfrac{1}{2}i \sum \left(A_n r^n + \frac{B_n}{r^{n+1}}\right) P_n(\cos\theta), \tag{2}$$

where r is the distance from the centre of the coil and θ is the polar angle measured from the coil axis. On the axis, where $r = z$ and $P_n(\cos\theta) = P_n(1) = 1$,

$$\Omega_{\text{axis}} = \tfrac{1}{2}i \sum \left(A_n z^n + \frac{B_n}{z^{n+1}}\right). \tag{3}$$

We determine the A's and B's by requiring that (3) be the expansion of (1) in powers of z. For $z < a$ we have from (1)

$$\Omega_{\text{axis}} = \tfrac{1}{2}i\left(1 - \frac{z}{a} + \frac{z^3}{2a^3} - \cdots\right).$$

Thus, for $r < a$, Ω is given by (2), with the B's equal to zero and

$$A_0 = 1, \quad A_2 = A_4 = \ldots = 0,$$

$$A_1 = -\frac{1}{a}, \quad A_{2n+1} = (-)^{n+1}\frac{(2n-1)!!}{2n!!}\frac{1}{a^{2n+1}},$$

where $(2n-1)!! = 1 \cdot 3 \cdots (2n-1)$ and $2n!! = 2 \cdot 4 \cdots 2n$. For $z > a$ we have from (1)

$$\Omega_{\text{axis}} = \tfrac{1}{2}i\left(\frac{a^2}{2z^2} - \frac{3a^4}{8z^4} + \cdots\right).$$

Thus, for $r > a$, Ω is given by (2), with the A's equal to zero and

$$B_0 = B_2 = B_4 = \ldots = 0,$$

$$B_1 = \frac{a^2}{2}, \quad B_{2n+1} = (-)^n \frac{(2n+1)!!}{(2n+2)!!} a^{2n+2}.$$

41.3. For any element $\mathbf{ds}$ the product $\mathbf{ds} \times \mathbf{r}$ has magnitude $a\,ds$ (moment of $\mathbf{ds}$ with respect to the centre of the coil) and is normal to the plane of the coil. Hence

$$H = \frac{i}{4\pi} \oint \frac{a\,ds}{r^3} = \frac{i}{4\pi} 4 \int_{-a}^{a} \frac{a\,ds}{(a^2 + s^2)^{\frac{3}{2}}} = \frac{i}{4\pi} \frac{4}{a} \left[\frac{s}{(a^2 + s^2)^{\frac{1}{2}}} \right]_{-a}^{a} = \frac{i\sqrt{2}}{\pi a}.$$

41.4.

$$\mathbf{B} = \nabla \times \mathbf{A} = \nabla \times \left(\tfrac{1}{2}\mathbf{B} \times \mathbf{x} \right)$$
$$= \tfrac{1}{2}\mathbf{B}(\nabla \cdot \mathbf{x}) - \tfrac{1}{2}(\mathbf{B} \cdot \nabla)\mathbf{x} = \tfrac{3}{2}\mathbf{B} - \tfrac{1}{2}\mathbf{B} = \mathbf{B}.$$

41.5. Very close to a wire carrying a current i the wire is indistinguishable from a straight line. The field lines are therefore circular and $B = \mu_0 i/(2\pi r)$, where r is the distance from the wire. The force on an element $\mathbf{ds}$ of a parallel loop carrying a current i' in the same direction is $i'\mathbf{ds} \times \mathbf{B}$. It is of magnitude $i'ds\,B = \mu_0 i i'/(2\pi r)$ and is directed towards the first loop.

41.6. The force (cf. (40.21)) is

$$\mathbf{f} = (\mathbf{m} \cdot \mathbf{grad})\mathbf{B} + \mathbf{m} \times \mathbf{curl}\,\mathbf{B},$$

but the second term is zero, because $\mathbf{B}$ is a vacuum field: $\mathbf{curl}\,\mathbf{B} = 0$. Since $B = \mu_0 i/(2\pi r)$ only depends on the distance r from the wire,

$$\mathbf{f} = m_r \frac{d}{dr}\mathbf{B} = -m_r \frac{\mathbf{B}}{r},$$

where m_r is the projection of $\mathbf{m}$ upon the shortest distance between the loop and the wire. The magnitude of the force is $\mu_0 i m_r/(2\pi r^2)$.

41.7.

$$\mathbf{f} = i_1 \oint \mathbf{ds}_1 \times \mathbf{B}_2 = i_1 \oint \mathbf{ds}_1 \times \frac{\mu_0 i_2}{4\pi} \oint \frac{\mathbf{ds}_2 \times \mathbf{r}}{r^3}$$
$$= \frac{\mu_0 i_1 i_2}{4\pi} \oint \oint \frac{\mathbf{ds}_1 \times (\mathbf{ds}_2 \times \mathbf{r})}{r^3}.$$

In the case of two parallel wires, the force on wire 1 is $i_1 \oint \mathbf{ds}_1 \times \mathbf{B}_2$, directed towards the other wire. Since the wires are parallel $\mathbf{B}_2$ is constant along wire 1, and the force is $i_1 B_2 = \mu_0 i_1 i_2/(2\pi r)$ per unit length, where r is the distance between the wires. The symmetry of the formula means that the same expression yields the force, per unit length, exerted by wire 1 on wire 2. We can therefore speak of 'the force between two parallel wires'.

41.8. We have

$$\frac{\mathbf{r}}{r^3} = -\operatorname{grad} \frac{1}{r},$$

$$\frac{\mathbf{ds}_1 \times (\mathbf{ds}_2 \times \mathbf{r})}{r^3} = -\mathbf{ds}_1 \times \left(\mathbf{ds}_2 \times \operatorname{grad} \frac{1}{r}\right)$$

$$= (\mathbf{ds}_1 \cdot \mathbf{ds}_2)\operatorname{grad}\frac{1}{r} - \left(\mathbf{ds}_1 \cdot \operatorname{grad}\frac{1}{r}\right)\mathbf{ds}_2.$$

Hence

$$\frac{\mu_0 i_1 i_2}{4\pi} \oint \oint \frac{\mathbf{ds}_1 \times (\mathbf{ds}_2 \times \mathbf{r})}{r^3}$$

$$= \frac{\mu_0 i_1 i_2}{4\pi} \oint \oint (\mathbf{ds}_1 \cdot \mathbf{ds}_2)\operatorname{grad}\frac{1}{r} - \frac{\mu_0 i_1 i_2}{4\pi} \oint \oint (\mathbf{ds}_1 \cdot \operatorname{grad}\frac{1}{r})\,\mathbf{ds}_2.$$

The second term vanishes because $\oint \mathbf{ds}_1 \cdot \operatorname{grad}(1/r) = 0$, which is true of the line integral of any gradient. The force is therefore given by the first term, and each pair of elements ds_1 and ds_2 at an angle θ may be regarded as contributing a force $\mu_0 i_1 i_2 \cos\theta ds_1 ds_2 |\operatorname{grad}(1/r)|/(4\pi) = \mu_0 i_1 i_2 \cos\theta ds_1 ds_2/(4\pi r^2)$.

41.9. The calculation is the same as that leading from (14.8) to (14.10):

$$\mathbf{B} = \mu_0 \mathbf{H} = -\mu_0 \operatorname{grad}\frac{1}{4\pi}\frac{\mathbf{m}\cdot\mathbf{r}}{r^3} = \frac{\mu_0}{4\pi}\left[\frac{3(\mathbf{m}\cdot\mathbf{r})\mathbf{r}}{r^5} - \frac{\mathbf{m}}{r^3}\right].$$

41.10.

$$\mathbf{t} = \mathbf{m} \times \mathbf{B} = \mathbf{m} \times \frac{\mu_0}{4\pi}\left[\frac{3(\mathbf{m}'\cdot\mathbf{r})\mathbf{r}}{r^5} - \frac{\mathbf{m}'}{r^3}\right]$$

$$= \frac{\mu_0}{4\pi}\left[\frac{3(\mathbf{m}\times\mathbf{r})(\mathbf{m}'\cdot\mathbf{r})}{r^5} - \frac{\mathbf{m}\times\mathbf{m}'}{r^3}\right].$$

41.11. The forces that two small loops exert on one another satisfy the law of action and reaction, but they are not directed along the line joining the loops.

41.12. Since the loops are small we may ignore the field inhomogeneities and use the torque formula of Exercise 41.10:

$$\mathbf{t} = \frac{\mu_0}{4\pi}\left[\frac{3(\mathbf{m}\times\mathbf{r})(\mathbf{m}'\cdot\mathbf{r})}{r^5} - \frac{\mathbf{m}\times\mathbf{m}'}{r^3}\right].$$

where $\mathbf{r}$ is the vector from $\mathbf{m}$ to $\mathbf{m}'$. Similarly, the torque on $\mathbf{m}'$ is

$$\mathbf{t}' = \frac{\mu_0}{4\pi}\left[\frac{3(\mathbf{m}'\times\mathbf{r})(\mathbf{m}\cdot\mathbf{r})}{r^5} - \frac{\mathbf{m}'\times\mathbf{m}}{r^3}\right].$$

Both torques must vanish, and it follows that both $\mathbf{m}$ and $\mathbf{m}'$ must point along the line joining them. The configuration of *stable* equilibrium must, further, be such that both torques be restoring. This is only possible if both moments are pointing in the same direction.

42.1. The field of the straight current line i, at a distance s from the line, is $B = \mu_0 i/(2\pi s)$. The flux of this field through the circle is

$$\frac{\mu_0 i}{2\pi} \int \frac{dS}{s} = L_{12} i.$$

The integral is the same as in Problem 42.1, if we replace b by d. Thus

$$L_{12} = \frac{\mu_0}{2\pi} 2\pi [d - \sqrt{(d^2 - a^2)}] = \mu_0 [d - \sqrt{(d^2 - a^2)}].$$

42.2. According to Fig. B.6 the magnetic field of a unit current flowing along the straight wire is $\mu_0/(2\pi x)$. In order to calculate its flux through the triangle, which is the sought coefficient of mutual inductance, we note that the strip in Fig. B.6 has widht dx and length $2h$, where $h = (a + b - x) \tan 30°$. Thus

$$L_{12} = \frac{\mu_0}{2\pi} \int_b^{b+a} \frac{2h \, dx}{x} = \frac{\mu_0}{\pi} \int_b^{b+a} \frac{(a + b - x) \tan 30° \, dx}{x}$$

$$= \frac{\mu_0}{3^{\frac{1}{2}} \pi} \int_b^{b+a} \frac{a + b - x}{x} \, dx = \frac{\mu_0}{3^{\frac{1}{2}} \pi} [(a + b) \ln(1 + a/b) - a].$$

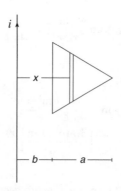

Fig. B.6

42.3. The magnetic flux through a loop is equal to the circulation of the vector potential around the loop (cf. (42.8)). According to Exercise 40.1 the vector potential of a circular current loop $r = a$, $z = 0$ in cylindrical coordinates is

$$A_r = A_z = 0, \qquad A_\phi = \frac{\mu_0}{4\pi} i \int_0^{2\pi} \frac{a \cos \psi \, d\psi}{(a^2 + r^2 + z^2 - 2ar \cos \psi)^{\frac{1}{2}}}.$$

Since $\mathbf{A}$ has only a ϕ component, which depends only on r and z, its circulation around a loop of radius a at $z = c$ is $2\pi a A_\phi(r = a, z = c)$. Thus

$$L_{12} = \frac{\mu_0}{4\pi} 2\pi a \int_0^{2\pi} \frac{a \cos \psi \, d\psi}{(c^2 + 2a^2 - 2a^2 \cos \psi)^{\frac{1}{2}}}.$$

When $c \gg a$ we have

$$(c^2 + 2a^2 - 2a^2 \cos \psi)^{-\frac{1}{2}} \approx \frac{1}{c}\left(1 + \frac{a^2}{c^2} \cos \psi\right),$$

and this leads to

$$L_{12} = \frac{\mu_0}{4\pi} \frac{2\pi^2 a^4}{c^3}.$$

43.1. According to $\mathbf{n} \times [\![\mathbf{H}]\!] = 0$, $H \sin \theta$ is continuous. According to $\mathbf{n} \cdot [\![\mathbf{B}]\!] = 0$, $\mu H \cos \theta$ is continuous. Hence $\mu \cot \theta$ is continuous.

43.2. Outside we assume

$$\mathbf{H}_e = \mathbf{H}_0 + \alpha\left(\frac{b}{r}\right)^3 [3(\mathbf{n} \cdot \mathbf{H}_0)\mathbf{n} - \mathbf{H}_0].$$

In the shell Ω must be of the form $\sum(A_n r^n + B_n r^{-n-1}) P_n(\cos \theta)$. Assuming only one term $(n = 1)$ of each kind, we set

$$\mathbf{H} = \beta\left(\frac{b}{r}\right)^3 [3(\mathbf{n} \cdot \mathbf{H}_0)\mathbf{n} - \mathbf{H}_0] + \gamma \mathbf{H}_0.$$

At $r = b$ the jump conditions $\mathbf{n} \times [\![\mathbf{H}]\!] = 0$ and $\mathbf{n} \cdot [\![\mathbf{B}]\!] = 0$ give

$$1 - \alpha = \gamma - \beta, \qquad \mu_0(1 + 2\alpha) = \mu(\gamma + 2\beta). \tag{1}$$

In the cavity we try

$$\mathbf{H}_i = \delta \mathbf{H}_0,$$

which corresponds to the first non-constant term of $\Omega = \sum C_n r^n P_n(\cos \theta)$. The two jump conditions at $r = a$ give

$$\gamma - \left(\frac{b}{a}\right)^3 \beta = \delta, \qquad \mu\left[\gamma + 2\left(\frac{b}{a}\right)^3 \beta\right] = \mu_0 \delta. \tag{2}$$

Solution of the two pairs of linear equations (1) and (2) for the constants α, β, γ and δ gives

$$\frac{H_i}{H_0} = \delta = \frac{9\mu\mu_0}{9\mu\mu_0 + 2(\mu - \mu_0)^2(1 - a^3/b^3)}.$$

43.3. Let P denote the position of the current line in the plane which is perpendicular to the line. The current potential $\mathbf{H}$ outside the material will be the one corresponding to the current line i through P and an image line i' through the image point P' inside the material. The field inside will be due to an image line i'' through P. From the jump condition $\mathbf{n} \times [\![\mathbf{H}]\!] = 0$ we obtain

$$i'' = i - i',$$

and from the jump condition $\mathbf{n} \cdot [\![\mathbf{B}]\!] = 0$,

$$\mu i'' = \mu_0(i + i').$$

Thus

$$i' = \frac{\mu - \mu_0}{\mu + \mu_0} i, \qquad i'' = \frac{2\mu_0}{\mu + \mu_0} i.$$

In the material the current potential lines are circles (around the current line i'' at P), and its intensity is $H = i''/(2\pi r)$, where r is the distance from P. The lines of the magnetic field $\mathbf{B} = \mu \mathbf{H}$ are the same circles, and

$$B = \frac{\mu i''}{2\pi r} = \frac{2\mu}{\mu + \mu_0} \frac{\mu_0 i}{2\pi r}.$$

The field is therefore $2\mu/(\mu + \mu_0)$ times the field $\mu_0 i/(2\pi r)$ in the absence of the material, and is in the same direction.

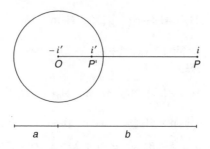

Fig. B.7

43.4. Let a be the radius of the cylinder and b be the distance of the current line i from the axis. In the plane which is perpendicular to the axis (Fig. B.7) we denote the position of the current line by P and the axis of the cylinder by O. Let P' be the image of P, that is, the point on OP at a distance a^2/b from O. The field outside the cylinder will be that of the current line i and current lines i' and $-i'$ through P' and O. (The image $-i'$ at the centre O ensures that the total image current inside the cylinder, which determines the integral $\oint H_s \, ds$ around the cylinder, be zero.) The field inside the cylinder will be that of a current i'' through P. The jump conditions are satisfied by

$$i' = \frac{\mu - \mu_0}{\mu + \mu_0} i, \qquad i'' = \frac{2\mu_0}{\mu + \mu_0} i.$$

Inside the cylinder the lines of the magnetic field $\mathbf{B} = \mu \mathbf{H}$ are again circular, just as those of $\mathbf{H}$ (which is due to i'' at P), and the magnetic field intensity is $2\mu/(\mu + \mu_0)$ times what it would have been in the absence of the cylinder.

We get the results of the previous exercise by taking the limit $a \to \infty$: the cylinder axis then recedes to infinity along with the image current $-i'$ flowing through it.

Chapter 12

44.1. This exercise is quite similar to the example treated in the text. At $t = 0$ the surface charge density jumps from zero to the constant $\sigma = Q/(4\pi a^2)$. Again, $V(r, t) = 0$ for

$t < (r - a)/c$. After that it is given by (cf. (44.28))

$$V = \frac{a\sigma}{2\epsilon_0 r} \int dR,$$

the integral having the lower limit $r - a$ and the upper limit $\min(ct, r + a)$. Thus, for $(r - a)/c < t < (r + a)/c$ the potential is

$$V(r, t) = \frac{Q}{4\pi\epsilon_0 r} \frac{1}{2}\left(1 - \frac{r - ct}{a}\right),$$

which is an outgoing spherical wave, of the form $f(r - ct)/r$. After $t = (r + a)/c$ the wave 'has passed' the point r and

$$V(r, t) = \frac{Q}{4\pi\epsilon_0 r}$$

is the static potential of a uniformly charged sphere.

44.2. Since there is only a current, $V = 0$. Using cylindrical coordinates, with the wire as the z axis and $j' \, d^3 x' = i \, dz$, the vector potential at a distance r from the wire is given by

$$A_z(r, t) = \frac{\mu_0}{4\pi} \int_{-\sqrt{c^2 t^2 - r^2}}^{\sqrt{c^2 t^2 - r^2}} \frac{i \, dz}{(r^2 + z^2)^{\frac{1}{2}}} = \frac{\mu_0 i}{2\pi} \int_0^{\sqrt{c^2 t^2 - r^2}} \frac{dz}{(r^2 + z^2)^{\frac{1}{2}}}$$

$$= \frac{\mu_0 i}{2\pi}\left[\ln\left(\sqrt{r^2 + z^2} + z\right)\right]_0^{\sqrt{c^2 t^2 - r^2}} = \frac{\mu_0 i}{2\pi} \ln\left(\frac{ct + \sqrt{c^2 t^2 - r^2}}{r}\right),$$

the other components being zero. The electric field $\mathbf{E} = -\mathbf{A}_t$ has only the component

$$E_z(r, t) = -\frac{\partial A_z}{\partial t} = -\frac{\mu_0 i c}{2\pi\sqrt{c^2 t^2 - r^2}}.$$

The magnetic field $\mathbf{B} = \mathbf{curl}\, \mathbf{A}$ has only the component

$$B_\phi(r, t) = -\frac{\partial A_z}{\partial r} = \frac{\mu_0 i}{2\pi r} \frac{ct}{\sqrt{c^2 t^2 - r^2}}.$$

After a long time $(ct \gg r)$ the electric field vanishes and the magnetic field becomes $\mu_0 i/(2\pi r)$.

45.1. Denoting by $\dot{\mathbf{d}}(t)$ the derivative of $\mathbf{d}(t)$ we have $\mathbf{d}_t(t - r/c) = \dot{\mathbf{d}}(t - r/c)$ and $\partial_i \mathbf{d}(t - r/c) = -c^{-1} n_i \dot{\mathbf{d}}(t - r/c)$. Hence

$$\operatorname{div}\mathbf{d} = \partial_i d_i = -\frac{1}{c}\mathbf{n} \cdot \dot{\mathbf{d}},$$

$$\operatorname{div} \frac{1}{r}\mathbf{d} = \frac{1}{r}\operatorname{div}\mathbf{d} + \left(\mathbf{grad}\,\frac{1}{r}\right) \cdot \mathbf{d} = -\frac{1}{cr}\mathbf{n} \cdot \dot{\mathbf{d}} - \frac{1}{r^2}\mathbf{n} \cdot \mathbf{d},$$

$$(\mathbf{curl}\,\mathbf{d})_i = \epsilon_{ijk}\partial_j d_k = -\epsilon_{ijk}\partial_j c^{-1} n_j \dot{d}_k = -\frac{1}{c}(\mathbf{n} \times \dot{\mathbf{d}})_i,$$

$$\mathbf{curl}\,\frac{1}{r}\mathbf{d} = \frac{1}{r}\mathbf{curl}\,\mathbf{d} + \left(\mathbf{grad}\,\frac{1}{r}\right) \times \mathbf{d} = -\frac{1}{cr}\mathbf{n} \times \dot{\mathbf{d}} - \frac{1}{r^2}\mathbf{n} \times \mathbf{d},$$

$$\operatorname{grad}_i(\mathbf{r} \cdot \mathbf{d}) = \partial_i x_j d_j = \delta_{ij} d_j - x_j c^{-1} n_i \dot{d}_j = \left[\mathbf{d} - \frac{r}{c}(\mathbf{n} \cdot \dot{\mathbf{d}})\mathbf{n}\right]_i.$$

For the Hertz superpotentials (45.9) we have, according to (45.2),

$$\mathbf{A} = \mu_0 \mathbf{Z}_t^e = \frac{\mu_0}{4\pi r} \dot{\mathbf{d}},$$

$$V = -\epsilon_0^{-1} \operatorname{div} \mathbf{Z}^e = -\frac{1}{4\pi\epsilon_0} \operatorname{div} \frac{1}{r}\mathbf{d} = \frac{1}{4\pi\epsilon_0}\left(\frac{1}{cr}\mathbf{n}\cdot\dot{\mathbf{d}} + \frac{1}{r^2}\mathbf{n}\cdot\mathbf{d}\right),$$

$$\mathbf{B} = \operatorname{curl}\mathbf{A} = \frac{\mu_0}{4\pi}\operatorname{curl}\frac{1}{r}\dot{\mathbf{d}} = -\frac{\mu_0}{4\pi}\left(\frac{1}{r^2}\mathbf{n}\times\dot{\mathbf{d}} + \frac{1}{cr}\mathbf{n}\times\ddot{\mathbf{d}}\right),$$

$$\mathbf{E} = -\mathbf{A}_t - \operatorname{grad}V = -\frac{\mu_0}{4\pi r}\ddot{\mathbf{d}} - \frac{1}{4\pi\epsilon_0}\operatorname{grad}\left(\frac{1}{cr}\mathbf{n}\cdot\dot{\mathbf{d}} + \frac{1}{r^2}\mathbf{n}\cdot\mathbf{d}\right)$$

$$= -\frac{\mu_0}{4\pi r}\ddot{\mathbf{d}} - \frac{1}{4\pi\epsilon_0}\operatorname{grad}\left(\frac{1}{cr^2}\mathbf{r}\cdot\dot{\mathbf{d}} + \frac{1}{r^3}\mathbf{r}\cdot\mathbf{d}\right)$$

$$= \frac{1}{4\pi\epsilon_0}\left\{\frac{1}{r^3}[3(\mathbf{n}\cdot\mathbf{d})\mathbf{n}-\mathbf{d}] + \frac{1}{cr^2}[3(\mathbf{n}\cdot\dot{\mathbf{d}})\mathbf{n}-\dot{\mathbf{d}}]\right\}$$

$$+ \frac{\mu_0}{4\pi r}[(\mathbf{n}\cdot\ddot{\mathbf{d}})\mathbf{n} - \ddot{\mathbf{d}}].$$

45.2. Substituting $\mathbf{m}(\mathbf{x}, t) = \mathbf{d}(t)\delta(\mathbf{x})$ in (45.6) we obtain the Hertz vectors

$$\mathbf{Z}^e = 0, \qquad \mathbf{Z}^m = -\frac{\mathbf{d}}{4\pi r}.$$

Substituting in (45.2) we find $V = 0$ and

$$\mathbf{A} = -\mu_0 \operatorname{curl}\mathbf{Z}^m = \frac{\mu_0}{4\pi}\operatorname{curl}\frac{1}{r}\mathbf{d} = -\frac{\mu_0}{4\pi}\left(\frac{1}{r^2}\mathbf{n}\times\mathbf{d} + \frac{1}{cr}\mathbf{n}\times\dot{\mathbf{d}}\right).$$

The identity (45.17) is proved as follows:

$$\operatorname{curl}(\mathbf{r}\times\mathbf{d}) = \nabla\times(\mathbf{r}\times\mathbf{d})$$

$$= \mathbf{r}(\nabla\cdot\mathbf{d}) + (\mathbf{d}\cdot\nabla)\mathbf{r} - \mathbf{d}(\nabla\cdot\mathbf{r}) - (\mathbf{r}\cdot\nabla)\mathbf{d}$$

$$= \mathbf{r}\operatorname{div}\mathbf{d} + (\mathbf{d}\cdot\nabla)\mathbf{r} - 3\mathbf{d} - (\mathbf{r}\cdot\nabla)\mathbf{d}$$

$$= -\mathbf{r}\frac{1}{c}\mathbf{n}\cdot\dot{\mathbf{d}} + \mathbf{d} - 3\mathbf{d} + \frac{r}{c}\dot{\mathbf{d}}$$

$$= -2\mathbf{d} - \frac{r}{c}\mathbf{n}\times(\mathbf{n}\times\dot{\mathbf{d}}).$$

We use this identity in the calculation of the magnetic field:

$$\mathbf{B} = \operatorname{curl}\mathbf{A} = -\frac{\mu_0}{4\pi}\left(\operatorname{curl}\frac{\mathbf{r}\times\mathbf{d}}{r^3} + \operatorname{curl}\frac{\mathbf{r}\times\dot{\mathbf{d}}}{cr^2}\right)$$

$$= -\frac{\mu_0}{4\pi}\left[\frac{1}{r^3}\operatorname{curl}(\mathbf{r}\times\mathbf{d}) + \operatorname{grad}\frac{1}{r^3}\times(\mathbf{r}\times\mathbf{d})\right.$$

$$\left. + \frac{1}{cr^2}\operatorname{curl}(\mathbf{r}\times\dot{\mathbf{d}}) + \operatorname{grad}\frac{1}{cr^2}\times(\mathbf{r}\times\dot{\mathbf{d}})\right]$$

$$= \frac{\mu_0}{4\pi}\left[\frac{3(\mathbf{n}\cdot\mathbf{d})\mathbf{n} - \mathbf{d}}{r^3} + \frac{3(\mathbf{n}\cdot\dot{\mathbf{d}})\mathbf{n} - \dot{\mathbf{d}}}{cr^2} + \frac{\mathbf{n}\times(\mathbf{n}\times\ddot{\mathbf{d}})}{c^2 r}\right].$$

The electric field is

$$\mathbf{E} = -\mathbf{A}_t = \frac{\mu_0}{4\pi}\left(\frac{\mathbf{n}\times\dot{\mathbf{d}}}{r^2} + \frac{\mathbf{n}\times\ddot{\mathbf{d}}}{cr}\right).$$

The fields in the far zone are

$$\mathbf{E} = \frac{\mu_0}{4\pi}\frac{\mathbf{n}\times\ddot{\mathbf{d}}}{cr}, \qquad \mathbf{B} = \frac{1}{c}\mathbf{n}\times\mathbf{E}.$$

The energy flux is

$$\mathbf{E}\times\mathbf{H} = \frac{\mathbf{E}\times\mathbf{B}}{\mu_0} = \frac{\mu_0}{4\pi}\frac{\ddot{d}^2\sin^2\theta}{4\pi c^3 r^2}\mathbf{n}.$$

Finally, the power radiated is

$$P = \int \mathbf{E}\times\mathbf{H}\cdot\mathbf{n}\,dS = \frac{\mu_0}{4\pi}\frac{2\ddot{d}^2}{3c^3}.$$

46.1. The function $t'(\mathbf{x}, t)$ is implicitly defined by $(46.9)_2$:

$$|\mathbf{x} - \mathbf{r}_0(t')| = c(t - t'). \tag{1}$$

Differentiation with respect to t gives

$$-\frac{(\mathbf{x} - \mathbf{r}_0)\cdot\dot{\mathbf{r}}_0}{|\mathbf{x} - \mathbf{r}_0(t')|}\frac{\partial t'}{\partial t} = -(\mathbf{n}\cdot\mathbf{v})\frac{\partial t'}{\partial t} = c\left(1 - \frac{\partial t'}{\partial t}\right),$$

or

$$\left(1 - \frac{\mathbf{n}\cdot\mathbf{v}}{c}\right)\frac{\partial t'}{\partial t} = 1.$$

Taking the gradient of (1) and noting that, for fixed t', $\mathbf{grad}\,|\mathbf{x} - \mathbf{r}_0(t')| = \mathbf{n}$ we obtain

$$\mathbf{n} - (\mathbf{n}\cdot\mathbf{v})\,\mathbf{grad}\,t' = -c\,\mathbf{grad}\,t',$$

or

$$\left(1 - \frac{\mathbf{n}\cdot\mathbf{v}}{c}\right)\mathbf{grad}\,t' = -\frac{1}{c}\mathbf{n}.$$

46.2. The fields derived from the Liénard–Wiechert potentials are

$$\mathbf{E} = -\mathbf{A}_t - \mathbf{grad}\,V = -\frac{\mu_0}{4\pi}\left(\frac{e\mathbf{v}}{r - \mathbf{r}\cdot\mathbf{v}/c}\right)_t - \frac{1}{4\pi\epsilon_0}\mathbf{grad}\,\frac{e}{r - \mathbf{r}\cdot\mathbf{v}/c},$$

$$\mathbf{B} = \mathbf{curl}\,\mathbf{A} = \frac{\mu_0}{4\pi}\mathbf{curl}\,\frac{e\mathbf{v}}{r - \mathbf{r}\cdot\mathbf{v}/c}$$

$$= \frac{\mu_0}{4\pi}\left(\frac{e}{r - \mathbf{r}\cdot\mathbf{v}/c}\mathbf{curl}\,\mathbf{v} + \mathbf{grad}\,\frac{e}{r - \mathbf{r}\cdot\mathbf{v}/c}\times\mathbf{v}\right).$$

Of course

$$\left(\frac{e\mathbf{v}}{r - \mathbf{r} \cdot \mathbf{v}/c}\right)_t = \frac{e\mathbf{v}_t}{r - \mathbf{r} \cdot \mathbf{v}/c} - \frac{e\mathbf{v}}{(r - \mathbf{r} \cdot \mathbf{v}/c)^2}(r - \mathbf{r} \cdot \mathbf{v}/c)_t,$$

$$\mathbf{grad}\,\frac{e}{r - \mathbf{r} \cdot \mathbf{v}/c} = -\frac{e}{(r - \mathbf{r} \cdot \mathbf{v}/c)^2}\,\mathbf{grad}(r - \mathbf{r} \cdot \mathbf{v}/c).$$

The vectors $\mathbf{r} = \mathbf{x} - \mathbf{r}_0(t')$ and $\mathbf{v} = \dot{\mathbf{r}}_0(t')$ must be differentiated with the help of (46.12). Thus

$$\mathbf{v}_t = \dot{\mathbf{v}}\frac{\partial t'}{\partial t} = \frac{\dot{\mathbf{v}}}{1 - \mathbf{n} \cdot \mathbf{v}/c},$$

$$(r - \mathbf{r} \cdot \mathbf{v}/c)_t = -\frac{(\mathbf{n} \cdot \mathbf{v}) + (v^2 + \mathbf{r} \cdot \dot{\mathbf{v}})/c}{1 - \mathbf{n} \cdot \mathbf{v}/c},$$

$$\mathbf{grad}(r - \mathbf{r} \cdot \mathbf{v}/c) = \frac{(1 - v^2/c^2)\mathbf{n} - (1 - \mathbf{n} \cdot \mathbf{v}/c)\mathbf{v}/c + (\mathbf{r} \cdot \dot{\mathbf{v}})\mathbf{n}/c^2}{1 - \mathbf{n} \cdot \mathbf{v}/c},$$

$$\mathbf{curl}\,\mathbf{v} = -\frac{\mathbf{n} \times \dot{\mathbf{v}}/c}{1 - \mathbf{n} \cdot \mathbf{v}/c}.$$

The remaining calculations require no more differentiations, only algebra.

Chapter 13

47.1. For $\epsilon'' = 0$, $\epsilon' < 0$ and $\mathbf{k}'$ parallel to $\mathbf{k}''$, eqn (47.7),

$$k'^2 - k''^2 + 2i\mathbf{k}' \cdot \mathbf{k}'' = \omega^2(\epsilon' + i\epsilon'')\mu,$$

gives $\mathbf{k}' = 0$. Thus $\mathbf{k} = i\mathbf{k}''$ is imaginary, and eqns (47.5) become

$$i\mathbf{k}'' \times \mathbf{H} = -\omega\epsilon\mathbf{E}, \qquad i\mathbf{k}'' \times \mathbf{E} = \omega\mu\mathbf{H}.$$

The time average of the Poynting vector is therefore

$$\tfrac{1}{2}\Re(\mathbf{E} \times \mathbf{H}^*) = \tfrac{1}{2}\Re\left(\mathbf{E} \times \frac{-i\mathbf{k}'' \times \mathbf{E}^*}{\omega\mu}\right)$$

$$= \frac{1}{2\omega\mu}\Re[-i\mathbf{k}''(\mathbf{E} \cdot \mathbf{E}^*) - \mathbf{E}^*(-i\mathbf{k}'' \cdot \mathbf{E})]$$

$$= \frac{1}{2\omega\mu}\Re(-i\mathbf{k}''|E|^2)$$

$$= 0.$$

48.1. According to $(48.12)_1$ and $(48.13)_1$ we have, when $\theta_t = \frac{1}{2}\pi$,

$$R_\perp = \frac{\sin^2(\theta_i - \frac{1}{2}\pi)}{\sin^2(\theta_i + \frac{1}{2}\pi)} = \frac{\cos^2\theta_i}{\cos^2\theta_i} = 1,$$

$$R_\parallel = \frac{\tan^2(\theta_i - \frac{1}{2}\pi)}{\tan^2(\theta_i + \frac{1}{2}\pi)} = \frac{\sin^2(\theta_i - \frac{1}{2}\pi)\cos^2(\theta_i + \frac{1}{2}\pi)}{\cos^2(\theta_i - \frac{1}{2}\pi)\sin^2(\theta_i + \frac{1}{2}\pi)}$$

$$= \frac{\cos^2\theta_i \sin^2\theta_i}{\sin^2\theta_i \cos^2\theta_i} = 1.$$

48.2. According to $(48.8)_1$ and $(48.10)_1$ we have, when $\sin\theta_i = 0$ and $\cos\theta_i = 1$,

$$E^{(r)} = \frac{n_1 - \sqrt{(\epsilon_2/\epsilon_0)}}{n_1 + \sqrt{(\epsilon_2/\epsilon_0)}} E^{(i)},$$

$$H^{(r)} = -\frac{n_1 - \sqrt{(\epsilon_2/\epsilon_0)}}{n_1 + \sqrt{(\epsilon_2/\epsilon_0)}} H^{(i)}.$$

Substituting $n_2 + i\kappa_2 = c\sqrt{\epsilon_2\mu_2} = \sqrt{\epsilon_2/\epsilon_0}\sqrt{\mu_2/\mu} = \sqrt{\epsilon_2/\epsilon_0}$, we have

$$E^{(r)} = \frac{n_1 - (n_2 + i\kappa_2)}{n_1 + (n_2 + i\kappa_2)} E^{(i)} = \frac{n - 1 - i\kappa}{n + 1 + i\kappa} E^{(i)},$$

where $n = n_2/n_1$ and $\kappa = \kappa_2/\kappa_1$. Similarly

$$H^{(r)} = -\frac{n - 1 - i\kappa}{n + 1 + i\kappa} H^{(i)}.$$

Hence

$$\mathbf{E}^{(r)} \times (\mathbf{H}^{(r)})^* = -\left|\frac{n - 1 - i\kappa}{n + 1 + i\kappa}\right|^2 \mathbf{E}^{(i)} \times (\mathbf{H}^{(i)})^*.$$

The time averaged incident and reflected energy fluxes $\frac{1}{2}\Re(\mathbf{E}^{(i)} \times (\mathbf{H}^{(i)})^*)$ and $\frac{1}{2}\Re(\mathbf{E}^{(i)} \times (\mathbf{H}^{(i)})^*)$ are oppositely directed. Thus the reflection coefficient is

$$R = \left|\frac{n - 1 - i\kappa}{n + 1 + i\kappa}\right|^2 = \frac{(n - 1)^2 + \kappa^2}{(n + 1)^2 + \kappa^2}.$$

48.3. For a light ray with angle of incidence θ_i we have, according to Snell's law,

$$\sin \theta_i = n \sin \theta_t.$$

Thus

$$\cos^2 \theta_t = 1 - \sin^2 \theta_t = 1 - \frac{1}{n^2} \sin^2 \theta_i.$$

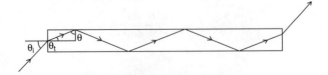

Fig. B.8

The transmitted ray will impinge on the wall at an angle of incidence θ, which must exceed the angle of total reflection, that is, $\sin \theta > 1/n$. But $\sin \theta = \cos \theta_t$ (Fig. B.8). Hence the requirement is

$$1 - \frac{1}{n^2} \sin^2 \theta_i > \frac{1}{n^2}$$

or

$$n^2 > 1 + \sin^2 \theta_i.$$

This condition will be met for any angle of incidence θ_i if $n^2 > 2$. Most kinds of glass have $n \approx 1.5 > \sqrt{2}$.

48.4. The required condition is that the incident and reflected rays at A be tangent to the tube, and that the reflection there be total. Thus the angle θ in Figure B.9 must exceed the angle of total reflection. Hence

$$\sin \theta = \frac{r}{r+d} > \frac{1}{n},$$

which results in $r > d/(n-1)$.

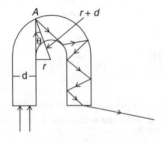

Fig. B.9

49.1. According to (49.8), $\mathbf{j} \cdot \mathbf{j}^*$ has the z dependence $e^{-2z/\delta}$. Hence the integral is $\int e^{-2z/\delta} \, dz = \frac{1}{2}\delta$ times the surface value of $\Re[\mathbf{j} \cdot \mathbf{j}^*/(2\sigma)]$, or

$$\frac{\sigma\delta}{2}\Re\left(\tfrac{1}{2}\mathbf{E} \cdot \mathbf{E}^*\right) = \frac{\sigma\delta}{2}\Re\left[\tfrac{1}{2}\mathbf{E} \cdot \left(\frac{i-1}{\sigma\delta}\mathbf{n} \times \mathbf{H}\right)^*\right]$$

$$= \Re\left(\frac{1+i}{4}\mathbf{n} \cdot \mathbf{E} \times \mathbf{H}^*\right)$$

$$= \Re\left(\frac{1+i}{4}\frac{1-i}{\sigma\delta}|H_s|^2\right)$$

$$= \frac{|H_s|^2}{2\sigma\delta}.$$

49.2. According to (49.8), $\mathbf{j}$ has the z dependence $e^{-(1-i)z/\delta}$. Hence the integral is $\delta/(1-i)$ times the surface value of $\mathbf{j} = \sigma\mathbf{E}$. According to (49.10), this is

$$\frac{\sigma\delta}{1-i}\mathbf{E} = -\mathbf{n} \times \mathbf{H} = \mathbf{H} \times \mathbf{n}.$$

50.1. We integrate the equation

$$\mathrm{div}_2(f \, \mathbf{grad}_2 \, g) = f\Delta_2 g + \mathbf{grad}_2 \, f \cdot \mathbf{grad}_2 \, g$$

over the cross-section. By using Green's theorem the integral of the divergence on the left-hand side is transformed to one of $f \, \partial g/\partial n$ over the circumference, which vanishes because either $f = 0$ or $\partial g/\partial n = 0$ there. Thus

$$\int \mathbf{grad}_2 \, f \, \mathbf{grad}_2 \, g \, dA = -\int f\Delta_2 g \, dA. \tag{1}$$

We first show that characteristic values κ^2 of $\Delta_2 f = -\kappa^2 f$, whether with $f = 0$ (a TM characteristic mode) or with $\partial f/\partial n = 0$ (a TE characteristic mode) on the circumference, are all positive. For let f_n be a solution of $\Delta_2 f_n = -\kappa_n^2 f_n$. Then, according to (1),

$$-\int f_n^* \Delta_2 f_n \, dA = \int f_n^* \kappa_n^2 f_n \, dA$$

$$= \kappa_n^2 \int |f_n|^2 \, dA$$

$$= \int |\mathbf{grad}_2 \, f_n|^2 \, dA. \tag{2}$$

Now, since f_n is a characteristic mode, $\int |f_n|^2 \, dA$ must be positive. Hence

$$\kappa_n^2 = \frac{\int |\mathbf{grad}_2 \, f_n|^2 \, dA}{\int |f_n|^2 \, dA},$$

so that $\kappa_n^2 > 0$. Finally, let f_m and f_n be characteristic modes belonging, respectively, to κ_m^2 and κ_n^2. Then, according to (1),

$$
\begin{aligned}
-\int f_m^* \Delta_2 f_n \, dA = \kappa_n^2 \int f_m^* f_n \, dA = \int \mathbf{grad}_2 \, f_n \cdot \mathbf{grad}_2 \, f_m^* \, dA, \\
-\int f_n (\Delta_2 f_m)^* \, dA = \kappa_m^2 \int f_n f_m^* \, dA = \int \mathbf{grad}_2 \, f_n \cdot \mathbf{grad}_2 \, f_m^* \, dA,
\end{aligned}
\tag{3}
$$

where, in the last equation, we have used the reality of κ_m^2. From these equations we deduce that

$$
(\kappa_m^2 - \kappa_n^2) \int f_m^* f_n \, dA = 0.
$$

Thus, if the characteristic values are different, $\int f_m^* f_n \, dA = 0$, and it follows from either of (3) that

$$
\int \mathbf{grad}_2 \, f_n \cdot \mathbf{grad}_2 \, f_m^* \, dA = 0.
$$

50.2. We seek separable solutions $f(x)g(y)$ of the characteristic value problem:

$$
-\Delta_2[f(x)g(y)] = -f''(x)g(y) - f(x)g''(y) = \kappa^2 f(x)g(y),
$$

or

$$
-f''(x)/f(x) - g''(y)/g(y) = \kappa^2.
\tag{4}
$$

Now $f''(x)/f(x)$ can only depend on x. But according to the last equation it is equal to $\kappa^2 + g''(y)/g(y)$, which does *not* depend on x. Hence f''/f equals a constant $-\lambda$ (say). By the same argument g''/g equals a constant $-\mu$.

The solutions of $f''(x) = -\lambda f(x)$ are $\cos(\lambda^{1/2}x)$ and $\sin(\lambda^{1/2}x)$ if $\lambda > 0$; $\cosh(-\lambda^{1/2}x)$ and $\sinh(-\lambda^{1/2}x)$ if $\lambda < 0$; and 1 and x (a linear function of x) if $\lambda = 0$. For a TM mode $f(x)$ must vanish at both $x = 0$ and $x = a_x$, and the only non-trivial solution is $\sin(n_x \pi x/a_x)$, with n_x a non-zero integer. Similarly, $g(y)$ must be $\sin(n_y \pi y/a_y)$, with n_y a non-zero integer.

For a TE mode $f'(x)$ must vanish at both $x = 0$ and $x = a_x$, and the only non-trivial solution is $\cos(n_x \pi x/a_x)$, with n_x an integer (possibly zero). Similarly, $g(y)$ must be $\cos(n_y \pi y/a_y)$, with n_y an integer. Thus we arrive at the characteristic modes (50.17) and characteristic values (50.18):

$$
\kappa^2 = \pi^2 \left(\frac{n_x^2}{a_x^2} + \frac{n_y^2}{a_y^2} \right).
$$

For TM characteristic modes n_x and n_y must both be non-zero. The smallest characteristic value is then $\pi^2(a_x^{-2} + a_y^{-2})$. For TE characteristic modes n_x and n_y cannot both vanish, because κ must be positive. The smallest characteristic value is then π^2, divided by the larger of a_x^2 and a_y^2.

Since we have restricted our search for characteristic modes to separable ones, of the form $f(x)g(y)$, the question (loosely put) is, have we indeed found all the characteristic modes? The answer is, we have found a sufficient number of them, for the sets (50.17) are complete: any TM wave can be represented as a superposition of the characteristic modes (50.17)$_1$, and any TE wave as a superposition of the modes (50.17)$_2$. These superpositions are nothing other than Fourier series.

50.3. Since the unit normal into the waveguide wall has no z component, the tangential components of **H** are given by

$$\mathbf{n} \times \mathbf{H} = (n_y H_z, -n_x H_z, n_x H_y - n_y H_x).$$

For TEM waves, according to (50.2)$_{2,4}$ with $E_z = 0$,

$$H_x = -\frac{k_z}{\omega\mu_0}E_y, \qquad H_y = \frac{k_z}{\omega\mu_0}E_x, \qquad H_z = 0.$$

Since $\omega = ck_z$ (cf. (50.3)$_1$), we obtain

$$\mathbf{n} \times \mathbf{H} = (0, 0, -c\epsilon_0\mathbf{n} \cdot \mathbf{E}).$$

Hence $|H_s|^2 = c^2\epsilon_0^2|\mathbf{n} \cdot \mathbf{grad}_2\, V|^2$. But since V is constant on the circumference of the cross-section, $|\mathbf{n} \cdot \mathbf{grad}_2\, V|^2 = |\mathbf{grad}_2\, V|^2$. Thus

$$\frac{1}{2\sigma\delta}\oint |H_s|^2\, dl = \frac{1}{2\mu_0\sigma\delta}\oint \epsilon_0|\nabla_2 V|^2\, dl.$$

For TM waves, according to (50.7)$_{3,4}$,

$$H_x = -\frac{i\omega\epsilon_0}{\kappa^2}\frac{\partial E_z}{\partial y}, \qquad H_y = \frac{i\omega\epsilon_0}{\kappa^2}\frac{\partial E_z}{\partial x}, \qquad H_z = 0.$$

Hence

$$\mathbf{n} \times \mathbf{H} = (0, 0, (i\omega\epsilon_0/\kappa^2)\mathbf{n} \cdot \mathbf{grad}\, E_z).$$

Thus

$$\frac{1}{2\sigma\delta}|H_s|^2 = \frac{\omega^2}{2\mu_0\sigma\delta c^2\kappa^4}\epsilon_0|\mathbf{n} \cdot \mathbf{grad}\, E_z|^2.$$

But for TM waves $E_z = 0$ in the walls, so that $\mathbf{grad}\, E_z$ is parallel to **n**. Hence

$$\frac{1}{2\sigma\delta}\oint |H_s|^2\, dl = \frac{\omega^2}{2\mu_0\sigma\delta c^2\kappa^4}\oint \epsilon_0|\mathbf{grad}_2\, E_z|^2\, dl.$$

For TE waves, according to (50.9)$_{1,2}$,

$$H_x = \frac{ik_z}{\kappa^2}\frac{\partial H_z}{\partial x}, \qquad H_y = \frac{ik_z}{\kappa^2}\frac{\partial H_z}{\partial y}.$$

Hence

$$\mathbf{n} \times \mathbf{H} = [n_y H_z, -n_x H_z, (ik_z/\kappa^2)(n_x \partial H_z/\partial y - n_y \partial H_z/\partial x)].$$

Thus

$$|H_s|^2 = n_y^2 |H_z|^2 + n_x^2 |H_z|^2 + \frac{k_z^2}{\kappa^4} \left| n_x \frac{\partial H_z}{\partial y} - n_y \frac{\partial H_z}{\partial x} \right|^2$$

$$= |H_z|^2 + \frac{k_z^2}{\kappa^4} \left| n_x \frac{\partial H_z}{\partial y} - n_y \frac{\partial H_z}{\partial x} \right|^2.$$

We have

$$\left| n_x \frac{\partial H_z}{\partial y} - n_y \frac{\partial H_z}{\partial x} \right|^2$$

$$= n_x^2 \left| \frac{\partial H_z}{\partial y} \right|^2 + n_y^2 \left| \frac{\partial H_z}{\partial x} \right|^2 - n_x n_y \left(\frac{\partial H_z}{\partial y} \frac{\partial H_z^*}{\partial x} + \frac{\partial H_z}{\partial x} \frac{\partial H_z^*}{\partial y} \right). \tag{1}$$

Now, for TE waves,

$$\frac{\partial H_z}{\partial n} = n_x \frac{\partial H_z}{\partial x} + n_y \frac{\partial H_z}{\partial y} = 0$$

on the circumference of the cross-section. Hence

$$0 = \left| \frac{\partial H_z}{\partial n} \right|^2$$

$$= n_x^2 \left| \frac{\partial H_z}{\partial x} \right|^2 + n_y^2 \left| \frac{\partial H_z}{\partial y} \right|^2 + n_x n_y \left(\frac{\partial H_z}{\partial y} \frac{\partial H_z^*}{\partial x} + \frac{\partial H_z}{\partial x} \frac{\partial H_z^*}{\partial y} \right). \tag{2}$$

Taking the sum of (1) and (2), we obtain

$$\left| n_x \frac{\partial H_z}{\partial y} - n_y \frac{\partial H_z}{\partial x} \right|^2 = \left(n_x^2 + n_y^2 \right) |\operatorname{grad}_2 H_z|^2 = |\operatorname{grad}_2 H_z|^2.$$

Hence

$$\frac{1}{2\sigma\delta} \oint |H_s|^2 \, dl = \frac{1}{2\mu_0\sigma\delta} \oint \mu_0 \left(|H_z|^2 + \frac{k_z^2}{\kappa^4} |\nabla_2 H_z|^2 \right) dl.$$

Chapter 14

51.1. To the first order, which is all we need in order to calculate a derivative, we have

$$\delta \ln \det A = \ln \det(A + \delta A) - \ln \det A = \ln \frac{\det(A + \delta A)}{\det A}$$

$$= \ln \det A^{-1}(A + \delta A) = \ln \det(I + A^{-1}\delta A)$$

$$\approx \ln(1 + \operatorname{tr} A^{-1}\delta A) \approx \operatorname{tr} A^{-1}\delta A.$$

Thus

$$\frac{d}{dt} \ln \det A = \operatorname{tr} A^{-1} \dot{A}.$$

51.2. Differentiation of (51.4) gives

$$\dot{\rho} \det F + \rho \det F \operatorname{tr} F^{-1} \dot{F} = 0,$$

which may be divided by the non-zero $\det F$.

51.3. Substituting $\dot{x}_i$ in Euler's equation (51.7), we obtain

$$\ddot{x}_i = (\dot{x}_i)_t + \dot{x}_j \partial_j \dot{x}_i = (\dot{x}_i)_t + (\operatorname{grad} \dot{\mathbf{x}})_{ij} \dot{x}_j = [\dot{\mathbf{x}}_t + (\operatorname{grad} \dot{\mathbf{x}}) \dot{\mathbf{x}}]_i.$$

51.4. From the definition $F_{ij} = \partial x_i/(\partial X_j)$ of F we have

$$\dot{F}_{ij} = \frac{\partial^2 x_i}{\partial t \partial X_j} = \frac{\partial \dot{x}_i}{\partial X_j} = \frac{\partial \dot{x}_i}{\partial x_k} \frac{\partial x_k}{\partial X_j}.$$

In matrix notation, this equation reads

$$\dot{F} = (\operatorname{grad} \dot{\mathbf{x}}) F.$$

Hence $\dot{F} F^{-1} = \operatorname{grad} \dot{\mathbf{x}}$.

51.5. From Euler's relation $\dot{\rho} = \rho_t + \dot{\mathbf{x}} \cdot \operatorname{grad} \rho$. Hence

$$\dot{\rho} + \rho \operatorname{div} \dot{\mathbf{x}} = \rho_t + \dot{\mathbf{x}} \cdot \operatorname{grad} \rho + \rho \operatorname{div} \dot{\mathbf{x}} = \rho_t + \operatorname{div} \rho \dot{\mathbf{x}} = 0.$$

51.6. In $\dot{\mathbf{G}} = \mathbf{F}$ the right-hand side is the resultant *applied*, or external, force. Consider two bodies 1 and 2. Let $\mathbf{F}_{21}$ be the resultant force exerted by 2 on 1, and let $\mathbf{F}_1$ be the resultant of the other forces acting on 1. Then

$$\dot{\mathbf{G}}_1 = \mathbf{F}_{21} + \mathbf{F}_1.$$

Similarly, let $\mathbf{F}_{12}$ be the resultant force exerted by 1 on 2, and $\mathbf{F}_2$ the resultant of the other forces acting on 2. Then

$$\dot{\mathbf{G}}_2 = \mathbf{F}_{12} + \mathbf{F}_2.$$

Now the composite body $1 + 2$ has momentum $\mathbf{G}_{12} = \int \dot{\mathbf{x}} \, dm = \mathbf{G}_1 + \mathbf{G}_2$, and is subject to the applied force $\mathbf{F}_1 + \mathbf{F}_2$. Hence

$$\dot{\mathbf{G}}_{12} = \dot{\mathbf{G}}_1 + \dot{\mathbf{G}}_2 = \mathbf{F}_1 + \mathbf{F}_2.$$

It follows that $\mathbf{F}_{21} + \mathbf{F}_{12} = 0$, which is the law of action and reaction.

In stating his 'new principle of mechanics', $m\ddot{\mathbf{x}} = \mathbf{F}$, Euler remarked that it applied to any mass, whether finite or infinitesimal, and that it included 'all the laws of mechanics'.

51.7. Let $\mathbf{F}_{21}$ be the force exerted by particle 2 on particle 1, and let $\mathbf{F}_1$ be the resultant of the other forces acting on particle 1. Then $\dot{\mathbf{L}}_O = \mathbf{M}_O$ for particle 1 reads

$$\frac{d}{dt}[(\mathbf{x}_1 - \mathbf{x}_O) \times m_1\dot{\mathbf{x}}_1] = (\mathbf{x}_1 - \mathbf{x}_O) \times (\mathbf{F}_{21} + \mathbf{F}_1).$$

Similarly, let $\mathbf{F}_{12}$ be the force exerted by particle 1 on particle 2, and let $\mathbf{F}_2$ be the resultant of the other forces acting on particle 2. Then

$$\frac{d}{dt}[(\mathbf{x}_2 - \mathbf{x}_O) \times m_2\dot{\mathbf{x}}_2] = (\mathbf{x}_2 - \mathbf{x}_O) \times (\mathbf{F}_{12} + \mathbf{F}_2).$$

For the composite body consisting of both particles the applied moment is $(\mathbf{x}_1 - \mathbf{x}_O) \times \mathbf{F}_1 + (\mathbf{x}_2 - \mathbf{x}_O) \times \mathbf{F}_2$. Thus

$$\frac{d}{dt}[(\mathbf{x}_1 - \mathbf{x}_O) \times m_1\dot{\mathbf{x}}_1 + (\mathbf{x}_2 - \mathbf{x}_O) \times m_2\dot{\mathbf{x}}_2]$$
$$= (\mathbf{x}_1 - \mathbf{x}_O) \times \mathbf{F}_1 + (\mathbf{x}_2 - \mathbf{x}_O) \times \mathbf{F}_2.$$

It follows that

$$(\mathbf{x}_1 - \mathbf{x}_O) \times \mathbf{F}_{21} + (\mathbf{x}_2 - \mathbf{x}_O) \times \mathbf{F}_{12} = 0.$$

From the law of action and reaction $\mathbf{F}_{21} = -\mathbf{F}_{12}$. Hence

$$(\mathbf{x}_2 - \mathbf{x}_1) \times \mathbf{F}_{12} = 0,$$

that is, $\mathbf{F}_{12}$ is along the line joining 1 and 2 (and so is $\mathbf{F}_{21} = -\mathbf{F}_{12}$).

51.8. We have

$$\dot{\mathbf{L}}_O = \frac{d}{dt}\int (\mathbf{x} - \mathbf{x}_O) \times \dot{\mathbf{x}}\, dm$$

$$= \int (\mathbf{x} - \mathbf{x}_O) \times \ddot{\mathbf{x}}\, dm$$

$$= \int (\mathbf{x} - \mathbf{x}_O) \times (\operatorname{div} T + \rho\mathbf{b})\, dV,$$

where we have used Cauchy's first equation (51.18). According to Euler's second law $\dot{\mathbf{L}}_O = \mathbf{M}_O$

$$\int (\mathbf{x} - \mathbf{x}_O) \times (\operatorname{div} T + \rho\mathbf{b})\, dV$$

$$= \oint (\mathbf{x} - \mathbf{x}_O) \times T\mathbf{n}\, dS + \int (\mathbf{x} - \mathbf{x}_O) \times \mathbf{b}\, dm.$$

Thus

$$\int (\mathbf{x} - \mathbf{x}_O) \times \operatorname{div} T\, dV = \oint (\mathbf{x} - \mathbf{x}_O) \times T\mathbf{n}\, dS, \tag{1}$$

which holds for any body, or for any part of a body. The ith component of the left-hand side is

$$\epsilon_{ijk} \int (\mathbf{x} - \mathbf{x}_O)_j \frac{\partial T_{kl}}{\partial x_l} dV$$

$$= \epsilon_{ijk} \int \frac{\partial}{\partial x_l} [(\mathbf{x} - \mathbf{x}_O)_j T_{kl}] dV - \int \epsilon_{ijk} \delta_{jl} T_{kl} dV$$

$$= \epsilon_{ijk} \oint (\mathbf{x} - \mathbf{x}_O)_j T_{kl} n_l \, dS - \int \epsilon_{ijk} T_{kj} \, dV.$$

The last surface integral is the ith component of the integral on the right-hand side of (1). Hence $\int \epsilon_{ijk} T_{kj} \, dV = 0$, and since this holds for any part of a body—however small—T must be symmetric.

Chapter 15

55.1.

$$\mathbf{b} \cdot \overset{*}{\mathbf{a}} = \mathbf{b} \cdot [\dot{\mathbf{a}} + \mathbf{a} \, \text{div} \, \dot{\mathbf{x}} - (\mathbf{a} \cdot \mathbf{grad}) \dot{\mathbf{x}}]$$

$$= \mathbf{b} \cdot \dot{\mathbf{a}} + (\mathbf{b} \cdot \mathbf{a}) I \cdot \text{grad} \, \dot{\mathbf{x}} - b_i a_j \frac{\partial \dot{x}_i}{\partial x_j}$$

$$= \mathbf{b} \cdot \dot{\mathbf{a}} + (\mathbf{b} \cdot \mathbf{a}) I \cdot \text{grad} \, \dot{\mathbf{x}} - (\mathbf{b} \otimes \mathbf{a})_{ij} (\text{grad} \, \dot{\mathbf{x}})_{ij}$$

$$= \mathbf{b} \cdot \dot{\mathbf{a}} + [(\mathbf{b} \cdot \mathbf{a}) I - \mathbf{b} \otimes \mathbf{a}] \cdot \text{grad} \, \dot{\mathbf{x}}.$$

55.2. Both sides of the equation are matrices. The $_{12}$ element of the left-hand side is

$$(a_2 b_3 - a_3 b_2) c_2 + (b_2 c_3 - b_3 c_2) a_2 + (c_2 a_3 - c_3 a_2) b_2 = 0$$

and so is I_{12}. All other non-diagonal terms can be checked in the same way. The $_{11}$ element of the left-hand side is

$$(a_2 b_3 - a_3 b_2) c_1 + (b_2 c_3 - b_3 c_2) a_1 + (c_2 a_3 - c_3 a_2) b_1,$$

and this equals $(\mathbf{a} \times \mathbf{b} \cdot \mathbf{c}) I_{11}$. The two other diagonal terms can be checked in the same way.

55.3. According to the formula of Exercise 55.1 we have

$$- \text{div} \, \mathcal{E} \times \mathcal{H} = \mathcal{J} \cdot \mathcal{E} + \mathcal{E} \cdot \overset{*}{\mathbf{D}} + \mathcal{H} \cdot \overset{*}{\mathbf{B}}$$

$$= \mathcal{J} \cdot \mathcal{E} + \mathcal{E} \cdot \dot{\mathbf{D}} + \mathcal{H} \cdot \dot{\mathbf{B}}$$

$$+ [(\mathcal{E} \cdot \mathbf{D} + \mathcal{H} \cdot \mathbf{B}) I - \mathcal{E} \otimes \mathbf{D} - \mathcal{H} \otimes \mathbf{B}] \cdot \text{grad} \, \dot{\mathbf{x}}. \qquad (1)$$

From the aether relation $\mathbf{D} = \epsilon_0 \mathbf{E} + \mathbf{P}$ and $\mathcal{E} = \mathbf{E} + \dot{\mathbf{x}} \times \mathbf{B}$ we obtain

$$\mathcal{E} \cdot \mathbf{D} = \mathcal{E} \cdot (\epsilon_0 \mathbf{E} + \mathbf{P}) = (\mathbf{E} + \dot{\mathbf{x}} \times \mathbf{B}) \cdot \epsilon_0 \mathbf{E} + \mathcal{E} \cdot \mathbf{P}$$

$$= \mathcal{E} \cdot \mathbf{P} + \epsilon_0 E^2 - \epsilon_0 \mathbf{E} \times \mathbf{B} \cdot \dot{\mathbf{x}}.$$

Similarly

$$\mathcal{E} \cdot \dot{\mathbf{D}} = \mathcal{E} \cdot (\epsilon_0 \dot{\mathbf{E}} + \dot{\mathbf{P}}) = (\mathbf{E} + \dot{\mathbf{x}} \times \mathbf{B}) \cdot \epsilon_0 \dot{\mathbf{E}} + \mathcal{E} \cdot \dot{\mathbf{P}}$$
$$= \mathcal{E} \cdot \dot{\mathbf{P}} + \left(\tfrac{1}{2}\epsilon_0 E^2\right)^{\cdot} - \epsilon_0 \dot{\mathbf{E}} \times \mathbf{B} \cdot \dot{\mathbf{x}}$$
$$= (\mathcal{E} \cdot \mathbf{P})^{\cdot} - \mathbf{P} \cdot \dot{\mathcal{E}} + \left(\tfrac{1}{2}\epsilon_0 E^2\right)^{\cdot} - \epsilon_0 \dot{\mathbf{E}} \times \mathbf{B} \cdot \dot{\mathbf{x}}$$

and

$$\mathcal{E} \otimes \mathbf{D} = \mathcal{E} \otimes (\epsilon_0 \mathbf{E} + \mathbf{P}) = \mathcal{E} \otimes \mathbf{P} + \epsilon_0 \mathbf{E} \otimes \mathbf{E} - \epsilon_0 \mathbf{B} \times \dot{\mathbf{x}} \otimes \mathbf{E}.$$

From $\mathcal{H} = \mathbf{B}/\mu_0 - \dot{\mathbf{x}} \times \epsilon_0 \mathbf{E} - \mathcal{M}$ we obtain

$$\mathcal{H} \cdot \mathbf{B} = \left(\frac{\mathbf{B}}{\mu_0} - \dot{\mathbf{x}} \times \epsilon_0 \mathbf{E} - \mathcal{M}\right) \cdot \mathbf{B}$$
$$= \frac{B^2}{\mu_0} - \epsilon_0 \mathbf{E} \times \mathbf{B} \cdot \dot{\mathbf{x}} - \mathcal{M} \cdot \mathbf{B}.$$

Similarly

$$\mathcal{H} \cdot \dot{\mathbf{B}} = \left(\frac{\mathbf{B}}{\mu_0} - \dot{\mathbf{x}} \times \epsilon_0 \mathbf{E} - \mathcal{M}\right) \cdot \dot{\mathbf{B}}$$
$$= \left(\frac{B^2}{2\mu_0}\right)^{\cdot} - \epsilon_0 \mathbf{E} \times \dot{\mathbf{B}} \cdot \dot{\mathbf{x}} - \mathcal{M} \cdot \dot{\mathbf{B}}$$

and

$$\mathcal{H} \otimes \mathbf{B} = \left(\frac{\mathbf{B}}{\mu_0} - \dot{\mathbf{x}} \times \epsilon_0 \mathbf{E} - \mathcal{M}\right) \otimes \mathbf{B}$$
$$= \frac{\mathbf{B} \otimes \mathbf{B}}{\mu_0} - \epsilon_0 \dot{\mathbf{x}} \times \mathbf{E} \otimes \mathbf{B} - \mathcal{M} \otimes \mathbf{B}.$$

Substituting the last six relations in (1), noting that

$$(\mathbf{E} \times \mathbf{B} \cdot \dot{\mathbf{x}})^{\cdot} = \dot{\mathbf{E}} \times \mathbf{B} \cdot \dot{\mathbf{x}} + \mathbf{E} \times \dot{\mathbf{B}} \cdot \dot{\mathbf{x}} + \mathbf{E} \times \mathbf{B} \cdot \ddot{\mathbf{x}}$$

and that, according to Exercise 55.2,

$$\mathbf{E} \times \mathbf{B} \otimes \dot{\mathbf{x}} + \dot{\mathbf{x}} \times \mathbf{E} \otimes \mathbf{B} + \mathbf{B} \times \dot{\mathbf{x}} \otimes \mathbf{E} = (\mathbf{E} \times \mathbf{B} \cdot \dot{\mathbf{x}})I,$$

we obtain the desired formula.

56.1. Since any transformation can be constructed from a succession of infinitesimal transformations it is sufficient to consider orthogonal matrices of the form $Q = I + A$, where A is infinitesimal. Working to the first order in A we have

$$I = Q^T Q = (I + A)^T (I + A) = I + A + A^T.$$

Thus A must be antisymmetric. We can write any antisymmetric matrix in the form

$$A_{ij} = -\epsilon_{ijk}\Omega_k$$

where Ω is a vector. We then have

$$(Q\mathbf{a})_i = Q_{ij}a_j = (I + A)_{ij}a_j = a_i - \epsilon_{ijk}\Omega_k a_j,$$

which is the ith component of $\mathbf{a} + \Omega \times \mathbf{a}$. We therefore require that, to the first order in Ω,

$$f(\mathbf{a} + \Omega \times \mathbf{a}) = f(\mathbf{a}).$$

The derivative of f with respect to any component of Ω must therefore vanish for $\Omega = 0$. This gives

$$a_i\frac{\partial f}{\partial a_j} - a_j\frac{\partial f}{\partial a_i} = 0$$

for all i and j.

56.2. Since $\mathrm{div}_i A = \partial_j A_{ij}$,

$$
\begin{aligned}
\mathrm{div}_i] \left(\mathbf{E} \otimes \mathbf{E} - \tfrac{1}{2}E^2 I\right) &= \partial_j\left(E_i E_j - \tfrac{1}{2}E^2\delta_{ij}\right) \\
&= E_i\partial_j E_j + E_j\partial_j E_i - \tfrac{1}{2}\partial_i(E_k E_k) \\
&= E_i\partial_j E_j + E_j(\partial_j E_i - \partial_i E_j).
\end{aligned}
$$

This is the ith component of $\mathbf{E}\,\mathrm{div}\,\mathbf{E} - \mathbf{E} \times \mathbf{curl}\,\mathbf{E}$.

56.3. Calculating as in the previous exercise, and proceeding to use the aether relations and Maxwell's equations, we obtain

$$
\begin{aligned}
\mathrm{div}[(\mathcal{M} \cdot \mathbf{B})I &- \mathcal{M} \otimes \mathbf{B}] \\
&= \mathcal{M} \times \mathbf{curl}\,\mathbf{B} - \mathcal{M}\,\mathrm{div}\,\mathbf{B} + (\mathcal{M} \cdot \mathbf{grad})\mathbf{B} + \mathbf{B} \times \mathbf{curl}\,\mathcal{M} \\
&= (\mathcal{M} \cdot \mathbf{grad})\mathbf{B} + \mathcal{M} \times \mathbf{curl}\,\mathbf{B} + \mathbf{B} \times \mathbf{curl}\,(\mathbf{B}/\mu_0 - \mathbf{H} + \dot{\mathbf{x}} \times \mathbf{P}) \\
&= (\mathcal{M} \cdot \mathbf{grad})\mathbf{B} + \mathcal{M} \times \mathbf{curl}\,\mathbf{B} + \mathbf{B} \times \mathbf{curl}\,\mathbf{B}/\mu_0 + (\mathbf{j} + \mathbf{D}_t - \mathbf{curl}\,\dot{\mathbf{x}} \times \mathbf{P}) \times \mathbf{B},
\end{aligned}
$$
$$\tag{1}$$

$$
\begin{aligned}
\mathrm{div}\,\mathcal{E} \otimes \mathbf{P} &= \mathcal{E}\,\mathrm{div}\,\mathbf{P} + (\mathbf{P} \cdot \mathbf{grad})\mathcal{E} \\
&= (\mathbf{P} \cdot \mathbf{grad})\mathcal{E} + (\mathbf{E} + \dot{\mathbf{x}} \times \mathbf{B})\,\mathrm{div}\,\mathbf{P} \\
&= (\mathbf{P} \cdot \mathbf{grad})\mathcal{E} + \mathbf{E}\,\mathrm{div}(\mathbf{D} - \epsilon_0\mathbf{E}) + \dot{\mathbf{x}} \times \mathbf{B}\,\mathrm{div}\,\mathbf{P} \\
&= (\mathbf{P} \cdot \mathbf{grad})\mathcal{E} + q\mathbf{E} - \epsilon_0\mathbf{E}\,\mathrm{div}\,\mathbf{E} + \dot{\mathbf{x}}\,\mathrm{div}\,\mathbf{P} \times \mathbf{B},
\end{aligned}
$$
$$\tag{2}$$

$$
\begin{aligned}
\mathrm{div}\,\mathbf{E} \times \mathbf{B} \otimes \dot{\mathbf{x}} &= \mathbf{E} \times \mathbf{B}\,\mathrm{div}\,\dot{\mathbf{x}} + (\dot{\mathbf{x}} \cdot \mathbf{grad})(\mathbf{E} \times \mathbf{B}) \\
&= \rho(\mathbf{E} \times \mathbf{B}/\rho)^{\cdot} - (\mathbf{E} \times \mathbf{B})_t \\
&= \rho(\mathbf{E} \times \mathbf{B}/\rho)^{\cdot} + \mathbf{E} \times \mathbf{curl}\,\mathbf{E} - \mathbf{E}_t \times \mathbf{B}.
\end{aligned}
$$

These equations, together with $\overset{*}{\mathbf{P}} = \mathbf{P}_t + \dot{\mathbf{x}}\,\mathrm{div}\,\mathbf{P} - \mathbf{curl}\,\dot{\mathbf{x}} \times \mathbf{P}$, yield (56.15).

56.4.

$$qE + j \times B = q(\mathcal{E} - \dot{x} \times B) + (\mathcal{J} + q\dot{x}) \times B = q\mathcal{E} + \mathcal{J} \times B.$$

56.5. According to (56.8), with $p = \mathbf{P} = \mathcal{M} = \dot{x} = 0$, the electromagnetic part of the stress is

$$T\mathbf{n} = -\left[\tfrac{1}{2}\epsilon_0 E^2 + B^2/(2\mu_0)\right]\mathbf{n} + ED_n + BH_n.$$

But in this formula $\mathbf{n}$ points out of the material. In (49.3) we have chosen $\mathbf{n}$ to point *into* the perfect conductor. With this choice for $\mathbf{n}$ the stress is

$$T\mathbf{n} = \tfrac{1}{2}(E_n D_n + B_s H_s)\mathbf{n} - ED_n = \tfrac{1}{2}(\mathbf{n}B_s H_s - ED_n),$$

where use has been made of (49.2) in writing $\epsilon_0 E^2 = \epsilon_0(E_n^2 + E_s^2) = E_n D_n$ and $B^2/\mu_0 = (B_n^2 + B_s^2)/\mu_0 = B_s H_s$, and in setting $\mathbf{n}E_n = \mathbf{E}$. Now, according to (49.2)–(49.3),

$$-ED_n = \tau\mathbf{E},$$
$$\mathbf{K} \times \mathbf{B} = (-\mathbf{n} \times \mathbf{H}) \times \mathbf{B} = \mathbf{n}(\mathbf{B} \cdot \mathbf{H}) - \mathbf{H}B_n = \mathbf{n}B_s H_s.$$

Thus we obtain for the stress

$$T\mathbf{n} = \tfrac{1}{2}(\tau\mathbf{E} + \mathbf{K} \times \mathbf{B}).$$

Since $\mathbf{n}$ is now directed into the perfect conductor, the first term, $-\tfrac{1}{2}D_n\mathbf{E} = -\tfrac{1}{2}D_n E_n\mathbf{n}$, is a tension, and the second term, $\tfrac{1}{2}B_s H_s\mathbf{n}$, a pressure.

56.6. After substitution of $\mathbf{b}$ from (47.9) into (47.11), the energy balance law takes the form (48.1):

$$\rho\dot{\varepsilon} = T \cdot \operatorname{grad}\dot{x} + \rho(\dot{g} \cdot \dot{x} + h) - \operatorname{div}(\mathbf{q} + \mathcal{E} \times \mathcal{H}).$$

In this equation we substitute ε from

$$\varepsilon = \varphi + \vartheta s + \mathbf{g} \cdot \dot{x} - \tfrac{1}{2}\dot{x}^2$$
$$+ [\epsilon_0 E^2/2 + B^2/(2\mu_0) + \mathcal{E} \cdot \mathbf{P} - \epsilon_0\mathbf{E} \times \mathbf{B} \cdot \dot{x}]/\rho,$$

(cf. (55.7)). Using

$$\rho\dot{\varphi} = \rho(\varphi_\rho\dot{\rho} + \varphi_\vartheta\dot{\vartheta} + \varphi_{\mathcal{E}} \cdot \dot{\mathcal{E}} + \varphi_{\mathbf{B}} \cdot \dot{\mathbf{B}})$$
$$= -p\operatorname{div}\dot{x} - \rho s\dot{\vartheta} - \mathbf{P} \cdot \dot{\mathcal{E}} - \mathcal{M} \cdot \dot{\mathbf{B}}$$

and equation (1) of Exercise 55.3, we arrive at the simple form of the energy balance equation,

$$\rho\vartheta\dot{s} = \mathcal{J} \cdot \mathcal{E} + \rho h - \operatorname{div}\mathbf{q}.$$

58.1. From the definition $A \cdot B = \operatorname{tr} AB^T$ of the matrix scalar product and the equation grad $\dot{\mathbf{x}} = \dot{F}F^{-1}$ (cf. (51.10)) we obtain

$$\tau \cdot \operatorname{grad} \dot{\mathbf{x}} = \operatorname{tr} \tau (\dot{F}F^{-1})^T = \operatorname{tr} \tau (F^T)^{-1}\dot{F}^T = \tau (F^T)^{-1} \cdot \dot{F}.$$

59.1.

$$(\mathbf{div}\, d)_i = \partial_j d_{ij} = \partial_j \tfrac{1}{2}(\partial_j \dot{x}_i + \partial_i \dot{x}_j) = \tfrac{1}{2}(\nabla^2 \dot{x}_i + \partial_i \operatorname{div} \dot{\mathbf{x}}).$$

But $\nabla^2 = \mathbf{grad}\operatorname{div} - \mathbf{curl}^2$. Hence

$$\mathbf{div}\, 2\lambda d = 2\lambda \operatorname{\mathbf{grad}}\operatorname{div} \dot{\mathbf{x}} - \lambda \operatorname{\mathbf{curl}}^2 \dot{\mathbf{x}}.$$

To this we add $\mathbf{div}\,(\zeta \operatorname{div} \dot{\mathbf{x}})I = \zeta \operatorname{\mathbf{grad}}\operatorname{div} \dot{\mathbf{x}}$.

59.2. Since the new, viscous terms in T are $2\lambda d + \zeta(\operatorname{div} \dot{\mathbf{x}})I$ (cf. (59.3)) and T enters the balance of energy (55.1) in the form $T \cdot \operatorname{grad} \dot{\mathbf{x}}$, the added terms are

$$[2\lambda d + \zeta(\operatorname{div} \dot{\mathbf{x}})I] \cdot \operatorname{grad} \dot{\mathbf{x}}$$
$$= [2\lambda d + \zeta(\operatorname{div} \dot{\mathbf{x}})I] \cdot \left(d + \tfrac{1}{2}[\operatorname{grad} \dot{\mathbf{x}} - (\operatorname{grad} \dot{\mathbf{x}})^T] \right).$$

The scalar product of a symmetric matrix and an antisymmetric one vanishes. Furthermore $\operatorname{div} \dot{\mathbf{x}} = I \cdot d = \operatorname{tr} d$. Hence

$$[2\lambda d + \zeta(\operatorname{div} \dot{\mathbf{x}})I] \cdot \operatorname{grad} \dot{\mathbf{x}} = 2\lambda(d \cdot d) + \zeta(\operatorname{tr} d)^2.$$

Chapter 16

61.1. The ith component of $(\operatorname{grad} \dot{\mathbf{x}})\dot{\mathbf{x}}$ is

$$(\operatorname{grad} \dot{\mathbf{x}})_{ij}\dot{x}_j = \dot{x}_j \partial_j \dot{x}_i = \dot{x}_j(\partial_j \dot{x}_i - \partial_i \dot{x}_j) + \partial_i \tfrac{1}{2}(\dot{x}_j \dot{x}_j).$$

This is the ith component of $(\operatorname{\mathbf{curl}} \dot{\mathbf{x}}) \times \dot{\mathbf{x}} + \operatorname{\mathbf{grad}}(\tfrac{1}{2}\dot{x}^2)$.

63.1. Omitting the viscous terms, the equation of motion in the form (61.7) is

$$\rho\left[\dot{\mathbf{x}}_t + (\operatorname{\mathbf{curl}} \dot{\mathbf{x}}) \times \dot{\mathbf{x}} + \operatorname{\mathbf{grad}} \tfrac{1}{2}\dot{x}^2\right] = -\operatorname{\mathbf{grad}} p + (\operatorname{\mathbf{curl}} \mathbf{B}) \times \mathbf{B}/\mu_0 + \rho\mathbf{b}.$$

Either of the motions $\dot{\mathbf{x}} = \pm\mathbf{B}/\sqrt{\mu_0\rho}$ will satisfy the requirement $\operatorname{div} \dot{\mathbf{x}} = 0$ of incompressibility; moreover, the terms involving the curls in the equation of motion will cancel. Since the fluid is perfectly conducting, the electric field set up will be $\mathbf{E} = -\dot{\mathbf{x}} \times \mathbf{B} = 0$, because these motions are everywhere parallel to the magnetic field. Thus $\mathbf{B}_t = -\operatorname{\mathbf{curl}} \mathbf{E}$ will vanish, and so will $\dot{\mathbf{x}}_t$. All that survives of the equation of motion is

$$0 = -\operatorname{\mathbf{grad}}\left(p + \tfrac{1}{2}\dot{x}^2 + \rho U\right)$$

(since $\rho\mathbf{b} = -\operatorname{\mathbf{grad}} \rho U$), and this is satisfied by choosing

$$p = \text{const.} - \tfrac{1}{2}\dot{x}^2 - \rho U.$$

63.2. Under the Galilean transformation $\mathbf{x}' = \mathbf{x} - \mathbf{u}t$ (and $t' = t$) the magnetic field and velocity transformation laws are $\mathbf{B}' = \mathbf{B}$ and $\mathbf{v}' = \mathbf{v} - \mathbf{u}$. If we express $\mathbf{B}'$ and $\mathbf{v}'$ in terms of the primed variables, we obtain

$$\mathbf{B}' = \mathbf{B}(\mathbf{x}) = \mathbf{B}(\mathbf{x}' + \mathbf{u}t),$$

$$\mathbf{v}' = \pm \frac{\mathbf{B}(\mathbf{x})}{\sqrt{\mu_0 \rho}} - \mathbf{u} = \pm \frac{\mathbf{B}(\mathbf{x}' + \mathbf{u}t) - \mathbf{B}_\infty}{\sqrt{\mu_0 \rho}}.$$

The wave velocity becomes zero at infinity.

64.1. Since $j_\phi = 0$, $B_z = 0$ according to $(64.6)_1$. Ohm's law $\mathbf{j} = \sigma \mathbf{E}$ results in $\mathbf{curl} \, \mathbf{j} = 0$, since $\mathbf{curl} \, \mathbf{E} = -\mathbf{B}_t = 0$. But this has the consequence that j_z is uniform over the cross section of the cylinder. Thus $i(r) = \pi r^2 j_z$, $B_\phi = \frac{1}{2}\mu_0 j_z r$, and the pinch equation becomes

$$\frac{dp}{dr} + \frac{1}{2}\mu_0 j_z^2 r = 0.$$

Hence

$$p + \frac{1}{4}\mu_0 j_z^2 r^2 = p + \frac{B_\phi^2}{\mu_0} = \text{const.}$$

64.2. If $j_z = 0$ then, according to $(64.6)_2$, B_ϕ must vanish; otherwise it would be singular on the axis. The pinch equation is

$$\frac{dp}{dr} + \frac{1}{2\mu_0} \frac{dB_z^2}{dr} = 0,$$

which integrates to

$$p + \frac{B_z^2}{2\mu_0} = \frac{B_0^2}{2\mu_0},$$

where B_0 is the field outside the cylinder.

65.1. Let $\mathbf{u}$ be the velocity of the plate and $\mathbf{n}$ the unit normal pointing into the plate. Since the fluid is inviscid there is no *a priori* restriction on its velocity component $\mathbf{n} \times \mathbf{v}$ parallel to the plate, but v_n must equal u_n because the plate is rigid. If the plate is perfectly conducting, however, its electric field is $\mathbf{E}_p = -\mathbf{u} \times \mathbf{B}_p$, just as in the fluid $\mathbf{E} = -\mathbf{v} \times \mathbf{B}$, and the jump condition $\mathbf{n} \times [\![\mathbf{E}]\!] - u_n [\![\mathbf{B}]\!] = 0$ reads

$$\mathbf{n} \times (\mathbf{E}_p - \mathbf{E}) - u_n(\mathbf{B}_p - \mathbf{B})$$
$$= -\mathbf{n} \times (\mathbf{u} \times \mathbf{B}_p - \mathbf{v} \times \mathbf{B}) - u_n(\mathbf{B}_p - \mathbf{B})$$
$$= -\mathbf{u}(\mathbf{n} \cdot \mathbf{B}_p) + \mathbf{B}_p u_n + \mathbf{v}(\mathbf{n} \cdot \mathbf{B}) - \mathbf{B}v_n - u_n(\mathbf{B}_p - \mathbf{B})$$
$$= (\mathbf{n} \cdot \mathbf{B})(\mathbf{v} - \mathbf{u}) = 0, \tag{1}$$

where use has been made of $v_n = u_n$ and $\mathbf{n} \cdot [\![\mathbf{B}]\!] = \mathbf{n} \cdot (\mathbf{B}_p - \mathbf{B}) = 0$. Since the plate is not parallel to $\mathbf{B}$, $\mathbf{n} \cdot \mathbf{B} \neq 0$, and we conclude that $\mathbf{v} = \mathbf{u}$, that is, equality of all components, not just the normal one.

The foregoing conclusion, that a non-zero $\mathbf{n \cdot B}$ forbids any slippage of a perfectly conducting fluid past a perfectly conducting rigid body, can be directly deduced from the fact that there is magnetic flux freezing on both sides.

Equations (65.12) are of the form $(A - \omega I)\mathbf{v} = 0$, with a symmetric A. Hence the characteristic vectors of the fast and slow modes are orthogonal. We obtain each mode alone by substituting the appropriate characteristic value ω_f or ω_s in (65.12). Let θ be the angle between $\mathbf{k}$ and $\mathbf{B}$. Since $\mathbf{B}$ is in the x direction, $k_x = k \cos \theta$ and $k_y = k \sin \theta$. The characteristic equation is

$$\omega^4 - (u_0^2 + u_A^2)k^2\omega^2 + u_0^2 u_A^2 k^4 \cos^2 \theta = 0.$$

In our fluid $u_0\sqrt{\mu_0\rho} = 2B$ or $u_0^2 = 4u_A^2$, and the characteristic equation becomes

$$\omega^4 - 5u_A^2 k^2\omega^2 + 4u_A^4 k^4 \cos^2 \theta = 0,$$

which has the fast and slow roots

$$\omega^2 = \tfrac{1}{2}k^2 \left[5u_A^2 \pm \sqrt{(25u_A^4 - 16u_A^4 \cos^2 \theta)} \right]$$

$$= \tfrac{5}{2}u_A^2 k^2 \left[1 \pm \sqrt{1 - \left(\tfrac{4}{5} \cos \theta\right)^2} \right].$$

The characteristic values are precisely those for which eqns (65.12) become dependent. For the modes, essentially the ratios u_y/u_x, we therefore use only one of the equations (65.12). Thus, for exciting a fast or slow magnetoacoustic mode, with a direction of propagation at an angle θ from $\mathbf{B}$, the plate velocity $\mathbf{u} = (u_x, u_y)$ must be such that

$$\frac{u_y}{u_x} = \frac{\tfrac{5}{8}\left[1 \pm \sqrt{1 - \left(\tfrac{4}{5}\cos \theta\right)^2}\right] - \cos^2 \theta}{\sin \theta \cos \theta}, \tag{2}$$

the plus sign referring to the fast mode, and the minus sign to the slow one.

If the plate is non-conducting the jump condition $\mathbf{n} \times [\![\mathbf{E}]\!] - u_n[\![\mathbf{B}]\!] = 0$ reads

$$\mathbf{n} \times \mathbf{E}_p - u_n \mathbf{B}_p = -(\mathbf{n} \cdot \mathbf{B})\mathbf{v},$$

instead of (1). It restricts $\mathbf{E}_p$ and $\mathbf{B}_p$, but does not prevent slippage of the fluid along the plate. In fact, vibration of the plate in its own plane cannot excite any motion in the inviscid fluid. It may create oscillating surface charge densities $\mathbf{n} \cdot [\![\mathbf{D}]\!]$ and current densities $\mathbf{n} \times [\![\mathbf{H}]\!]$ at the boundary, and excite electromagnetic waves (Chapter 12), but these are suppressed by the magnetohydrodynamic approximation.

If a non-conducting plate vibrates normal to its own plane, that is, $\mathbf{u}$ parallel to $\mathbf{n}$, it can still excite magnetoacoustic waves, but only with propagation directions that are given by (2), with $u_y/u_x = n_y/n_x = \pm 1$ (the sign depending on whether the angle between $\mathbf{n}$ and $\mathbf{B}$ is, respectively, 135° or 45°).

Chapter 17

66.1. Here $\mathbf{D} = Q/(4\pi r^2)\mathbf{n}$ and $\mathbf{E} = \mathbf{D}/\epsilon$ are radial. On the surface of the sphere the stress

$$\mathbf{t} = ED_n - \tfrac{1}{2}\mathbf{E} \cdot \mathbf{D} = \tfrac{1}{2}ED_n = \tfrac{1}{2}\frac{Q}{4\pi\epsilon a^2}\frac{Q}{4\pi a^2}\mathbf{n}$$

is radial too. In order to get the force we integrate the component $t \cos \theta$ over a half-sphere:

$$\int_0^{\frac{\pi}{2}} t \cos \theta \, 2\pi a^2 \sin \theta \, d\theta = t \frac{2\pi a^2}{2} = \frac{Q^2}{32\pi \epsilon a^2}.$$

66.2. The height to which the liquid rises is given by (66.13):

$$\rho g z(r) = \tfrac{1}{2} (\epsilon - \epsilon_0) E^2(r).$$

Multiplying by $2\pi r$ and integrating, we obtain

$$\rho g h (b^2 - a^2) = (\epsilon - \epsilon_0) \int_a^b E^2(r) r \, dr.$$

Now in a cylindrical capacitor $E(r) \propto r^{-1}$ and $V = \int_a^b E(r) \, dr$. Hence

$$\int_a^b E^2(r) r \, dr = \frac{V^2}{\ln (b/a)}.$$

67.1. (a) Since $\mathbf{P}$ is uniform, the bound charge density $- \operatorname{div} \mathbf{P}$ vanishes. But there are uniform bound surface charge densities $P_n = \pm P$ on the slab faces. Thus the electric field is like that of a parallel plate capacitor. Except for edge effects—which disappear when the slab is infinite—$\mathbf{E}$ will vanish outside. So will $\mathbf{D}$. From $\mathbf{n} \cdot [\![\mathbf{D}]\!] = 0$ it follows that $\mathbf{D} = 0$ inside the slab as well. Thus inside the slab $\epsilon_0 \mathbf{E} + \mathbf{P} = 0$ and $N = 1$. In case (b) $\mathbf{E}$ must vanish (and therefore $\mathbf{D} = \mathbf{P}$), otherwise there will be a non-vanishing field outside, in accordance with $\mathbf{n} \times [\![\mathbf{E}]\!] = 0$. Thus $N = 0$. Case (c) is similar to case (b), and $N = 0$ again. Case (d) is the two-dimensional analogue of the sphere discussed in Section 67. We try a two-dimensional dipole field (cf. Problem 29.1) outside the cylinder,

$$\epsilon_0 \mathbf{E}_e = \alpha \frac{a^2}{r^2} [2(\mathbf{n} \cdot \mathbf{P})\mathbf{n} - \mathbf{P}],$$

and a uniform field

$$\epsilon_0 \mathbf{E} = \beta \mathbf{P}$$

inside. The jump condition $\mathbf{n} \times [\![\mathbf{E}]\!] = 0$ gives $\beta = -\alpha$, and the jump condition $\mathbf{n} \cdot [\![\mathbf{D}]\!] = 0$, $\beta + 1 = \alpha$. Thus $\alpha = \tfrac{1}{2}$ and $\beta = -\tfrac{1}{2}$. Hence $N = -\beta = \tfrac{1}{2}$.

68.1. Let a be the radius of the sphere. We attempt a solution of the form

$$\mathbf{D}_e = \alpha \left(\frac{a}{r}\right)^3 [3(\mathbf{n} \cdot \mathbf{P}_0)\mathbf{P}_0 - \mathbf{P}_0]$$

$$\mathbf{D} = \beta \mathbf{P}_0$$

for the electric displacement outside and inside the sphere. The jump condition $\mathbf{n} \cdot [\![\mathbf{D}]\!] = 0$ yields the relation $2\alpha = \beta$. The electric field inside the sphere is determined by the aether relation $\mathbf{D} = \epsilon_0 \mathbf{E} + \mathbf{P} = \epsilon \mathbf{E} + \mathbf{P}_0$. Thus $\mathbf{E} = (\mathbf{D} - \mathbf{P}_0)/\epsilon$. Outside we have $\mathbf{E}_e = \mathbf{D}_e/\epsilon_0$. The jump condition $\mathbf{n} \times [\![\mathbf{E}]\!] = 0$ yields $-\alpha/\epsilon_0 = (\beta - 1)/\epsilon$. After solving for α and β we obtain

$$\beta = \frac{2\epsilon_0}{2\epsilon_0 + \epsilon}, \qquad \mathbf{E} = -\frac{\mathbf{P}_0}{2\epsilon_0 + \epsilon}.$$

Chapter 18

70.1. (a) Since $\mathbf{n} \cdot [\![\mathbf{B}]\!] = 0$, $\mathbf{B}$ must vanish inside, otherwise it will not vanish outside. Thus $\mathbf{H} = -\mathcal{M}$ and $N = 1$. (b) In this case, since $\mathbf{n} \times [\![\mathbf{H}]\!] = 0$, $\mathbf{H}$ must vanish (but there will be a magnetic field $\mathbf{B} = \mu_0 \mathcal{M}$) in the slab. Thus $N = 0$. Case (c) is similar to case (b). In case (d) we try a two-dimensional dipole field outside the cylinder, and a uniform field inside (Exercise 67.1):

$$\mathbf{H}_e = \alpha \left(\frac{a}{r}\right)^2 [2(\mathbf{n} \cdot \mathcal{M})\mathbf{n} - \mathcal{M}],$$
$$\mathbf{H} = \beta \mathcal{M}.$$

From $\mathbf{n} \times [\![\mathbf{H}]\!] = 0$ we obtain $-\alpha = \beta$, and from $\mathbf{n} \cdot [\![\mathbf{B}]\!] = 0$, $\alpha = 1 + \beta$. The solutions are $\alpha = \frac{1}{2}$ and $\beta = -\frac{1}{2}$. Thus $N = -\beta = \frac{1}{2}$. (e) This problem is three-dimensional. We try

$$\mathbf{H}_e = \alpha \left(\frac{a}{r}\right)^3 [3(\mathbf{n} \cdot \mathcal{M})\mathbf{n} - \mathcal{M}],$$
$$\mathbf{H} = \beta \mathcal{M}.$$

The two jump conditions lead to the equations $-\alpha = \beta$ and $2\alpha = 1 + \beta$. The solutions are $\alpha = \frac{1}{3}$ and $\beta = -\frac{1}{3}$. Thus $N = -\beta = \frac{1}{3}$.

70.2. According to the last exercise, the field outside the earth would be

$$\mathbf{H}_e = \alpha \left(\frac{a}{r}\right)^3 [3(\mathbf{n} \cdot \mathcal{M})\mathbf{n} - \mathcal{M}].$$

The vector $3(\mathbf{n} \cdot \mathcal{M})\mathbf{n}$ is radial (i.e. vertical), of magnitude $3(\mathbf{n} \cdot \mathcal{M}) = 3\mathcal{M} \sin \lambda$, where λ is the magnetic latitude. The vector $-\mathcal{M}$ has a vertical component $-\mathcal{M} \sin \lambda$ and a horizontal component $\mathcal{M} \cos \lambda$. Thus, to within a common factor, the magnetic field has vertical and horizontal components $2 \sin \lambda$ and $\cos \lambda$. It follows that the tangent of the dip angle δ is

$$\tan \delta = \frac{2 \sin \lambda}{\cos \lambda} = 2 \tan \lambda.$$

70.3. Any continuous toroidal $\mathbf{B}$ which depends only on the distance from the torus axis satisfies div $\mathbf{B} = 0$ and the jump condition $\mathbf{n} \cdot [\![\mathbf{B}]\!] = 0$. The current potential $\mathbf{H}$ must satisfy Ampère's law $\oint \mathbf{H} \cdot d\mathbf{s} = ni$. Since $H = B/\mu - \mathcal{M}_0$ in the material and $H = B/\mu_0$ in the gap, Ampère's law requires

$$(2\pi - \alpha)r(B/\mu - \mathcal{M}_0) + \alpha r B/\mu_0 = ni,$$

where r is the distance from the axis of the torus. Thus

$$B = \frac{\mu ni/r + (2\pi - \alpha)\mu \mathcal{M}_0}{2\pi + (\mu/\mu_0 - 1)\alpha}.$$

70.4. The magnetic field of $\mathbf{m}_1$ is

$$\mathbf{B}_1 = -\,\mathbf{grad}\,\Omega_1, \qquad \Omega_1 = \frac{\mu_0}{4\pi}\frac{\mathbf{m}_1 \cdot \mathbf{r}}{r^3},$$

where $\mathbf{r}$ is the position relative to $\mathbf{m}_1$. The magnetic force on $\mathbf{m}_2$ is

$$\mathbf{F}_2 = (\mathbf{m}_2 \cdot \mathbf{grad})\mathbf{B}_1 = -(\mathbf{m}_2 \cdot \mathbf{grad})\,\mathbf{grad}\,\Omega_1 = -\,\mathbf{grad}\,U,$$

where

$$U = \mathbf{m}_2 \cdot \mathbf{grad}\,\Omega_1 = \frac{\mu_0}{4\pi}\left[\frac{\mathbf{m}_1 \cdot \mathbf{m}_2}{r^3} - \frac{3(\mathbf{m}_1 \cdot \mathbf{r})(\mathbf{m}_2 \cdot \mathbf{r})}{r^5}\right]$$

may be regarded as a magnetic potential energy. For a magnet $\mathbf{m}_2 = -\mathbf{m}_1$ at a height h above $\mathbf{m}_1$ the total potential energy

$$U + Mgh = -\frac{\mu_0}{4\pi}\frac{2(\mathbf{m}_1 \cdot \mathbf{m}_2)}{h^3} + Mgh = \frac{\mu_0 m^2}{2\pi h^3} + Mgh$$

must be at a minimum. Thus we obtain for h the equation

$$-\frac{3\mu_0 m^2}{2\pi h^4} + Mg = 0, \qquad \text{or} \qquad h^4 = \frac{3\mu_0 m^2}{2\pi Mg}.$$

With three magnets we seek a minimum (with respect to h and k) of

$$\frac{\mu_0 m^2}{2\pi h^3} - \frac{\mu_0 m^2}{2\pi (h+k)^3} + \frac{\mu_0 m^2}{2\pi k^3} + Mgh + Mg(h+k),$$

which yields the equations

$$-\frac{3\mu_0 m^2}{2\pi h^4} + \frac{3\mu_0 m^2}{2\pi (h+k)^4} + 2Mg = 0,$$

$$\frac{3\mu_0 m^2}{2\pi (h+k)^4} - \frac{3\mu_0 m^2}{2\pi k^4} + Mg = 0. \tag{1}$$

Eliminating Mg we obtain

$$\frac{2}{k^4} - \frac{1}{h^4} - \frac{1}{(h+k)^4} = 0.$$

Thus for $x = h/k$ we have

$$2 - x^{-4} - (1+x)^{-4} = 0.$$

The root $x = 0.850115$ must be found by numerical means. Returning now to (1) we find

$$h^4 = (1 - x^4)\frac{3\mu_0 m^2}{2\pi Mg} = 0.831^4 \frac{3\mu_0 m^2}{2\pi Mg}, \qquad k = \frac{h}{x} = 1.18h.$$

72.1. We attempt a superposition of $\mathbf{B}_0$ and a dipole field:

$$\mathbf{B} = \mathbf{B}_0 + \alpha \left(\frac{a}{r}\right)^3 [3(\mathbf{n} \cdot \mathbf{B}_0)\mathbf{n} - \mathbf{B}_0].$$

The boundary condition $\mathbf{n} \cdot \mathbf{B} = 0$ on $r = a$ requires $(1 + 2\alpha)(\mathbf{n} \cdot \mathbf{B}_0) = 0$. Thus $\alpha = -\frac{1}{2}$ and the field is

$$\mathbf{B} = \left(1 + \frac{a^3}{2r^3}\right)\mathbf{B}_0 - \frac{3a^3}{2r^3}(\mathbf{n} \cdot \mathbf{B}_0)\mathbf{n}.$$

On the sphere

$$\mathbf{B} = \frac{3}{2}[\mathbf{B}_0 - (\mathbf{n} \cdot \mathbf{B}_0)\mathbf{n}],$$

which exerts the magnetic pressure

$$\frac{B^2}{2\mu_0} = \frac{9}{8\mu_0}[\mathbf{B}_0 - (\mathbf{n} \cdot \mathbf{B}_0)\mathbf{n}]^2 = \frac{9}{8\mu_0}[B_0^2 - (\mathbf{n} \cdot \mathbf{B}_0)^2] = \frac{9B_0^2 \sin^2 \theta}{8\mu_0}.$$

72.2. With $\mathbf{K} = \mu_0^{-1}\mathbf{n} \times \mathbf{B}$ we obtain

$$\tfrac{1}{2}\mathbf{K} \times \mathbf{B} = \tfrac{1}{2}\mu_0^{-1}(\mathbf{n} \times \mathbf{B}) \times \mathbf{B} = \tfrac{1}{2}\mu_0^{-1}[(\mathbf{n} \cdot \mathbf{B})\mathbf{B} - B^2\mathbf{n}] = -\frac{B^2}{2\mu_0}\mathbf{n},$$

since $\mathbf{n} \cdot \mathbf{B} = 0$ on the surface of the superconductor. This is precisely (72.1).

72.3. We have

$$\gamma(p, \vartheta) = \varphi[\rho(p, \vartheta), \vartheta] + \frac{p}{\rho(p, \vartheta)},$$

where $\rho(p, \vartheta)$ is (implicitly) defined by $p = \rho^2 \varphi_\rho(\rho, \vartheta)$. The partial derivatives of $\gamma(p, \vartheta)$ are

$$\gamma_\vartheta(p, \vartheta) = \varphi_\rho(\rho, \vartheta)\rho_\vartheta(p, \vartheta) + \varphi_\vartheta(\rho, \vartheta) - \frac{p}{\rho^2}\rho_\vartheta(p, \vartheta)$$

$$= \varphi_\vartheta(\rho, \vartheta) = -s$$

and

$$\gamma_p(p, \vartheta) = \varphi_\rho(\rho, \vartheta)\rho_p(p, \vartheta) + \frac{1}{\rho} - \frac{p}{\rho^2}\rho_p(p, \vartheta) = \frac{1}{\rho}.$$

72.4. The critical field $H_c(p, \vartheta)$ is (implicitly) defined by

$$\gamma_s \left[p + \tfrac{1}{2}\mu_0 H_c^2(p, \vartheta), \vartheta\right] = \gamma_n(p, \vartheta).$$

Differentiating with respect to p, and recalling $\partial\gamma(p, \vartheta)/\partial p = \rho^{-1}$, we obtain

$$\frac{1}{\rho_s}\left[1 + \frac{\partial}{\partial p}\tfrac{1}{2}\mu_0 H_c^2(p, \vartheta)\right] = \frac{1}{\rho_n},$$

which is (72.18).

72.5. Differentiating eqn (72.13) with respect to ϑ at fixed H_c and using (72.11) we obtain

$$\rho_s^{-1}\frac{\partial p}{\partial \vartheta} - s_s = \rho_n^{-1}\frac{\partial p}{\partial \vartheta} - s_n.$$

Rearranging, we find

$$\vartheta \frac{\partial p}{\partial \vartheta} = \frac{\vartheta(s_n - s_s)}{\rho_n^{-1} - \rho_s^{-1}} = \frac{\mathcal{L}}{\rho_n^{-1} - \rho_s^{-1}}.$$

72.6. It is clear that, in order to satisfy the condition $\mathbf{n} \cdot \mathbf{B} = 0$ on the surface of the superconductor, $\mathbf{m}$ and its image $\mathbf{m}'$ must have the same component along the surface, but their components normal to the surface must have opposite signs. Two such magnets repel each other.

Bibliographical notes

Below is a short list of sources for various topics covered in this book, and of books for further reading.

Whittaker, E. T. (1960). *A history of the theories of the aether and electricity*. Harper, New York. An authoritative history of electromagnetism. Whittaker's allocation of credit for special relativity is regarded by many as eccentric, because it appears to belittle Einstein's contribution.

Truesdell, C. and Toupin, R. (1960). *The classical field theories*, Vol. III/1 of Flügge's *Handbuch der Physik*. Springer-Verlag, Berlin. This encyclopedic article provides an excellent presentation of the principles of electromagnetism, and of continuum mechanics. It is also noteworthy for its historical references, and for its comprehensive bibliography. Truesdell and Toupin's seemingly innocent introduction of body heating played an essential role in Coleman and Noll's work.

Coleman, B. D. and Noll, W. (1963). The thermodynamics of elastic materials with heat conduction and viscosity. *Archive for Rational Mechanics and Analysis*, **13**, 167–178. The pioneering work, from which modern thermodynamics has developed.

Truesdell, C. (1969). *Rational thermodynamics*. McGraw Hill, New York. A very readable introduction to modern thermodynamics. A second, much enlarged edition, with appendices contributed by 23 other experts, was published by Springer-Verlag, New York, in 1984.

Landau, L. D. and Lifshitz, E. M. (1975). *The classical theory of fields*, Vol. 2 (4th edn) of *A course of theoretical physics*. Pergamon Press, Oxford. The first half of this volume presents the electrodynamics of charged particles. The second half is devoted to general relativity. Landau and Lifshitz's *Course of theoretical physics*, begun in the late 1930s, has set an unsurpassed standard of elegance.

Landau, L. D., Lifshitz, E. M. and Pitaevskii, L. P. (1984). *Electrodynamics of continuous media*, Vol. 8 (2nd edn) of *A course of theoretical physics*. Pergamon Press, Oxford. Covers all aspects of the electromagnetism of macroscopic bodies. Relies heavily on quantum mechanics. The treatment of moving bodies in Section 76 is unfortunately based on Minkowski's (1908) relativistic extension of Hertz's ill-fated aether relations.

Griffiths, D. J. (1989). *Introduction to electrodynamics* (2nd edn). Prentice Hall, London. A modern textbook, noteworthy for its wealth of ingenious examples and exercises. The chapter on special relativity is especially commendable.

Jackson, D. J. (1975). *Classical electrodynamics* (2nd edn). Wiley, New York. A classical book on the electrodynamics of charged particles. Contains an exceptionally thorough treatment of potential problems, and of Liénard–Wiechert potentials.

Index

Page numbers are <u>underlined</u> when they represent the main source of information, e.g. a definition.
Page numbers in *italic* give instructive examples.